Advances in Experimental Medicine and Biology

Volume 1038

More information about this series at http://www.springer.com/series/5584

Hongzhi Sun • Xiangdong Wang
Editors

Mitochondrial DNA and Diseases

Springer

Editors
Hongzhi Sun
Liaoning Province Key Laboratory of Cancer Metabolomics
Jinzhou Hospital of Jinzhou Medical University
Jinzhou, Liaoning, China

Xiangdong Wang
Zhongshan Hospital Institute of Clinical Science
Fudan University, Shanghai Medical College
Shanghai, China

ISSN 0065-2598 ISSN 2214-8019 (electronic)
Advances in Experimental Medicine and Biology
ISBN 978-981-10-6673-3 ISBN 978-981-10-6674-0 (eBook)
https://doi.org/10.1007/978-981-10-6674-0

Library of Congress Control Number: 2017958161

Printed on acid-free paper

This Springer imprint is published by Springer Nature
The registered company is Springer Nature Singapore Pte Ltd.
The registered company address is: 152 Beach Road, #21-01/04 Gateway East, Singapore 189721, Singapore

Contents

About the Editors

Hongzhi Sun, MD, PhD, is a famous surgical specialist in China and the director of the First Affiliated Hospital of Liaoning Medical University. He is the chairman of Surgical Biomaterials Specialized Committee, the vice chairman of Tumor Cell Biology Specialized Committee and Surgical Biological Cell Technology Specialized Committee in Liaoning Cell Biology Society, a member of the Breast Sciences Section of Liaoning Medical Association, the executive director of Liaoning Anticancer Association, the associate editor of the international journal *Molecular and Cellular Therapies*, the editorial board member of *Chinese Journal of Tissue Engineering Research* and *Clinical Rehabilitation* and Liaoning Medical University newspaper. He hosted seven fund projects and published more than ten scientific papers with the impact factor over 20. His major research is focused on the mechanism of tumorigenesis and progression, invasion and metastasis, and the basic and clinical research of biotherapy of tumor.

Xiangdong Wang, MD, PhD, is a distinguished professor of medicine, director of Shanghai Institute of Clinical Bioinformatics, executive director of Clinical Science Institute of Fudan University Zhongshan Hospital, director of Fudan University Center of Clinical Bioinformatics, deputy director of Shanghai Respiratory Research Institute, and visiting professor of King's College London. His main research is focused on clinical bioinformatics, disease-specific biomarkers, lung chronic diseases, cancer immunology, and molecular and cellular therapies. He is the author of more than 200 scientific publications with the impact factor about 900, citation number about 6918, h-index 46, i10-index 181, and cited journal impact factor about 7000.

Chapter 1
How Far Can Mitochondrial DNA Drive the Disease?

Hongzhi Sun, Weibin Shi, and Xiangdong Wang

Abstract Mitochondria are one of the dominant drivers for producing cellular energy to meet a large number of biological functions, of which the mitochondrial DNA (mtDNA) is the control center of energetic driving force and the dominant driver of mitochondrial molecular diversification. mtDNA transcription generates the necessary RNAs to regulate the extent and nature of mtRNA post-transcriptional modifications and the activity of nucleus-encoded enzymes. With a special focus on mtDNA, the current volume aims to overview the biology and structures of mtDNA, regulatory roles of mtDNA in lung diseases, or involvement of mtDNA in metabolism. We explore the significance of mtDNA sequencing, methylation, stability, and mutation in the pathogenesis of the diseases. Molecular mechanisms by which mtDNA contribute to the regulation of mitochondrial homeostasis and drug resistance are also discussed. We also point out the importance of mitochondrial ribosome, single cell biology, and gene editing in the understanding of the development of mitochondrial dysfunction in lung disease.

Keywords Mitochondria • mtDNA • Diseases • Energy • Metabolism

---

H. Sun (✉)
Liaoning Province Key Laboratory of Cancer Metabolomics, Jinzhou Hospital of Jinzhou Medical University, Jinzhou, Liaoning, China
e-mail: cmushz@163.com

W. Shi (✉)
Medical Examination Center of Zhongshan Hospital, Fudan University, Shanghai, China
e-mail: shi.weibin@zs-hospital.sh.cn

X. Wang (✉)
Zhongshan Hospital Institute of Clinical Science, Fudan University, Shanghai Medical College, Shanghai, China
e-mail: xiangdong.wang@clintransmed.org

H. Sun, X. Wang (eds.), *Mitochondrial DNA and Diseases*, Advances in Experimental Medicine and Biology 1038,
https://doi.org/10.1007/978-981-10-6674-0_1

1 Why We Focus on mtDNA

Mitochondria are the dominant driver of cellular energy for a large number of biological functions, while mitochondrial DNA (mtDNA) is the control center of energetic driving force and the dominant driver of mitochondrial molecular diversification. mtDNA can act as the part of genome to transcript genetic messages to maintain cell's own function, as the genetic copy for other cells which have the capacity of the sensibility to mtDNA and as the toxicant to "poison" owners intra- or extramitochondrially. Macrophages have strong uptake and weak digestion capacities of tumor mtDNA, while dendritic cells have strong sensitivity and digestion capacities [1]. Both cells belong to the family of native immunity but have different senses to mtDNA, due to the variation of integrin-associated protein-signal regulatory protein-α axis function between cells. Elements of mtDNA or elements associated with mtDNA function are considered as diagnostic biomarkers or therapeutic targets. With the development of biotechnology, mapping mtDNA genomes can provide more details of the genetic information for the understanding molecular mechanisms mtDNA is involved in. For example, a new strategy of PacBio full-length transcriptome sequencing was applied for the first full-length human mitochondrial transcriptome, and two novel long non-coding RNAs of ATP-binding cassette permease MDL1 and MDL1AS were discovered in human mitochondrial DNA from animal mitochondrial genomes [2]. mtDNA epigenetics is also a new merging area to be explored. mtDNA methylation and histone modifications were detected and suspected to regulate the mitochondrial unfolded protein response, although the exact mechanisms are not unclear yet. The oxidative phosphorylation system and RNA granules ribosome are the major players in the posttranscriptional processes of mitochondrial gene expression, mtDNA perturbation, and the coordination of mitochondrial and cytosolic translation [3]. mtDNA transcription generates the necessary RNAs to regulate the extent and nature of mtRNA posttranscriptional modifications and the activity of nucleus-encoded enzymes. With a special focus on mtDNA, the current volume aims to overview the biology and structures of mtDNA, regulatory roles of mtDNA in lung diseases, or involvement of mtDNA in metabolism. We explore the significance of mtDNA sequencing, methylation, stability, and mutation in the pathogenesis of the diseases. Molecular mechanisms by mtDNA that contribute to the regulation of mitochondrial homeostasis and drug resistance are also discussed. We also point out the importance of mitochondrial ribosome, single-cell biology, and gene editing in the understanding of the development of mitochondrial dysfunction in lung disease.

2 Why We Start with the Relationship Between mtDNA and Lung Diseases

Lung diseases include acute and chronic lung injury and failure, infections, interstitial inflammation, vascular pathology, allergic immune disorders, and cancer. The incidence of lung diseases is one of the highest ones in the world, with a continuous increase. For example, the incidence and mortality of lung cancers become the top one in all kinds of cancers, and chronic lung diseases become the leading one in all kinds of chronic diseases, especially in the developing countries with air pollution [4–6]. Even though, lung diseases are relatively ignored, gain less attentions than they should have, or are underestimated, due to a large compensatory capacity the lung has, less incidence of sudden death, and wide distribution of clinical category.

Lu et al. collects experimental and clinical evidence to show the potential links between mtDNA and lung cancer and demonstrates the fact that alterations of mtDNA copy number and sequence mutations may contribute to carcinogenesis and development of lung cancer [7]. One of the exciting points from the chapter is that variations of mtDNA sequencing, mutation, and epigenetics are suggested to be potential biomarkers to monitor severity, duration, stage, response to therapy, and prognosis in patients with lung cancer. The significance of mtDNA mutations and altered copy numbers in lung cancer and clinical relevance was comprehensively overviewed [8–10], although a number of issues should be considered before mtDNA changes can be an effective strategies for lung cancer diagnosis and therapy. One of the most challenges to develop mtDNA-dependent biomarkers is to ensure the disease specificity and sensitivity, since mitochondria are the power and metabolism driver of the cells and have the function and responsibility more generally.

Qian et al. bring a critical question how we can properly identify disease-specific values from mtDNA sequencing, mutations, molecular interactions and networks, signaling pathways, and heteroplasmy in mitochondrial genome [11]. From practical points, this chapter presents what the proper and optimal way to prepare the library preparation, sequencing, data analysis, and professional interpretations and call the special attentions to recognize advantages and disadvantages of mtDNA. One of the critical issues the chapter delivers is how and why we should choose the most appropriate mtDNA sequencing in the real application, and we should seriously take care of bioethics in multi-aspects. Another important issue the chapter bring on the table at the first time is the urgent need to evaluate the potential cytotoxicity and genotoxicity of gene editing, especially the long-term effects, before mtDNA editing marches into clinic.

3 How mtDNA Works in the Disease

It is easier to understand the importance of mtDNA in mitochondrial function, cellular metabolism, and organ function, while it is hard to describe how mtDNA works and regulates the element interaction intra- or extramitochondrially. The mechanism by which mtDNA regulates and controls the occurrence of the disease is more complex than we can image. Of those, the epigenetic mechanisms are recently overviewed and believed to be important in bidirectional mitonuclear communication between mitochondria and the nuclear epigenome represents [12]. It seems that the mitochondria can generate a number of the intermediate or terminal metabolites to mediate or participate in the processes of acetylation or methylation to regulate and alter nDNA epigenetic phenotypes. The modified epigenetic messages in the nuclear are transferred into the mitochondria to mediate mtDNA gene expression of mitochondria-produced proteins or mtDNA acetylation or methylation. The mitochondrial proteins can participate and regulate the production of bioenergy, oxygenation, and metabolism in mitochondria, which in turn can feedback the signaling to the nuclear or mitochondria. It is possible or necessary for nDNA and mtDNA to have such crosstalk and interactions.

Wang et al. furthermore explore the intercommunication and interaction of the molecules within mitochondria during immune responses or the development of immune-related diseases [13]. This chapter collects scientific evidence to show the roles of mtDNA in immune responses and lead an initial discussion on the main function and mechanisms of mtDNA in immune responses. mtDNA per se can interact with native immune cells and influence their capacities, dependently upon the sensitivity of cells. mtDNA can also contribute to the immune response by regulating mtDNA-dependent TLR9 signaling pathways, autophagy pathways, or cGAS-STING signaling pathways. The chapter also uncovers the potentials of mtDNA-oriented therapeutic targets and mtDNA gene editing in immune diseases and inflammation, although it is still at the early stage of the discovery and development.

Huo et al. describe potential mechanisms by which mutations or sequence aberrations of nDNA or mtDNA occur in the disease and specially focus on USP30. USP30 is a mitochondrial deubiquitinase to contribute to mitochondrial fusion by mediating the deubiquitination of ubiquitylated forms of mitofusins [14]. USP30 can regulate the removal of Ub-chains from Lys 6 and Lys 11 on mitochondria-derived proteins and of the ubiquitin chains added by Parkin, a cytoplasmic E3 ubiquitin-ligase. Mitochondrial fission and mitophagy as well as mtDNA genetic alterations can be regulated and controlled by autophagy-related genes, mitochondrial-quality control pathways, or USP30-triggered mitochondrial dynamic signaling.

4 Whether mtDNA Can Influence Cell Metabolism

Lv et al. demonstrate the fact that metabolic disorders, oxidative stress, or inflammation can alter mtDNA, resulting in the occurrence of mitochondrial dysfunction [15]. It is questioned how mtDNA control and influence energy metabolism, signaling pathways, and cell death. The intra- or extracellular factors can induce DNA single- or double-strain breaks and provoke DNA-dependent protein kinase, e.g., aging [16]. The activated DNA-dependent protein kinase can phosphorylate threonines 5 and 7 of HSP90α and downregulate the activity of AMP-activated protein kinase, leading to the dysfunction of mitochondrial biogenesis, energy metabolism, and physical fitness. Drugs and toxicants could alter mitochondrial motility, structure, and cell injury [17–19], probably through mtDNA-associated mitochondrial dysfunction. Intra- or extracellular toxicants could induce overproductions of oxidative stress, dysfunction of mitochondrial respiration, alterations of mitochondrial membrane potential, and apoptosis through the downregulation of intracellular adenosine triphosphate activities and the down-activation of mitochondrial enzymes, e.g., mitochondrial respiratory enzymes, NADH dehydrogenase, cytochrome c oxidase, and oxidative stress-sensitive aconitase. It is critical that toxins-induced compromises and fragmentation of nDNA and mtDNA can deliver "wrong" messages to activate the apoptosis signal pathways, e.g., apoptosis-initiator caspase-9 and apoptosis-effector caspase-3/caspase-7. This is one of complex mechanisms to reach mtDNA-associated or mitochondria-dependent cytotoxicity and metabolic disorders. Song et al. at first time describe potential mechanisms of mtDNA in telocytes and foresee the importance to explore telocyte mitochondrial function and mtDNA regulations in this volume [20].

Pan et al. extend the discussion from mtDNA-regulated mitochondrial function to mitochondria-dependent metabolism in cancer and mitochondria-mediated metabolic regulation in cancer drug resistance [21]. This chapter systemically overviews the historical development of mitochondrial metabolism and characteristic phenotypes of mitochondria-dependent metabolism from various aspects, e.g., glycolysis and the Warburg effect, pentose phosphate pathway, glutamine reductive carboxylation lipid, and amino acid metabolism. Pan et al. combine their own with published knowledge and experience to propose a new therapeutic strategy for different categories of cancer on the basis of the principle of mitochondrial metabolism. Firstly, it is critical to select precision methodologies to carry out the personalized diagnosis tools by measuring mutations of nDNA and mtDNA, mass of mtDNA and mitochondria, quality of mitochondrial OXPHOS, or levels of mitochondrial metabolites and RO. Secondly, novel therapeutics have been developed and optimized, e.g., drug repositioning, specific pathways of inhibitors/activators of genetic engineering techniques, and mitochondria-targeting drug resistance. Finally, the mitochondria-targeting delivery systems for cancer should be designed and validated to meet the clinical needs, including targeting peptides, lipophilic cations, nanoparticles, nanovesicles, or nano-liposomes. The chapter clearly points out the future development of mitochondria-based therapies and explore the potential of clinical application.

Lu emphasizes the importance of mitochondrial ATP-dependent Lon protease on the pathogenesis of diseases, especially in cancer, as the potential biomarker for cancer diagnosis and novel target for drug discovery and development [22]. nDNA can deliver the messages for the encoding ATP-dependent Lon proteases in the mitochondria in various cells, which are the microenvironment for evolution and quality control of proteins by the signaling pathways and mediating factors. The balance of mitochondrial genome (mtDNA) and folding of mitochondria proteins can be maintained by the selective degradation of misfolded, oxidized, and short-lived regulatory proteins within mitochondria. A large number of intra- or extracellular factors, e.g., ROS, inflammatory mediators, or kinases, can influence the expressions and activities of ATP-dependent Lon proteases, leading mtDNA mutations, mitochondrial dysfunction, or cell injury. Disordered activities of ATP-dependent Lon proteases also change the sensitivity and susceptibility of cells to drugs and increase drug resistance and tolerance [23–27]. Mitochondrial ATP-dependent Lon protease was suggested as a potential biomarker for cancer diagnosis and therapeutic target for drug discovery and development, even though the mechanism by proteases that interact with mtDNA remains unclear.

5 Whether mtDNA Mutations and Genetic Changes Make the Sense

Zhang et al. describe that the mitochondria are sensitive to both endogenous and exogenous factors and become structural alterations of mitochondria and mtDNA [28]. One of the most important mechanisms is mtDNA repair pathways to maintain the mtDNA integrity and mitochondrial injury as well as the occurrence of diseases. The base excision repair pathway may contribute to mtDNA mutations and genetic changes and can be an alternative to treat related diseases. Potential mechanisms include a large number of enzymes, e.g., DNA glycosylases, polymerase γ, ligase, AP endonuclease, and mtDNA repair pathways, e.g., SP-BER, LP-BER, mtDNA degradation, direct DNA repair, nucleotide excision repair, mismatch repair, DNA strand break repair, or nonhomologous or microhomology-mediated end joining. Zhang et al. also discuss about roles of mitochondrial ribosomes, as oxidative phosphorylation, in the production of ATP during the development of diseases [29]. A particular example of epithelial mitochondria and mtDNA is illustrated to understand the relationship among mtDNA, mitochondria, cell, organ, and diseases [30].

Gao et al. arise another important roles of mtDNA epigenetics, since there is a little discussion on the methylation of mtDNA, and question whether mtDNA methylation occurs in biology and pathology [31]. More evidence suggests that mtDNA methylation plays an important role in mitochondrial gene regulation and the development of diseases and is affected by a large number of intra- or extracellular factors. mtDNA methylation is proposed as a useful biomarker of disease diagnosis, although those studies on mtDNA methylation are just initial. There are great needs to develop methodologies to detect mtDNA methylation-orientated signal pathways, interaction, and communications among regions of mtDNA sequences and between mtDNA and nDNA. It is also perspective to consider that mtDNA heterogeneity epigenetics should be investigated by measuring single-cell DNA sequencing, comprehensive characterizations of mtDNA, and bidirectional effects between mtDNA and 3D genome, instability, and gene editing. It would be more help to combine the single-cell biology with CRIPRS to mtDNA function [32–35]. Wang and Fang overview the sensitivity and vulnerability of mitochondria to drugs and drug candidates or compromised effects of drugs on mitochondrial function and provide the comprehensive understanding regarding the mechanisms underlying mitochondrial dysfunction to uncover therapeutic targets in multiple pathologies [36].

Acknowledgments The work was supported by Zhongshan Distinguished Professor Grant (XDW), the National Nature Science Foundation of China (91230204, 81270099, 81320108001, 81270131, 81300010), the Shanghai Committee of Science and Technology (12JC1402200, 12431900207, 11410708600, 14431905100), Operation funding of Shanghai Institute of Clinical Bioinformatics, Ministry of Education for Academic Special Science and Research Foundation for PhD Education (20130071110043), and National Key Research and Development Program (2016YFC0902400, 2017YFSF090207).

References

1. Xu MM, Pu Y, Han D, Shi Y, Cao X, Liang H, et al. Dendritic cells but not macrophages sense tumor mitochondrial DNA for cross-priming through signal regulatory protein α signaling. Immunity. 2017;47(2):363–73. https://doi.org/10.1016/j.immuni.2017.07.016. PMID: 28801234
2. Gao S, Tian X, Chang H, Sun Y, Wu Z, Cheng Z, et al. Two novel lncRNAs discovered in human mitochondrial DNA using PacBio full-length transcriptome data. Mitochondrion. 2017. https://doi.org/10.1016/j.mito.2017.08.002. PMID: 28802668

3. Pearce SF, Rebelo-Guiomar P, D'Souza AR, Powell CA, Van Haute L, Minczuk M. Regulation of mammalian mitochondrial gene expression: recent advances. Trends Biochem Sci. 2017;42(8):625–39. PMID: 28285835
4. Fang XC, Wang XD, Danderson M, Bai CX. Chronic obstructive pulmonary disease in China: the burden and the importance of proper management. Chest. 2011;139(4):920–9. PMID: 21467059
5. Chen C, Shi L, Li Y, Wang X, Yang S. Disease-specific dynamic biomarkers selected by integrating inflammatory mediators with clinical informatics in ARDS patients with severe pneumonia. Cell Biol Toxicol. 2016;32(3):169–84. PMID: 27095254
6. Shi L, Zhu B, Xu M, Wang X. Selection of AECOPD-specific immunomodulatory biomarkers by integrating genomics and proteomics with clinical informatics. Cell Biol Toxicol. 2017. https://doi.org/10.1007/s10565-017-9405-x. PMID: 28779230
7. Liu F, Sanin DE, Wang X. Mitochondrial DNA in lung cancer. Adv Exp Med Biol. 2017;1038:2
8. Wang X. New biomarkers and therapeutics can be discovered during COPD-lung cancer transition. Cell Biol Toxicol. 2016;32(5):359–61. PMID: 27405768
9. Gu J, Wang X. New future of cell biology and toxicology: thinking deeper. Cell Biol Toxicol. 2016;32(1):1–3. PMID: 26874518
10. Wang X. CBT profiles of cabozantinib approved for advanced renal cell carcinomas. Cell Biol Toxicol. 2016;32:259–61. PMID:27383755
11. Qian M, Spada C, Wang X. Approach, application and bioethics of mtDNA sequencing in cancer. Adv Exp Med Biol. 2017;1038:3
12. Matilainen O, Quirós PM, Auwerx J. Mitochondria and epigenetics – crosstalk in homeostasis and stress. Trends Cell Biol. 2017;27(6):453–63. PMID: 28274652
13. Wang L, Liebmen M, Wang X. Roles of mitochondrial DNA signaling in immune responses. Adv Exp Med Biol. 2017;1038:4
14. Hou J, Eldeeb MA, Wang X. Beyond deubiquitylation: USP30-mediated regulation of mitochondrial homeostasis. Adv Exp Med Biol. 2017;1038:5
15. Lv J, Bhatia M, Wang X. Roles of mitochondrial DNA in energy metabolism. Adv Exp Med Biol. 2017;1038:6
16. Park SJ, Gavrilova O, Brown AL, Soto JE, Bremner S, Kim J, et al. DNA-PK promotes the mitochondrial, metabolic, and physical decline that occurs during aging. Cell Metab. 2017;25(5):1135–1146.e7. PMID: 28467930
17. Kikuchi S, Ninomiya T, Kohno T, Kojima T, Tatsumi H. Cobalt inhibits motility of axonal mitochondria and induces axonal degeneration in cultured dorsal root ganglion cells of rat. Cell Biol Toxicol. 2017. https://doi.org/10.1007/s10565-017-9402-0. PMID: 28656345
18. Zerin T, Kim JS, Gil HW, Song HY, Hong SY. Effects of formaldehyde on mitochondrial dysfunction and apoptosis in SK-N-SH neuroblastoma cells. Cell Biol Toxicol. 2015;31(6):261–72. PMID: 26728267
19. Vedi M, Sabina EP. Assessment of hepatoprotective and nephroprotective potential of withaferin a on bromobenzene-induced injury in Swiss albino mice: possible involvement of mitochondrial dysfunction and inflammation. Cell Biol Toxicol. 2016;32(5):373–90. PMID: 27250656
20. Song D, Cretoiu D, Wang X. Mitochondrial DNA in telocytes. Adv Exp Med Biol. 2017;1038:7
21. Pan Y, Cao M, Yang Q, Go VG, Lee PWN, Xiao GG. Metabolic regulation in mitochondria and drug resistance. Adv Exp Med Biol. 2017;1038:8
22. Lu B. Mitochondrial Lon protease and cancer. Adv Exp Med Biol. 2017;1038:9
23. Guney Eskiler G, Cecener G, Tunca B, Egeli U. An in vitro model for the development of acquired tamoxifen resistance. Cell Biol Toxicol. 2016;32(6):563–81. PMID: 27585693
24. Peng WX, Han X, Zhang CL, Ge L, FY D, Jin J, Gong AH. FoxM1-mediated RFC5 expression promotes temozolomide resistance. Cell Biol Toxicol. 2017. https://doi.org/10.1007/s10565-017-9381-1. PMID: 28185110

25. Gu W, Dong N, Wang P, Shi C, Yang J, Wang J. Tamoxifen resistance and metastasis of human breast cancer cells were mediated by the membrane-associated estrogen receptor ER-α36 signaling in vitro. Cell Biol Toxicol. 2017;33(2):183–95. PMID: 27837347
26. Sidor-Kaczmarek J, Cichorek M, Spodnik JH, Wójcik S, Moryś J. Proteasome inhibitors against amelanotic melanoma. Cell Biol Toxicol. 2017. https://doi.org/10.1007/s10565-017-9390-0. PMID: 28281027
27. Ge S, Li T, Yao Q, Yan H, Huiyun Z, Zheng Y, Zhang B, He S. Expression of proteinase-activated receptor (PAR)-2 in monocytes from allergic patients and potential molecular mechanism. Cell Biol Toxicol. 2016;32(6):529–42. PMID: 27423452
28. Zhang L, Reyes A, Wang X. The role of base excision repair in maintaining mitochondrial DNA stability. Adv Exp Med Biol. 2017;1038:10
29. Zhang L, Wang W, Zhu B, Wang X. Regulatory roles of mitochondrial ribosome in lung diseases and single cell biology. Adv Exp Med Biol. 2017;1038:11
30. Zhang L, Wang W, Zhu B, Wang X. Epithelial mitochondrial dysfunction in lung disease. Adv Exp Med Biol. 2017; 1038: 12
31. Gao D, Zhu B, Sun H, Wang X. Mitochondrial DNA methylation and related disease. Adv Exp Med Biol. 2017;1038:13
32. Chu MP, Kriangkum J, Venner CP, Sandhu I, Hewitt J, Belch AR, Pilarski LM. Addressing heterogeneity of individual blood cancers: the need for single cell analysis. Cell Biol Toxicol. 2017;33(2):83–97. PMID: 27761761
33. Wang W, Gao D, Wang X. Can single-cell RNA sequencing crack the mystery of cells? Cell Biol Toxicol. 2017. https://doi.org/10.1007/s10565-017-9404-y. PMID: 28733864
34. Wang W, Zhu B, Wang X. Dynamic phenotypes: illustrating a single-cell odyssey. Cell Biol Toxicol. 2017.; PMID: 28638956
35. Wang W, Wang X. Single-cell CRISPR screening in drug resistance. Cell Biol Toxicol. 2017;33(3):207–10. https://doi.org/10.1007/s10565-017-9396-7. PMID: 28474250
36. Wang W, Hou J, Zhu Z, Fang H. Is mitochondrial cell fragility a cell weakness? Adv Exp Med Biol. 2017;1038:14

Weibin Shi, who graduated from Shanghai Jiao Tong University, is now working as the supervising technologist and director of the Health Control Center in Zhong Shan Hospital. Since joining Zhong Shan Hospital in 1989, he has been working for nearly 30 years. And during the period, he has acquired Radiology Technology Qualification (Intermediate). His publication record includes three research papers in core journals. He is a member of the committee of Shanghai Medical Association, Shanghai Health Administration Association, and Chinese Medical Association Shanghai branch.

Chapter 2
Mitochondrial DNA in Lung Cancer

Fangming Liu, David E. Sanin, and Xiangdong Wang

Abstract Mitochondrial DNA (mtDNA) variations are increasingly discovered and expected to be potential biomarkers to monitor severity, duration, stage, response to therapy, and prognosis in patients with lung cancer. The present article illustrates alterations of mtDNA in lung cancer, including alterations of mtDNA copy number and sequence mutations, as well as their possible mechanisms for carcinogenesis and development of lung cancer. The clear and comprehensive relationships between mtDNA variations and lung cancer are to be further confirmed to benefit effective strategies for lung cancer diagnosis and therapy.

Keywords Mitochondria • DNA • Mutation • Lung cancer • Diagnosis

1 Introduction

Lung cancer is an increasingly severe disease, accounting for 30% of all cancer-related deaths [1], and has the highest occurrence and mortality among malignancies. This will continue to increase if there are still no effective measures and therapies against lung cancer. Although the mechanism of lung cancer carcinogenesis remains unclear, tumor development is associated with mitochondrial DNA (mtDNA) mutations and alterations in mitochondrial genomic function [2, 3]. mtDNA mutations are found in various kinds of tumors, such as lung cancer [1, 4], bladder cancer [4], breast cancer [5], thyroid cancer [6], and prostate cancer [7], since mtDNA homogeneity mutation has been identified in colorectal cancer [8].

F. Liu
Zhongshan Hospital Clinical Science Institute, Fudan University, Shanghai, China

D.E. Sanin
Department of Immunometabolism, Max Planck Institute of Immunobiology and Epigenetics, Freiburg im Breisgau, Germany

X. Wang (✉)
Zhongshan Hospital Institute of Clinical Science, Fudan University, Shanghai Medical College, Shanghai, China
e-mail: xiangdong.wang@clintransmed.org

H. Sun, X. Wang (eds.), *Mitochondrial DNA and Diseases*, Advances in Experimental Medicine and Biology 1038,
https://doi.org/10.1007/978-981-10-6674-0_2

Characterizations of mtDNA genes vary among lung cancer subtypes, stages, and severity [9, 10]. The present article aims at overviewing the understanding of mtDNA biology, the significance of mtDNA mutations and altered copy numbers in lung cancer, the potential of mtDNA-associated biomarkers, and the comparison between mtDNA and nDNA biomarkers.

2 mtDNA Biology

mtDNAs are double-stranded closed circular DNA with 16,596 base pairs, composed of coding regions without introns and noncoding region. The coding region contains 13 polypeptides, 22 transfer RNAs, and 2 ribosomal RNAs [11]. The noncoding region (also called D-loop region) is responsible for mtDNA replication and transcription [12, 13], which has significant regulatory elements, such as L-strand promoter, H-strand promoter, and H-strand replication origin. In coding regions, almost every base plays a role in gene assembly [14]. As a result, once mutations occur, DNA sequence will probably be altered and followed by the alteration of coding proteins and mitochondrial dysfunction, as an important cause of human pathology.

Compared with nuclear DNA (nDNA), mtDNAs have higher mutation rates, and there are hundreds of thousands of mtDNA copies per cell. Mitochondrion is a major source to generate adenosine triphosphate (ATP) for cells through oxidative phosphorylation [15], during which reactive oxygen species (ROS) are generated as by-products and lead to DNA damage. mtDNAs are in close physical proximity to the electron transport chain and superoxide anions generated in oxidative phosphorylation and easier to acquire oxidative damage [11] (Fig. 2.1). mtDNAs contain

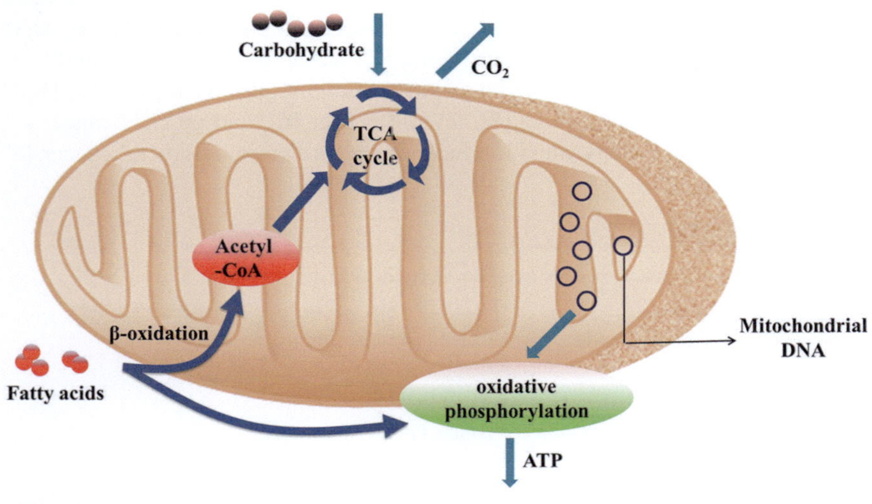

Fig. 2.1 An illustration of mtDNA distribution in mitochondria. mtDNAs are double-stranded closed circular DNA; they are in close physical proximity to the electron transport chain and reactive oxygen species produced in oxidative phosphorylation. It's more likely for mtDNAs to acquire oxidative damage

naked and lacked protective histones, more susceptible to the damage of carcinogens. Once mtDNA is compromised, mtDNA is hardly self-repaired due to the lack of DNA repair system and excision repair capacity in the mitochondria [14]. There are a large number of intra- and extracellular factors to contribute to the DNA damage of base pair mismatch and high mutation rates of mtDNAs, responsible for the development of mitochondrial dysfunction and cell susceptibility [15].

High mutations and copies of mtDNA as distinct features are involved in the carcinogenesis and are expected to be potential biomarkers for cancer detection and monitoring. The frequency of sequence mutations may increase during mtDNA copying and sequencing. The higher the number of mtDNA copies, the more susceptible mtDNAs are for mutation. The small size and closed circular structure of mtDNAs are more resistant to the DNA damages [11]. The appearance of the same mutation in mtDNAs and tumors indicates the clear relationship between [4], since it is questioned whether the gene mutation is from mtDNA to tumor tissue or reversely, genetically regulated from the source, and represent the targeting similarity between. It seems clear that mtDNAs can impact tumor cell growth and sensitivity to treatments [4, 16, 17].

3 The Significance of mtDNA Mutations

mtDNA may undergo mutations, e.g., insertion, deletion, or point mutations, in response to carcinogenetic factors [18, 19]. The mutation rate of mtDNA is much higher than that of nDNA due to its sensitivity to cancerogens. There are an amount of mutations in mitochondrial D-loop in lung cancer [19, 20]. mtDNA noncoding region is important in the transcription and duplication of mtDNA. D-loop is 1124 bp in size and a replicate origin where heavy and light chains of mtDNA are located [18]. There are two high-variable regions HVI and HVII located between 16,024–16,383 and 57–372 in D-loop, respectively. Alterations of D-loop region are possibly associated with mitochondrial dysfunction and mutation of nuclear genome. Single nucleotide polymorphisms of minor alleles of nucleotides 235A/G and 324A/G in D-loop region can be a risk factor of lung cancers. Of single nucleotide polymorphisms, the minor alleles 151C/T, 200A/G, 524C/CA, and 16274G/A are associated with the induction of squamous cell carcinoma, while the mutation of the minor allele 16,298 T/C may trigger the occurrence of small cell lung cancer [1]. In another sequencing of mtDNA, 8 of 27 primary lung tumors had mutations in the D-loop region, including a C-to-G transversion in nucleotide 16,114 and deletions/insertions in polyinosinic acid-polycytidylic acid [19]. In addition, 12 SCLC and 16 NSCLC cell lines of lung tumors have high frequency of homopolymeric C tract or single nucleotide polymorphisms in the D-loop region [18].

Mutations in mtDNA coding region are correlated with clinical characteristics in patients with lung cancer. Of those,36 mutations were detected in mtDNA coding region and 19 missense mtDNA mutations in 9 mtDNA coding region, including COI, COII, COIII, ATPase 6, ND2, ND3, ND4, ND5, and Cyt b [3]. These alterations of mtDNA may play an important role in tumorigenesis and metastasis of lung can-

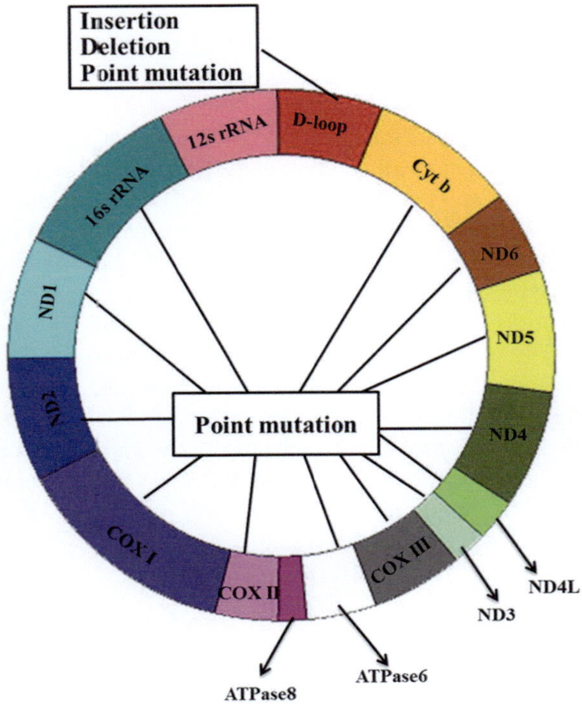

Fig. 2.2 Mutations in different regions of mtDNA. mtDNAs are composed of 16 coding and noncoding regions without introns. Several mutations of mtDNA in lung cancer were showed in this figure (insertion, deletion, or point mutations), according to the literature. The noncoding region D-loop is the "hot spot" of mutations

cer, as shown in Fig. 2.2. Mitochondrial DNA is vulnerable to alterations. When a mutation takes place, the altered mtDNAs can coexist with wild-type mtDNA in a heteroplasmy state [21, 22] . During cell replication, mtDNAs were divided into daughter cells randomly, and the percentage of variant mtDNAs in daughter cells is uncertain. After lots of passages of those heteroplasmic cells, the wild-type or mutation mtDNAs will transform into a predominant rate called homoplasmy. The heteroplasmic mutations result in highly variable inheritance and biological impact [22] due to the random distribution of genetic materials (Fig. 2.3). Thus, there are still several challenges to identify the association between mtDNA mutations and clinical phenotypes.

mtDNA mutations may contribute to carcinogenesis by decreasing cellular energy capacities, although the exact mechanism remains unclear. The mutation rate of mtDNA at a certain proportion can reduce the cellular energy capacity and cause the lack of cell and tissue bioenergetics to maintain normal functions. Or it is possible that tumors are induced by increasing mitochondrial oxidative stress [23]. The mitochondria are the primary source of endogenous ROS, including unpaired electrons and oxygen free radical. mtDNA encodes 13 polypeptides involved in electron transport chain, vulnerable to mutagens due to the lack of introns and protective histones. It is also possible that mtDNA mutations altered proteins and the function of respiratory chain is damaged [24]. The increased amount of electrons could result in more free radical production and interact with genome as tumor initiators, mutagenizing proto-oncogenes into oncogenes [25]. mtDNA mutations induce carci-

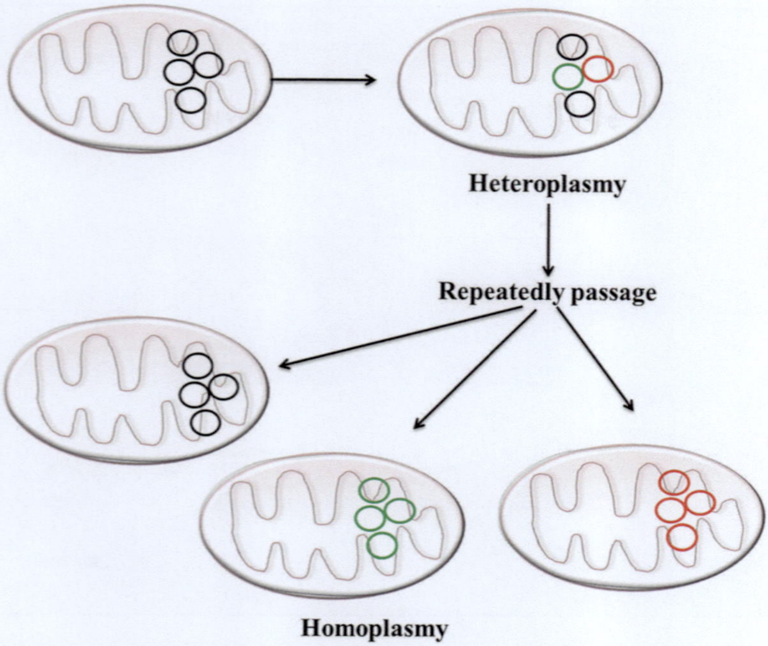

Fig. 2.3 The random genetic mutations of mtDNA. The altered mtDNAs can coexist with wild-type mtDNA in a heteroplasmy state. During cell replication, mtDNAs were divided into daughter cells randomly. After lots of passages of those heteroplasmic cells, the wild-type or mutation mtDNAs will transform into a predominant rate called homoplasmy that causes highly variable inheritance and biological impact

noma through modulating apoptosis controlled by electrochemical gradients. The reaction through mitochondrial electron transport chain can reduce electrochemical gradients, opening the mitochondrial permeability transition pore [22] and releasing cell death-promoting factors to destruct cytoplasm and chromatin. In cancer cells with mtDNA mutations, the electrochemical gradients increase more than in wild-type cells [26], possibly making carcinoma cells resistant to apoptosis (Fig. 2.4).

4 Altered Copy Number of mtDNAs

Each human cell contains an average of 100–500 mitochondria of which each has two to ten copies of mtDNAs [27]. The normal combination and operation of the respiratory chain require a complete and functional mitochondrial genome. mtDNA function depends not only on the integrity of mtDNAs' molecular structure but also on the copy number of mtDNAs in the cell. In addition to somatic mtDNA mutations including point mutation, deletion, and insertion, alterations of mtDNA copy number vary among human cancers. For instance, the copy number of mtDNAs

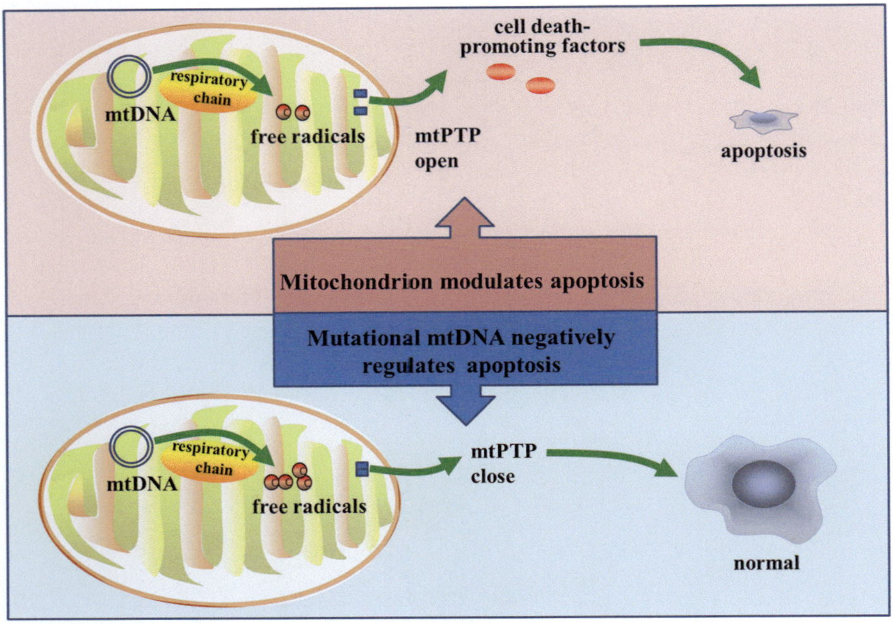

Fig. 2.4 Hypotheses about mtDNA mutations contributing to carcinogenesis. Proteins of electron transport chain are altered by mtDNA mutation, and the function of respiratory chain is damaged. The increased amount of electrons could result in more free radical production. In addition, mitochondria control cell apoptosis by electrochemical gradient ($\triangle\Psi$), produced by mitochondrial electron transport chain. Loss of $\triangle\Psi$ results in opening of the mitochondrial permeability transition pore (mtPTP). In cancer cells with mtDNA mutations, the $\triangle\Psi$ may increase, and mtPTP is closed; thus, the release of cell death-promoting factors was held back. By this way, mtDNA mutations possibly make carcinoma cells resistant to apoptosis

increased in prostate cancer cells [28] or in saliva of head and neck cancer [29] or serum levels of cell-free mtDNAs in testicular cancer [30]. Those data show that mtDNAs are altered not only in cancer cells but also in other body fluids. On the other hand, the contents of mtDNA reduced in 82% of cancerous breast tissues [31] and in hepatocellular carcinoma patients associated with poorer prognosis and shorter 5-year overall survival rates [32]. It indicates that the regulation of mtDNA copy number is out of control, with cancer site specificity [2, 33].

Alterations of mtDNA content also exist in patients with lung cancer. Lee et al. [34] found increased mtDNA content in about 50% of patients with lung cancers, while decreased mtDNA content was found in about 23%. Multiple factors influence alterations of mtDNA content, e.g., cancer type, DNA damage type, or sample size. The reduction of mtDNA copy number seems more common in lung cancer. For example, the copy number of mtDNAs is reduced in lung carcinoma tissues [35] and in advanced stages of the disease correlated with a shorter survival time [36]. There was no correlation between mtDNA content and gender, age, or smoking status [11] [36, 37]. Lee et al. reported that mtDNA contents initially increased with pack-years of smoking and then declined to levels below nonsmokers [38].

Different changes of mtDNA content in cancer may be caused by diverse mechanisms. D-loop region has regulatory elements containing promoters and replication origin and controls mtDNA replication and transcription. Once mutations occurred in D-loop region especially in regulatory elements, the replication and transcription rate of mtDNAs would be changed. The mutation in D-loop region was about 40% hepatocellular cancer, while decreased mtDNA content was more than 70% cancer with D-loop region mutation [34]. It is possible that the mutation in D-loop region decreases mtDNA content or a point mutation in coding region reduces mtDNA copy number. The introduction of mtDNA with mutant A3243G into mtDNA-deficient teratocarcinoma cells results in the depletion of mtDNA content [39].

Another possibility is that mtDNA contents are altered through ROS-mediated pathway. High levels of ROS would lead to mtDNA damage (oxidative damage), and 8-hydroxy-2-deoxyguanosine (8-oxo-G) as an indicator of oxidative damage increased in lung tissues of heavy smokers [27]. The 8-oxoguanine-DNA glycosylase 1 protein and polymerase γ are two key enzymes to repair 8-oxo-G damage in mtDNAs [40, 41]. Devitalized 8-oxoguanine-DNA glycosylase 1 and deficient polymerase γ lead to accumulation of mtDNA mutations responsible for changes of mtDNA content. The deficiency of repair system probably is responsible for the decrease of mtDNA content in ROS-mediated damage. ROS-mediated damage can also be a compensatory mechanism by which the process of mtDNA content to copy increases an oxidative stress [27]. When it comes to oxidative damage, the combination of mtDNA mutations and compensatory mechanism should be considered to estimate the changes of mtDNA copy number (Fig. 2.5).

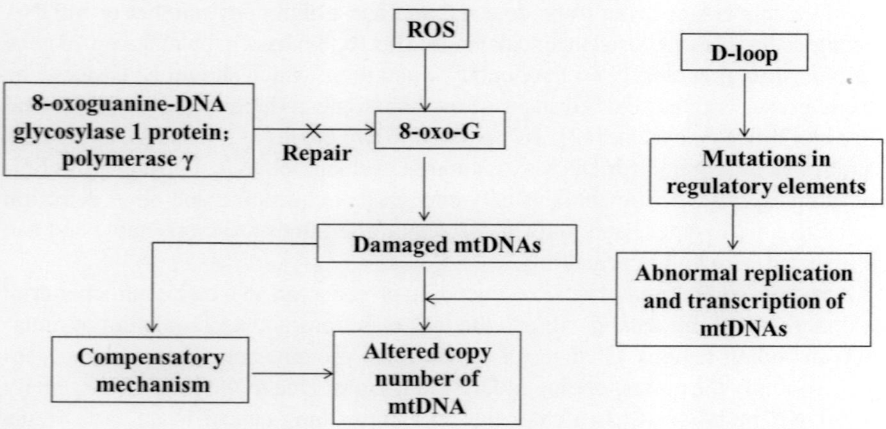

Fig. 2.5 Possible mechanisms of altered copy number of mtDNAs. When mutations occurred in regulatory elements in D-loop region, the replication and transcription rate of mtDNAs would be changed, decreasing mtDNA copy number. Another possibility is that high levels of ROS would lead to production of 8-hydroxy-2-deoxyguanosine (8-oxo-G) as an indicator of oxidative damage. Meanwhile, devitalized 8-oxoguanine-DNA glycosylase 1 and deficient polymerase γ lead to a failure of repairing 8-oxo-G damage in mtDNAs, which is responsible for changes of mtDNA content. ROS-mediated damage can also be a compensatory mechanism to estimate changes of mtDNA copy number

Loss of p53 protein is another reason to reduce mtDNA content through increased accumulation of mtDNA damage [42], since p53 protein regulates the response to DNA damage and can maintain the stability of mtDNAs [43]. In coping with mtDNA damage, p53 protein can move to mitochondria and interact with polymerase γ to strengthen the expression of polymerase γ to repair oxidative damage [42]. The alteration of mtDNA copy number in lung cancer is subject to the mtDNA mutations and oxidative damage. Although there are several causes to interpret the changes of mtDNA content, a clear and comprehensive explanation for increased or decreased mtDNA content of each case remains unknown. More research is required to determine the exact relationship between mtDNA content and lung cancer development.

5 Potential Value of mtDNA as Biomarkers for Lung Cancer Diagnosis

Although nuclear genome detection plays a significant role in cancer screening, the biological role of mtDNA alteration in tumor formation has drawn more and more attention in recent years. The alteration of mtDNA content or sequence mutations has evitable relationship with oncogenesis and becomes potential biomarkers for certain types of cancers [4, 44]. It is widely accepted that mtDNA is more vulnerable to mutagens by genotoxin, oxidative press, or other factors compared with nuclear genome [45, 46]. mtDNA has potential implications for lung cancer diagnosis, since mtDNA is easier to be detected and has a high copy number or mtDNA mutations have strong resistance to damage. The 16,596 base pairs make up 16 gene regions, arranging closely on the double-strand loop, which should be easier to be characterized with current advances of methodologies. There are several thousand copies of mtDNA in a cell [47], as compared to two copies of nuclear genome. Less samples were required for DNA extraction to evaluate mtDNA. For precious clinical samples, mtDNA content is usually enough for sequencing and other detection methods. On the other hand, mtDNA sequence alterations are more stable and can be detected with high repeatability.

Abnormal expression of mtDNA encoding proteins can also cause mitochondrial dysfunction and cell damage, due to the lack of introns and accumulation of mutations in coding regions. Changes of downstream proteins and clinical phenotypes can also imply the corresponding mtDNA mutations. Due to the high heterogeneity of mtDNA mutations, it is a challenge to identify lung cancer in an early stage. Sanger sequencing is a common type of mitochondrial sequencing technique, but as to heterogeneity detection, the next generation of sequencing will be more competitive. In addition, there are quantitative PCR techniques, array chips [48], and other measurement techniques, which can also achieve good detection results.

It is necessary for further study to identify and validate the function of mtDNA mutations in initiation and progression of lung cancer in order to be helpful to develop lung cancer detection and treatment strategies. It is also a challenge to

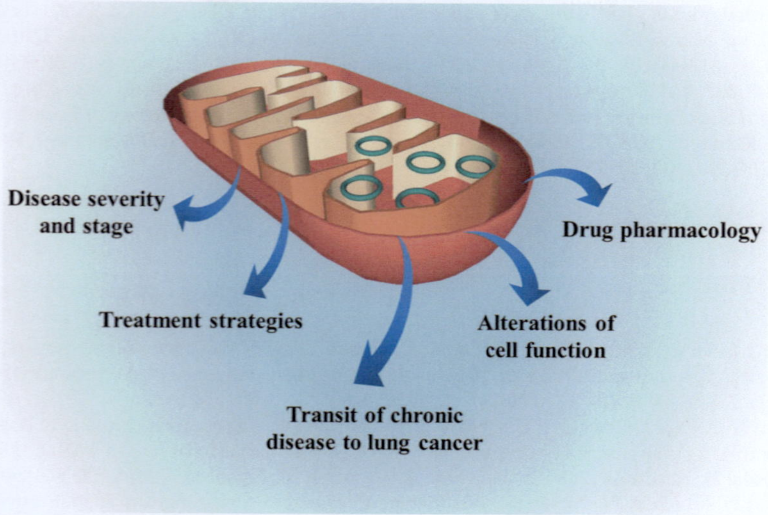

Fig. 2.6 Potential value of mtDNA as biomarkers. mtDNA has a prospect of becoming biomarkers to develop lung cancer detection and treatment strategies, to monitor disease severity and stage, to monitor the transit of chronic diseases especially chronic obstructive pulmonary disease to lung cancer, and to describe alterations of cell functions induced by mtDNA mutations. In addition, it will be a power if mtDNA alterations can act as a biomarker to measure the cell behavior in drug pharmacology

apply mtDNA changes for clinical application as biomarkers, especially to monitor the transit of chronic obstructive pulmonary disease to lung cancer [49–51], to integrate clinical phenotype mtDNA mutations for dynamic monitoring of disease severity and stage [52–54], and to describe alterations of cell functions induced by mtDNA mutations [55–58]. Mitochondrion is one of the most important organelles to regulate and control drug pharmacological profiles. It will be a power if mtDNA alterations can act as a biomarker to measure the cell behavior in drug metabolism [59–61], pharmacokinetics [62–64], and sensitivity to drug toxicity [65–68] (Fig. 2.6).

6 Conclusion

Alterations of mtDNA possibly are related to carcinogenesis of lung cancer, as well as their mechanisms for promoting the development of lung cancer, including alteration of mtDNA copy number and sequence mutations. Although the definite mechanism by which mtDNA is altered is to be further confirmed, it will be helpful in developing diagnostic and therapeutic strategies for lung cancer.

Acknowledgments The work was supported by Zhongshan Distinguished Professor Grant (XDW), the National Nature Science Foundation of China (91230204, 81270099, 81320108001, 81270131, 81300010), the Shanghai Committee of Science and Technology (12JC1402200, 12431900207, 11410708600, 14431905100). Operation funding of Shanghai Institute of Clinical Bioinformatics, Ministry of Education for Academic Special Science and Research Foundation for PhD Education (20130071110043), and National Key Research and Development Program (2016YFC0902400, 2017YFSF090207).

References

1. Ding C, Li R, Wang P, Jin P, Li S, Guo Z. Identification of sequence polymorphisms in the D-loop region of mitochondrial DNA as a risk factor for lung cancer. Mitochondrial DNA. 2012;23:251–4. PMID:22708867
2. Hsu H-C, Li S-J, Chen C-Y, Chen M-F. Eicosapentaenoic acid protects cardiomyoblasts from lipotoxicity in an autophagy-dependent manner. Cell Biol Toxicol. 2017:1–13. PMID:28741157
3. Kazdal D, Harms A, Endris V, Penzel R, Kriegsmann M, Eichhorn F, Muley T, Stenzinger A, Pfarr N, Weichert W. Prevalence of somatic mitochondrial mutations and spatial distribution of mitochondria in non-small cell lung cancer. Br J Cancer. 2017.; PMID:28557978
4. Fliss MS, Usadel H, Caballero OL, Wu L, Buta MR, Eleff SM, Jen J, Sidransky D. Facile detection of mitochondrial DNA mutations in tumors and bodily fluids. Science. 2000;287:2017–9. PMID:10720328
5. Richard SM, Bailliet G, Páez GL, Bianchi MS, Peltomäki P, Bianchi NO. Nuclear and mitochondrial genome instability in human breast cancer. Cancer Res. 2000;60:4231–7. PMID:10945635
6. Witte J, Lehmann S, Wulfert M, Yang Q, Röher HD. Mitochondrial DNA mutations in differentiated thyroid cancer with respect to the age factor. World J Surg. 2007;31:51–9. PMID:17171498
7. Yuan Y, Wang W, Li H, Yu Y, Tao J, Huang S, Zeng Z. Nonsense and missense mutation of mitochondrial ND6 gene promotes cell migration and invasion in human lung adenocarcinoma. BMC Cancer. 2015;15:346. PMID:25934296
8. Pagani IS, Kok CH, Saunders VA, Van der Hoek MB, Heatley SL, Schwarer AP, Hahn CN, Hughes TP, White DL, Ross DMA. Method for next-generation sequencing of paired diagnostic and remission samples to detect mitochondrial DNA mutations associated with leukemia. J Mol Diagn. 2017.; PMID:28732215
9. Bao L, Zhang Y, Wang J, Wang H, Dong N, Su X, Xu M, Wang X. Variations of chromosome 2 gene expressions among patients with lung cancer or non-cancer. Cell Biol Toxicol. 2016;32:419–35. PMID:27301951
10. Bao L, Diao H, Dong N, Su X, Wang B, Mo Q, Yu H, Wang X, Chen C. Histone deacetylase inhibitor induces cell apoptosis and cycle arrest in lung cancer cells via mitochondrial injury and p53 up-acetylation. Cell Biol Toxicol. 2016;32:469–82. PMID:27423454
11. Fang X, Netzer M, Baumgartner C, Bai C, Wang X. Genetic network and gene set enrichment analysis to identify biomarkers related to cigarette smoking and lung cancer. Cancer Treat Rev. 2013;39:77–88. PMID:22789435
12. Anderson S, Bankier AT, Barrell BG, de Bruijn MH, Coulson AR, Drouin J, Eperon IC, Nierlich DP, Roe BA, Sanger F. Sequence and organization of the human mitochondrial genome. Nature. 1981;290:457–65. PMID:7219534
13. Medvedev R, Hildt E, Ploen D. Look who's talking—the crosstalk between oxidative stress and autophagy supports exosomal-dependent release of HCV particles. Cell Biol Toxicol. 2017:1–21. PMID:27987184

14. Kang D, Hamasaki N. Mitochondrial oxidative stress and mitochondrial DNA. Clin Chem Lab Med. 2003;41:1281–8. PMID:14580153
15. Zhu L-Z, Hou Y-J, Zhao M, Yang M-F, Fu X-T, Sun J-Y, Fu X-Y, Shao L-R, Zhang H-F, Fan C-D. Caudatin induces caspase-dependent apoptosis in human glioma cells with involvement of mitochondrial dysfunction and reactive oxygen species generation. Cell Biol Toxicol. 2016;32:333–45. PMID:27184666
16. Yeh JJ, Lunetta KL, van Orsouw NJ, Jr MF, Mutter GL, Vijg J, Dahia PL, Eng C. Somatic mitochondrial DNA (mtDNA) mutations in papillary thyroid carcinomas and differential mtDNA sequence variants in cases with thyroid tumours. Oncogene. 2000;19:2060–6. PMID:10803467
17. Morais R, Desjardins P, Turmel C, Zinkewichpéotti K. Development and characterization of continuous avian cell lines depleted of mitochondrial DNA. Vitro Cell & Dev Biol J Tissue Cult Assoc. 1988;24:649–58. PMID:2840430
18. Suzuki M, Toyooka S, Miyajima K, Iizasa T, Fujisawa T, Bekele NB, Gazdar AF. Alterations in the mitochondrial displacement loop in lung cancers. Clin Cancer Res. 2003;9:5636–41. PMID:14654546
19. Kim W-H, Shen H, Jung D-W, Williams DR. Some leopards can change their spots: potential repositioning of stem cell reprogramming compounds as anti-cancer agents. Cell Biol Toxicol. 2016;32:157–68. PMID:27156576
20. Nishikawa M, Nishiguchi S, Shiomi S, Tamori A, Koh N, Takeda T, Kubo S, Hirohashi K, Kinoshita H, Sato E. Somatic mutation of mitochondrial DNA in cancerous and noncancerous liver tissue in individuals with hepatocellular carcinoma. Cancer Res. 2001;61:1843–5. PMID:11280735
21. Chatterjee A, Mambo E, Sidransky D. Mitochondrial DNA mutations in human cancer. Oncogene. 2006;25:4663–74. PMID:16892080
22. Wallace DC. Mitochondrial diseases in man and mouse. Science. 1999;283:1482–8. PMID:10066162
23. Khansari N, Shakiba Y, Mahmoudi M. Chronic inflammation and oxidative stress as a major cause of age- related diseases and cancer. Recent Patents Inflamm Allergy Drug Discov. 2009;3:73–80. PMID:19149749
24. Pelicano H, Carney D, Huang P. ROS stress in cancer cells and therapeutic implications. Drug Resist Updat. 2004;7:97–110. PMID:15158766
25. Brandon M, Baldi P, Wallace DC. Mitochondrial mutations in cancer. Oncogene. 2006;25:4647–62. PMID:16892079
26. Kulawiec M, Owens KM, Singh KK. Cancer cell mitochondria confer apoptosis resistance and promote metastasis. Cancer Biol Ther. 2014;8:1378–85. PMID:19556849
27. Iii HDH, Liu CS, Rothman N, Weinstein SJ, Bonner MR, Shen M, Lim U, Virtamo J, Cheng W, Albanes D. Mitochondrial DNA copy number and lung cancer risk in a prospective cohort study. Carcinogenesis. 2010;31:847–9. PMID:20176654
28. Mizumachi T, Muskhelishvili L, Naito A, Furusawa J, Fan CY, Siegel ER, Kadlubar FF, Kumar U, Higuchi M. Increased distributional variance of mitochondrial DNA content associated with prostate cancer cells as compared with normal prostate cells. Prostate. 2008;68:408–17. PMID:18196528
29. Jiang WW, Masayesva B, Zahurak M, Carvalho AL, Rosenbaum E, Mambo E, Zhou S, Minhas K, Benoit N, Westra WH. Increased mitochondrial DNA content in saliva associated with head and neck cancer. Clin Cancer Res. 2005;11:2486–91. PMID: 15814624
30. Ellinger J, Albers P, Müller SC, Von RA, Bastian PJ. Circulating mitochondrial DNA in the serum of patients with testicular germ cell cancer as a novel noninvasive diagnostic biomarker. BJU Int. 2009;104:48–52. PMID:19154496
31. Fan AX, Radpour R, Haghighi MM, Kohler C, Xia P, Hahn S, Holzgreve W, Zhong XY. Mitochondrial DNA content in paired normal and cancerous breast tissue samples from patients with breast cancer. J Cancer Res Clin. 2009;135:983–9. PMID:19125299

32. Datta S, Majumder M, Biswas NK, Sikdar N, Roy B. Increased risk of oral cancer in relation to common Indian mitochondrial polymorphisms and autosomal GSTP1 locus. Cancer. 2007;110:1991–9. PMID:17886251
33. Malik AN, Czajka A. Is mitochondrial DNA content a potential biomarker of mitochondrial dysfunction? Mitochondrion. 2013;13:481–92. PMID:23085537
34. Lee HC, Li SH, Lin JC, CC W, Yeh DC, Wei YH. Somatic mutations in the D-loop and decrease in the copy number of mitochondrial DNA in human hepatocellular carcinoma. Mutat Res. 2004;547:71–8. PMID:15013701
35. Wang Y, Liu VWS, Xue WC, Cheung ANY, Ngan HYS. Association of decreased mitochondrial DNA content with ovarian cancer progression. Br J Cancer. 2006;95:1087–91. PMID:17047655
36. Xu H, He W, Jiang HG, Zhao H, Peng XH, Wei YH, Wei JN, Xie CH, Liang C, Zhong YH. Prognostic value of mitochondrial DNA content and G10398A polymorphism in non-small cell lung cancer. Oncol Rep. 2013;30:3006–12. PMID:24101028
37. Wang H, Dai J. Changes on mitochondrial DNA content in non-small cell lung cancer. Chin J Lung Cancer. 2011;14:141. PMID:21342645
38. Lee HC, CY L, Fahn HJ, Wei YH. Aging- and smoking-associated alteration in the relative content of mitochondrial DNA in human lung. FEBS Lett. 1998;441:292. PMID:9883902
39. Turner CJ, Granycome C, Hurst R, Pohler E, Juhola MK, Juhola MI, Jacobs HT, Sutherland L, Holt IJ. Systematic segregation to mutant mitochondrial DNA and accompanying loss of mitochondrial DNA in human NT2 teratocarcinoma cybrids. Genetics. 2005;170:1879–85. PMID:15944344
40. Singh KK, Sigala B, Sikder HA, Schwimmer C. Inactivation of Saccharomyces cerevisiae OGG1 DNA repair gene leads to an increased frequency of mitochondrial mutants. Nucleic Acids Res. 2001;29:1381–8. PMID:11239005
41. Zhang D, Mott JL, Chang SW, Denniger G, Feng Z, Zassenhaus HP. Construction of transgenic mice with tissue-specific acceleration of mitochondrial DNA mutagenesis. Genomics. 2000;69:151–61. PMID:11031098
42. Higuchi M. Regulation of mitochondrial DNA content and cancer. Mitochondrion. 2007;7:53–7. PMID:17320491
43. Sánchez AB, González-Sistal A. Novel role of p53 in maintaining mitochondrial genetic stability through interaction with DNA Pol γ. EMBO J. 2005;24:3482–92. PMID:16163384
44. Isaacs C, Cavalli LR, Cohen Y, Pennanen M, Shankar LK, Freedman M, Singh B, Liu M, Gallagher A, Rone JD. Detection of LOH and mitochondrial DNA alterations in ductal lavage and nipple aspirate fluids from high-risk patients. Breast Cancer Res Treat. 2004;84:99–105. PMID:14999140
45. Bandy B, Davison AJ. Mitochondrial mutations may increase oxidative stress: implications for carcinogenesis and aging? Free Radic Biol Med. 1990;8:523–39. PMID:2193852
46. Shay JW, Werbin H. Are mitochondrial DNA mutations involved in the carcinogenic process? Mutat Res. 1987;186:149–60. PMID:3306359
47. Chiu RW, Chan LY, Lam NY, Tsui NB, Ng EK, Rainer TH, Lo YM. Quantitative analysis of circulating mitochondrial DNA in plasma. Clin Chem. 2003;49:719–26. PMID:12709361
48. Sui G, Zhou S, Wang J, Canto M, Lee EE, Eshleman JR, Montgomery EA, Sidransky D, Califano JA, Maitra A. Mitochondrial DNA mutations in preneoplastic lesions of the gastrointestinal tract: a biomarker for the early detection of cancer. Mol Cancer. 2006;5:73. PMID:17166268
49. Wang X. New biomarkers and therapeutics can be discovered during COPD-lung cancer. Transition. 2016.; PMID:27405768
50. Gu J, Wang X. New future of cell biology and toxicology: thinking deeper. 2016; PMID:26874518
51. Wang X. CBT profiles of cabozantinib approved for advanced renal cell carcinomas. Cell Biol Toxicol. 2016;32:259–61. PMID:27383755

52. Chen C, Shi L, Li Y, Wang X, Yang S. Disease-specific dynamic biomarkers selected by integrating inflammatory mediators with clinical informatics in ARDS patients with severe pneumonia. Cell Biol Toxicol. 2016;32:169–84. PMID:27095254
53. Shi L, Zhu B, Xu M, Wang X. Selection of AECOPD-specific immunomodulatory biomarkers by integrating genomics and proteomics with clinical informatics. Cell Biol Toxicol. 2017:1–15. PMID:28779230
54. Arumugam P, Samson A, Ki J, Song JM. Knockdown of clusterin alters mitochondrial dynamics, facilitates necrosis in camptothecin-induced cancer stem cells. Cell Biol Toxicol. 2017;33:307–21. PMID:28064403
55. Mossoba ME, Flynn TJ, Vohra S, Wiesenfeld PL, Sprando RL. Human kidney proximal tubule cells are vulnerable to the effects of Rauwolfia serpentina. Cell Biol Toxicol. 2015;31:285–93. PMID:26838987
56. Wu Y, Geng X-C, Wang J-F, Miao Y-F, Lu Y-L, Li B. The HepaRG cell line, a superior in vitro model to L-02, HepG2 and hiHeps cell lines for assessing drug-induced liver injury. Cell Biol Toxicol. 2016;32:37–59. PMID:27027780
57. Fathi H, Ebrahimzadeh MA, Ziar A, Mohammadi H. Oxidative damage induced by retching; antiemetic and neuroprotective role of Sambucus ebulus L. Cell Biol Toxicol. 2015;31:231–9. PMID:26493312
58. Zhu D, Liu Z, Pan Z, Qian M, Wang L, Zhu T, Xue Y, Wu D. A new method for classifying different phenotypes of kidney transplantation. Cell Biol Toxicol. 2016;32:323–32. PMID:27278387
59. Kikuchi S, Ninomiya T, Kohno T, Kojima T, Tatsumi H. Cobalt inhibits motility of axonal mitochondria and induces axonal degeneration in cultured dorsal root ganglion cells of rat. Cell Biol Toxicol. 2017:1–15. PMID:28656345
60. Kaarniranta K, Tokarz P, Koskela A, Paterno J, Blasiak J. Autophagy regulates death of retinal pigment epithelium cells in age-related macular degeneration. Cell Biol Toxicol. 2017:1–16. PMID:27900566
61. Chiang H-C, Wang C-H, Yeh S-C, Lin Y-H, Kuo Y-T, Liao C-W, Tsai F-Y, Lin W-Y, Chuang W-H, Tsou T-C. Comparative microarray analyses of mono (2-ethylhexyl) phthalate impacts on fat cell bioenergetics and adipokine network. Cell Biol Toxicol. 2017:1–16. PMID:28083810
62. Sanyal S, Das P, Law S. Effect of chronic pesticide exposure on murine cornea: a histopathological, cytological and flow cytometric approach to study ocular damage by xenobiotics. Cell Biol Toxicol. 2016;32:7–22. PMID:26897134
63. Soltani B, Ghaemi N, Sadeghizadeh M, Najafi F. Curcumin confers protection to irradiated THP-1 cells while its nanoformulation sensitizes these cells via apoptosis induction. Cell Biol Toxicol. 2016;32:543–61. PMID:27473378
64. Keta O, Bulat T, Golić I, Incerti S, Korać A, Petrović I, Ristić-Fira A. The impact of autophagy on cell death modalities in CRL-5876 lung adenocarcinoma cells after their exposure to γ-rays and/or erlotinib. Cell Biol Toxicol. 2016;32:83–101. PMID:27026538
65. Ganta KK, Mandal A, Chaubey B. Depolarization of mitochondrial membrane potential is the initial event in non-nucleoside reverse transcriptase inhibitor efavirenz induced cytotoxicity. Cell Biol Toxicol. 2017;33:69–82. PMID:27639578
66. Giromini C, Rebucci R, Fusi E, Rossi L, Saccone F, Baldi A. Cytotoxicity, apoptosis, DNA damage and methylation in mammary and kidney epithelial cell lines exposed to Ochratoxin A. Cell Biol Toxicol. 2016;32:249–58. PMID:27184667
67. Tlotleng N, Vetten MA, Keter FK, Skepu A, Tshikhudo R, Gulumian M. Cytotoxicity, intracellular localization and exocytosis of citrate capped and PEG functionalized gold nanoparticles in human hepatocyte and kidney cells. Cell Biol Toxicol. 2016;32:305–21. PMID:27184667
68. Divolis G, Mavroeidi P, Mavrofrydi O, Papazafiri P. Differential effects of calcium on PI3K-Akt and HIF-1α survival pathways. Cell Biol Toxicol. 2016;32:437–49. PMID:27344565

Fangming Liu, A member of Clinical Science Institute of Fudan University Zhongshan Hospital. She got a master's degree in Fuzhou University and completed scientific research projects in Shanghai Institute of Microsystem and Information Technology Chinese Academy of Sciences during graduate student period. Her previous studies are focused on early diagnosis of lung cancer, through exploring relationship between promoter methylation and early-stage lung cancer and establishing the technology platform for tumor marker protein detection in body fluids based on biological nanosensor. Her current study is focused on lung cancer metabolism and tumor immunology.

David E. Sanin, PhD, Postdoctoral Research Associate, Max Planck Institute of Immunobiology and Epigenetics, Freiburg in Breisgau, Germany. After receiving a scholarship granted by the National Institute of Science and Technology of Colombia (COLCIENCIAS), Dr. Sanin did his PhD in the University of York, with Dr. Adrian Mountford. His doctoral work focused on understanding the regulation and role of IL-10 production during the early stages of helminth skin infection.

Dr. Sanin has received prizes for the quality of his work awarded by the University of York and the American Society of Tropical Medicine and Hygiene. In addition to presenting his works in multiple international conferences, in some of which he has been a guest speaker, Dr. Sanin has published in several international journals such as *PLOS Pathogens* and the *Journal of Immunology*. Currently Dr. Sanin works on the metabolic regulation of antigen-presenting cells with Dr. Edward Pearce.

Xiangdong Wang, MD, PhD, is a distinguished professor of Medicine, director of Shanghai Institute of Clinical Bioinformatics, executive director of Clinical Science Institute of Fudan University Zhongshan Hospital, director of Fudan University Center of Clinical Bioinformatics, deputy director of Shanghai Respiratory Research Institute, and visiting professor of King's College London. His main research is focused on clinical bioinformatics, disease-specific biomarkers, chronic lung diseases, cancer immunology, and molecular and cellular therapies. He is the author of more than 200 scientific publications with the impact factor about 900, citation number about 6918, h-index 46, i10-index 181, and cited journal impact factor about 7000.

Chapter 3
Approach, Application, and Bioethics of mtDNA Sequencing in Cancer

Mengjia Qian, Claudio Spada, and Xiangdong Wang

Abstract Mitochondrial DNA (mtDNA) is more vulnerable to mutations and associated with many solid tumors. Through mtDNA sequencing, we can find useful information on the mutations implicated in diseases and can better define the impact of mitochondrial dysfunction on the process of carcinogenesis. In current article, we will discuss the current approaches of mtDNA sequencing and the challenges we should overcome, their applications in various cancers, and the potential bioethics problems we should face in the application of mtDNA sequencing in clinical diagnosis and treatment.

Keywords mtDNA sequencing • Cancer • Bioethics

1 Introduction

Mitochondrial diseases (MDs) is caused by a dysfunction of the mitochondrial respiratory chain which is associated with the mutations of genes encoding either nuclear DNA (nDNA) or mtDNA (mtDNA) [1]. mtDNA mutations not only lead to the mitochondrial diseases but also affect a wide range of cancers. mtDNA is a 16.319 bp closed-circle genome that encodes 13 respiratory chain subunits, 2 rRNAs and 22 tRNAs [2]. Compared with nDNA, mtDNA lacks repair capacity and is three to ten times more vulnerable to mutations [3]. Through mtDNA sequencing, we can

M. Qian
Zhongshan Hospital Institute of Clinical Science, Fudan University, Shanghai Medical College, Shanghai, China

Shanghai Institute of Clinical Bioinformatics, Shanghai, China

X. Wang (✉)
Zhongshan Hospital Institute of Clinical Science, Fudan University, Shanghai Medical College, Shanghai, China
e-mail: xiangdong.wang@clintransmed.org

C. Spada
Shanghai Institute of Clinical Bioinformatics, Shanghai, China

H. Sun, X. Wang (eds.), *Mitochondrial DNA and Diseases*, Advances in Experimental Medicine and Biology 1038,
https://doi.org/10.1007/978-981-10-6674-0_3

find useful information on those mutations and define the impact of mitochondrial dysfunction on the process of carcinogenesis, to improve the diagnosis and treatment in cancers. Here, we will discuss potential applications of mtDNA sequencing in the diagnosis and treatment in cancers including lung cancer.

mtDNA sequencing is measured with the next-generation sequencing (NGS). Currently, NGS is used in the detection of mitochondrial genome. There are many different sequencing platforms and methods to approve the sequencing, although a number of challenges need to be faced on the data analysis and library preparation. In the current article, we will summarize main approaches of mtDNA sequencing and discuss proper ways to avoid potential problems we may face in the process of mtDNA sequencing. With the development of the mtDNA sequencing in clinic, we should be aware of the ethics problems they brought. To some extent, more and more mutations will be found to be related with cancer, people will have the idea of editing their genes or even design babies. Another problem is with more genome be sequenced, we have to promise the safety of these large volume of data and information they may tell. The present article will discuss the bioethics of mtDNA sequencing and try to find proper ways to handle this issue (Fig. 3.1).

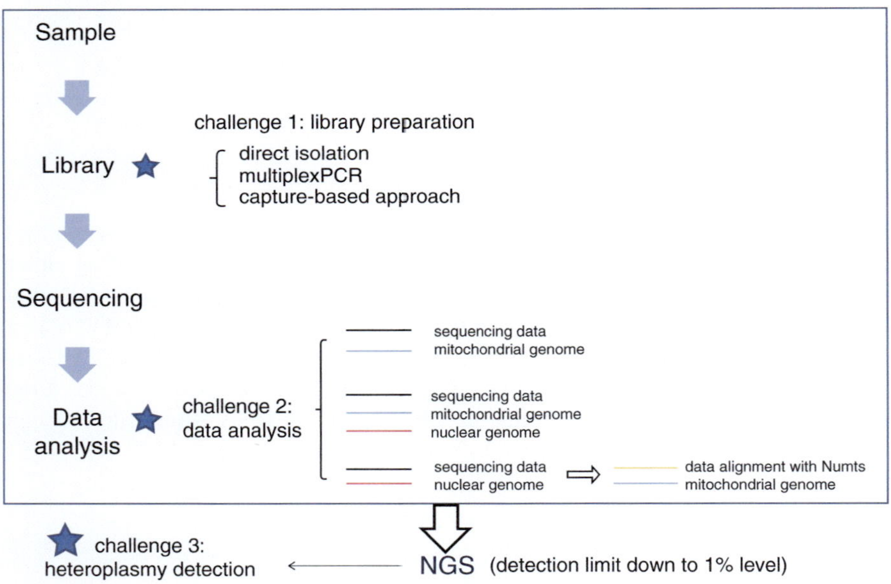

Fig. 3.1 Challenges of current mtDNA sequencing approaches
The main process of mtDNA sequencing includes sample acquisition, library preparation, sequencing, and data analysis. However, there are still some challenges we should overcome in our current sequencing approaches. The first one is how to prepare high-quality libraries; we listed several ways, and each one has its own advantages and disadvantages. The second one is how to identify mitochondrial genome from nuclear genome through data analysis. We summarized three ways. One is align the raw data directly against the mitochondrial genome; second one is against both the nuclear genome and mitochondrial genome at the same time; third one against the nuclear genome first, then against the mitochondrial genome. All the methods are based on NGS, which can make us overcome the third challenge, how to detect low level heteroplasmy. NGS can reduce the detection limit as low as 1% level

2 Approaches of mtDNA Sequencing: Opportunities and Challenges

How to Perfectly Prepare Sequencing Libraries? The next-generation sequencing assay for mtDNA detection has become available recently [4]. For example, duplex sequencing is one of the methodologies of NGS to detect very-low-frequency genetic alterations [5]. It has been successfully applied in mtDNA sequencing [6]. The main process of mtDNA sequencing in clinic mainly includes sample acquisition, library preparation, sequencing, data generation and analysis, variant alleles report, and interpretation of clinical significance to the doctors [7]. Whole mtDNA genome sequencing can describe all known and potentially novel mutations related with cancer in a single sequencing run [8]. However, there are still many challenges exist in mtDNA sequencing that we should overcome and make it more suitable in clinic.

There are mainly three methods to derive mtDNA fragments for sequencing, including direct isolation of mitochondrion [9], multiplex PCR [10], and capture-based approach with mtDNA-specific oligonucleotides [11]. Isolation of mitochondria to derive mtDNA is only suitable to small number of samples, while the other two methods cannot get rid of the affection of nuclear encoded mitochondrial pseudogenes (Numts) [12], especially in the samples that mtDNA occurring at a low frequency [13]. Moreover, the process of library construction in capture-based approach will easily cause high false positives in mutation callings [14]. In order to avoid false variant reports, the optimal method is differential centrifugation isolation followed by exonuclease digest. Such method can get more than 35% mtDNA reads and yielded hundreds-fold enrichment over baseline, especially suitable for small amounts of samples [15].

MitoRCA-seq is a new library construction approach suitable for small amounts of samples and is a PCR-free procedure with only 1 ng of total DNA and rolling circle amplification. The method can amplify full-length mtDNA and reduce Numt contamination simultaneously, to easily detect low-frequency point mutations of mtDNA [16]. However, the methods of library construction are various with advantages and disadvantages. More studies should be done to explore a perfect way to prepare sequencing libraries that not only disregard the Numts but also have low false positives. Importantly, it can be done with small amounts of samples and little artificial time.

How to Identify Mitochondrial Genome from Nuclear Genome? The current sequencing approaches mainly focus on resequencing both the targeted mtDNA and candidate nuclear genes [17]. Co-amplification of nDNA in the process of sequencing will easily affect the analysis of mtDNA sequencing, especially in low-cellularity samples [18]. So, the difficulty in dealing with the sequencing data is how to identify mitochondrial genome from nuclear copies of the mitochondrial genomes. Three main methods to handle this issue are to align the raw reads directly against the mitochondrial genome, against the nuclear and mitochondrial genomes at the same time, and against Numts first and then align only the nonaligned reads to the mitochondrial genome, while those approaches will largely ignore the reads that really derive from the Numts which actually introduce false heteroplasmic variability, will easily cause incorrect alignment, and will largely decrease the coverage of the mitochondrial genome and reduce the chance of detecting true variants [19].

How to Detect Heteroplasmic Mutations? In mitochondrial genome sequencing, there is a challenge to detect low-abundance point mutations and heteroplasmic mutations [20]. A typical human cell contains about 100 mitochondria, and 1 mitochondrion has an average of 5 mtDNA molecules; thus, 1 human cell totally has about 500 copies of mtDNA [21]. The heteroplasmy describes a phenomenon that some mitochondrial genomes have mutations within a single cell, while others not. Heteroplasmy has the ability to promote tumor growth [22] and predict whether mutant phenotypes will be expressed [23]. One of the challenges is to detect heteroplasmic mutations. NGS can be used to detect the heteroplasmy to a higher level of detail and compare with Sanger sequencing. The results are that the limit of Sanger sequencing is 10%, while NGS is 1% [24]. When the mean depth reached 5000×, NGS is able to have enough number of reads to detect true heteroplasmy even in badly covered regions [25].

However, data analysis through different software will generate different results, e.g., mtDNA-Server offers the analysis of mtDNA studies. The workflow of the analysis mainly includes parallel read alignment, heteroplasmy detection, artifact or contamination identification, variant annotation, and quality control metrics. The specificity and sensitivity are evaluated to detect heteroplasmy down to 1% level [26].

Future Directions of Method Development mtDNA sequence can be obtained from whole exome sequencing (WES), whole genome sequencing (WGS), and high-throughput RNA sequencing technology (RNA-seq) data [19]. Mitochondrial genome found in exome sequencing data is for the detection of pathogenic point mutations possible through comprehensive single diagnostic test [27]. Due to the high copy number of mtDNA per cell, the average coverage of the mitochondrial genome from exome sequencing can reach nearly full [19]. Despite the repeated poly-C sequence errors, the exome sequencing error is much lower when the sequencing depth reaches >20-fold as compared to Sanger sequencing [28].The detection limit of heteroplasmy obtained from exome sequencing is only 1% [29]. If deeper sequencing is performed, it is likely to detect lower levels of heteroplasmy in mitochondrial genome [28]. mtDNA sequencing data obtained from WGS are useful in the detection of copy number variants and intronic mutations [30]. However, this approach is restricted by the high cost.

mtDNA copy number had a significant correlation with mitochondrial gene expression levels measured by RNA-seq [31]. Genome sequencing has been performed in a large number of mitochondrial studies, of which there are few studies focusing on mitochondrial transcriptomes. This will be changed with the development of the technology of RNA-seq. Transcriptomes of intergenic regions of mitochondrial genomes can also be active [32]. More mitochondrial transcriptomes will be sequenced due to the improvement of operation and cost efficiency [33].

Furthermore, the development of data analysis of RNA-seq should also be accelerated. Recently developed package of method and software called MINTmap could analyze nuclear and mitochondrial RNA fragments from RNA-seq data. It can calculate both raw and normalized abundance of the RNAs and report the false positives of them [34]. mtDNA heteroplasmy can be found in tissues and acted on some somatic mtDNA mutations [35]. So, mtDNA sequencing at single-cell level becomes much more important in future direction to detect mitochondrial genome through WES, WGS, and RNA-seq, especially at single-cell level.

3 Applications of mtDNA Sequencing in Cancer

The mitochondrial medicine includes nDNA and mtDNA control, level of heteroplasmy, tissue energy demand, maternal inheritance, and mitotic segregation [36]. Of those, the detection of mitochondrial genome has better clinical sensitivity and impact [37]. NGS approach is quicker and cheaper in diagnosis of mitochondrial mutations. Through analyzing thousands of NGS mitochondrial genome data, we can find revolutionary diagnostic process in mitochondrial medicine and understand unresolved cases and previously unknown genetic and pathogenic mechanisms [38].

The overproduction of reactive oxygen species (ROS) from mitochondria in tumor cells can induce genomic instability and modify gene expression and signaling pathways contributing to tumorigenesis [39] and leading to the development and cancer recurrence and therapy resistance [40]. For example, mitochondrial D310 mutation was reported to serve as an ideal marker for tumor clonality in many solid tumors [41]. Somatic mtDNA mutations and altered mtDNA copy number exist in nearly all ovarian, breast, kidney, urinary bladder, pancreas, colorectal, and lung tumors [42]. It can be examined to identify tumor stage and metastasis stage [43], especially for bladder, breast, and kidney cancer which have significant depletion of mtDNA [40].

Lung Cancer In the beginning of neoplastic transformation stage, mitochondrial dysfunction results in the excessive production of ROS, leading to mtDNA damage and mutations [44]. mtDNA mutation is a marker of poor survival in cancer prognosis which was proved by comparing the frequency of mtDNA mutations in patients with non-small cell lung cancer (NSCLC) at different stages. Currently, most reported mutations have been focused on the D-loop region of mtDNA [45], which is a 1124 bp noncoding region related with mtDNA expression, replication, and transcription [46] and may play a key role in carcinogenesis [47]. Mutations in this region may lead to the change on mtDNA copy number and gene expression [48]. D-loop region had a relatively higher frequency of somatic mutations among the whole mtDNA genomes in tissue samples from patients with lung cancer were collected and sequenced the whole mtDNA genomes [49]. D-loop mutation rate of mtDNA in the lung cancer increased by measuring mtDNA sequencing in the exhaled breath condensate of patients. Those results suggest that mtDNA sequencing can be an invasive method to identify specific markers and mtDNA changes could be the specific markers for the carcinogenesis of lung cancer [50].

Moreover, mtDNA mutations are highly associated with epidermal growth factor receptor gene mutations in patients with lung cancer [51], suggesting that mtDNA and nDNA mutations contribute to the progression of lung cancer. mtDNA could be selected as markers in cancer when measuring a complete mitochondrial genome sequencing in squamous cell carcinoma [52]. Study on whole mitochondrial genome sequencing of 26 pair of tumor and non-neoplastic tissue samples was extended by reviewing published data of 326 cases to analyze the potential role of mitochondrial mutations and frequency in lung cancer [53]. They demonstrate that mtDNA mutations found in NSCLC are rather passenger mutations and their suitability as relevant targets for anticancer therapy should be questioned and furthermore validated.

Other Cancers mtDNA D-loop sequencing and its reliability were assessed in head-and-neck tumors [54]. mtDNA sequencing is suggested as a useful and precise method to identify different tumor clones and evaluate the relationship between primary cancer and lymph node metastasis [55]. In order to evaluate a new potential therapy for renal carcinoma, 2 mutated mitochondrial molecules, COX1 and ND5, were selected after 13 coding genes were sequenced and aligned in kidney cancer [56]. Derived peptides from these mutated genes were measured and synthesized to develop a tumor mitochondria vaccine to treat cancer [57]. A chemical compound, called organometallic half-sandwich Os (II), was tested to treat ovarian cancer both in vitro and in vivo. Through whole-transcriptome sequencing, three mutations in the mitochondrial genome, p.1257 V, p.N447S, and p.L517P, were identified for coding ND5, a proton pump to maintain the coupling gradient in mitochondria. Consequently, induced overproduction of ROS led to downstream DNA damage, defining a promising candidate for ovarian cancer treatment [56].

mtDNA mutations in breast cancer are caused by a deficiency in the electron transport chain function, and change of ROS levels [58] might be correlated with increased mtDNA copy number [59] or decreased mtDNA copy number [43]. High copy number of mtDNA was detected in body fluids which are noninvasive and very accessible, especially in urinary bladder carcinoma [60]. Whole mtDNA sequencing using body fluids is recommended for both initial tumor screening and tumor surveillance of urinary bladder carcinoma, better than other current cytological and molecular methods in the patient follow-up. D310 mutation in other cancers has a hot spot, e.g., head-and-neck tumor (37%), breast cancer (29%), or rectal cancer (28%) [61], while not found in bladder cancer [62]. Mitochondrial C16069T polymorphism was considered as bladder cancer-specific mtDNA mutation.

Colorectal Cancer Higher levels of heteroplasmy and decreased mtDNA copy number are correlated with colorectal adenocarcinoma through NGS approach [63]. The mitochondrial genome and nuclear genome have cross talk in the regulation of oncogenic pathways, since both simultaneously had mutations on the genes involved in the same function [64]. For example, MUTYH-associated polyposis is susceptibly related with colorectal cancer carcinogenesis and was driven by both mutations at nuclear gene KRAS and mitochondrial genes [65]. These results indicated that mtDNA instability can be recommended as a cancer hallmark.

Recently, the mutation of mitochondrial tRNA genes was suggested as a biomarker in many cancers. tRNA genes have many functions including processing and translation which are related with mitochondrial protein synthesis. For example, A12308G, a polymorphic mutation in V-loop tRNA, was reported highly associated with many cancers [66–68], as a useful detection spot in colorectal cancer diagnostic [69]. Mitochondrial tRNA gene mutation is associated with tumor differentiation in oral squamous cell carcinoma [70]. Though the application of mtDNA sequencing in cancer has been discussed, both in diagnosis and treatment and in various cancers, we still have more to do to make it suitable in real clinical stage (Fig. 3.2).

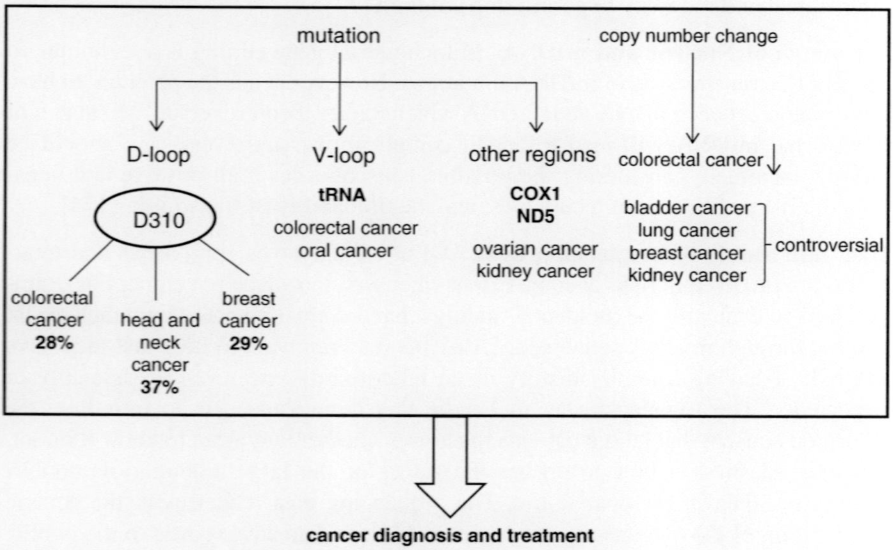

Fig. 3.2 Application of mtDNA sequencing in cancer diagnosis and treatment
MtDNA changes and copy number change can be biomarkers for cancer diagnosis and treatment. D310 mutation which located in D-loop region of the mitochondrial is a hot spot in many solid tumors, while mutation in V-loop tRNA is also a novel target. Copy number change can also be regarded as biomarker in cancer diagnosis and treatment. Despite decreased mtDNA copy number is related with colorectal cancer, whether high copy number or low copy number is related with other cancers still need to be further studied

4 Bioethics in Gene Editing

The Growing Need of Baby Design With the development of the sequencing technology, more cancer-related mitochondrial genes are discovered and needed for editing genes to get rid of those unwanted genes. mtDNA mutations can arise from de novo or are maternal inherited [71]. For example, point mutations are always inherited while large deletions usually appear from de novo. Preimplantation genetic diagnosis (PGD) is a technique to test genetic disorders in embryos and prevent the inherited mtDNA diseases before intrauterine transfer as the tool. However, PGD still has many limitations, of which PGD cannot 100% eliminate the risk of transmitting mtDNA diseases to offspring but only can reduce the percentage. Another limitation is that PGD may lead to random and rapid changes in mtDNA heteroplasmy levels to their offspring [72]. In recent years, mtDNA replacement therapy as a novel method has raised the attention of the public because it has so many ethical problems. The most controversial one is that this technique will allow us to design babies. Without knowing the effects for the future of the designed babies, it is unethical and too risky to permit this program [73].

Mismatch of Nuclear and mtDNA Mitochondrial gene editing is a technique to prevent the transmission of mtDNA mutations. However, it has the potential to have a mismatch between nDNA and mtDNA which makes it controversial. Mismatch of nDNA and mtDNA will lead to health complications, and every case should be strictly examined, considering the possible consequences both positive and negative, the risks of treatment techniques, and the probability of transmitting [74].

Risk and Safety of Sequencing Data All novel diagnosis approaches and treatments are involved in risks at some extent when first introduced in clinic. For example, how to deal with the incidental findings that are not associated with the current disease through mtDNA sequencing? Patients may not want to hear that they have the risk of having a family history of an inherited disease, even occasionally or expectedly. The suggested way to handle this circumstance is to introduce the informed consent that let the patients to choose whether they want to know it or not. Another question is how to ensure the safety of the large amounts of mtDNA sequencing data in the near future. The sequencing data related with the genetic information of the patients are very private. If these data are exposed to the public, people will easily find them in the public databases which will seriously affect the privacy of the patients. It would be very important to let the people be aware of the risks and sign the informed consent before they enroll [75] (Fig. 3.3).

Potential Cytotoxicity Gene editing as an alternative of clinical therapies goes closer to the clinical practice and becomes more reliable and realistic for patients, with the rapid development of CRISPR. Cytotoxicity and genotoxicity become the priority of gene editing to be considered as other drug development [76–78]. It is questioned if mtDNA gene editing can change more mitochondrial functions as designed, which often occurs when the cell responses to drug, even those nature herbs [79–81]. It is important to clarify the difference of mitochondrial

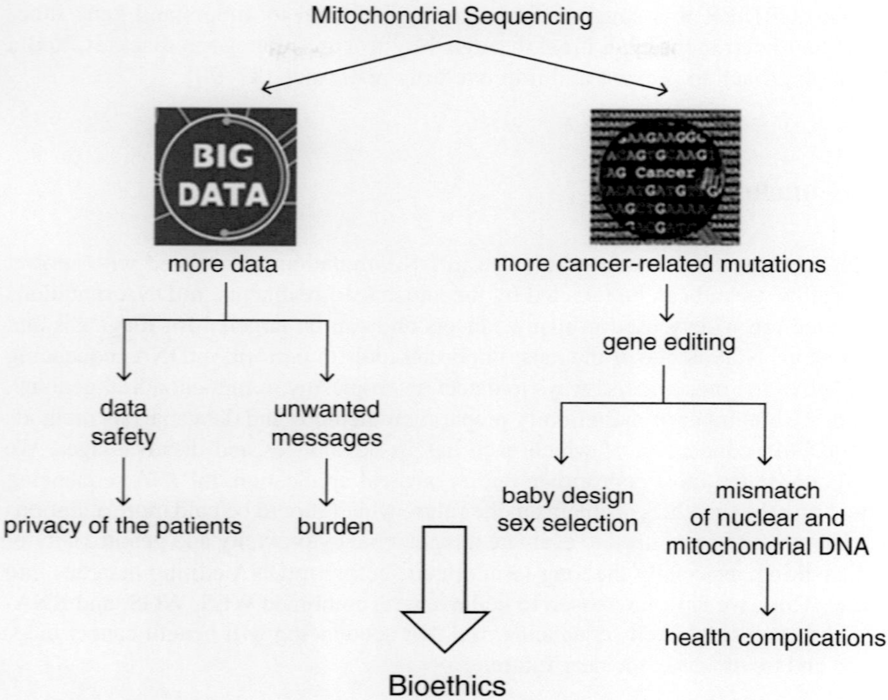

Fig. 3.3 Bioethics problems
With the development of mtDNA sequencing applied in clinic, there are many bioethics problems appeared that we should discuss and evaluate. On one hand, mtDNA sequencing will produce large volume of data, if we cannot ensure the safety of these data, the privacy of the patients will be affected. Moreover, from analysis of these data, we will get more information unrelated with current disease which may add the burden of the patients. On the other hand, mtDNA sequencing will explore more cancer-related mutations which will make people have the idea of gene editing. Gene editing will make people have the choice to select sex of the baby or even design a baby, and the mismatch of nuclear and mtDNA may bring health complications. These bioethics problems we should further discuss and evaluate

dysfunctions induced by chemicals from those by mtDNA gene editing [38, 40, 82–84]. Genotoxicity is another issue for mtDNA editing to be faced, since it can be caused by a large number of influencing factors [85], which may be generated directly or indirectly during mtDNA editing or post-editing modification. It should be clearly understood that manipulations of nDNA or mtDNA mutations in the human genome for gene therapy and homology-directed repair and editing must be fully specific and exact and have the lowest off-target effects to protect the human genotoxicity. Lin et al. developed a new method with the monomeric endonuclease to recognize at least 33 bp by fusing the DNA-recognizing domain of TALEN and to reengineer homing endonuclease I-SceI. Using such method, homology-directed repair and editing can be performed using the reengineered I-SceI to increase target specificity and low genotoxicity [86]. Recently, the single-cell CRISPR was suggested as a strong potential to understand gene functions and heterogeneity, a breakthrough way to cure monogene diseases, and a future approach to prevent and improve drug resistance [87–90].

5 Conclusion

mtDNA sequencing demonstrates that mtDNA mutations are related with several cell pathways and can be targeted by the anticancer treatments. mtDNA mutations exist and are widely used in many cancers and can be targeted for diagnosis and treatments. NGS is one of the most important tools to perform mtDNA sequencing and one of the most precise ways to detect heteroplasmy in mitochondrial genome. There are a number of main library preparation methods and data analysis methods of mtDNA sequencing, of which each has its advantages and disadvantages. We shall choose the most appropriate one in our real application. mtDNA sequencing may bring the bioethics problems in the future which should be paid more attentions to. There is an urgent need to evaluate the potential cytotoxicity and genotoxicity of gene editing, especially the long-term effects, before mtDNA editing marches into clinic. Thus, we have the reason to believe, with combined WES, WGS, and RNA-seq and even single-cell sequencing, mtDNA sequencing will benefit cancer diagnosis and treatment in the near future.

Acknowledgments The work was supported by Zhongshan Distinguished Professor Grant (XDW), The National Nature Science Foundation of China (91230204, 81270099, 81320108001, 81270131, 81300010), The Shanghai Committee of Science and Technology (12JC1402200, 12431900207, 11410708600, 14431905100), Operation funding of Shanghai Institute of Clinical Bioinformatics, Ministry of Education for Academic Special Science and Research Foundation for PhD Education (20130071110043), and National Key Research and Development Program (2016YFC0902400, 2017YFSF090207).

References

1. Filosto M, Mancuso M. Mitochondrial diseases: a nosological update. Acta Neurol Scand. 2007;115:211–21. PMID:17376118
2. Anderson S, Bankier AT, Barrell BG, de Bruijn MH, Coulson AR, Drouin J, et al. Sequence and organization of the human mitochondrial genome. Nature. 1981;290:457–65. PMID:7219534
3. Wallace DC. A mitochondrial paradigm of metabolic and degenerative diseases, aging, and cancer: a dawn for evolutionary medicine. Annu Rev Genet. 2005;39:359–407. PMID:16285865
4. Wong LJ. Next generation molecular diagnosis of mitochondrial disorders. Mitochondrion. 2013;13:379–87. PMID:23473862
5. Kennedy SR, Schmitt MW, Fox EJ, Kohrn BF, Salk JJ, Ahn EH, et al. Detecting ultralow-frequency mutations by Duplex Sequencing. Nat Protoc. 2014;9:2586–606. PMID:25299156
6. Stoler N, Arbeithuber B, Guiblet W, Makova KD, Nekrutenko A. Streamlined analysis of duplex sequencing data with Du Novo. Genome Biol. 2016;17:180. PMID:27566673

7. Watt S, Jiao W, Brown AM, Petrocelli T, Tran B, Zhang T, et al. Clinical genomics information management software linking cancer genome sequence and clinical decisions. Genomics. 2013;102:140–7. PMID:23603536
8. McCormick E, Place E, Falk MJ. Molecular genetic testing for mitochondrial disease: from one generation to the next. Neurotherapeutics J Am Soc Exp Neurother. 2013;10:251–61. PMID:23269497
9. Ye F, Samuels DC, Clark T, Guo Y. High-throughput sequencing in mtDNA research. Mitochondrion. 2014;17:157–63. PMID:24859348
10. Li M, Schroeder R, Ko A, Stoneking M. Fidelity of capture-enrichment for mtDNA genome sequencing: influence of NUMTs. Nucleic Acids Res. 2012;40:e137. PMID:22649055
11. Mason VC, Li G, Helgen KM, Murphy WJ. Efficient cross-species capture hybridization and next-generation sequencing of mitochondrial genomes from noninvasively sampled museum specimens. Genome Res. 2011;21:1695–704. PMID:21880778
12. Wolff JN, Shearman DC, Brooks RC, Ballard JW. Selective enrichment and sequencing of whole mitochondrial genomes in the presence of nuclear encoded mitochondrial pseudogenes (numts). PloS One. 2012;7:e37142. PMID:22606342
13. Petruzzella V, Carrozzo R, Calabrese C, Dell'Aglio R, Trentadue R, Piredda R, et al. Deep sequencing unearths nuclear mitochondrial sequences under Leber's hereditary optic neuropathy-associated false heteroplasmic mtDNA variants. Hum Mol Genet. 2012;21:3753–64. PMID:22589247
14. Schonberg A, Theunert C, Li M, Stoneking M, Nasidze I. High-throughput sequencing of complete human mtDNA genomes from the Caucasus and West Asia: high diversity and demographic inferences. Eur J Hum Genet. 2011;19:988–94. PMID:21487439
15. Gould MP, Bosworth CM, McMahon S, Grandhi S, Grimerg BT, LaFramboise T. PCR-free enrichment of MtDNA from human blood and cell lines for high quality next-generation DNA sequencing. PloS One. 2015;10:e0139253. PMID:26488301
16. Ni T, Wei G, Shen T, Han M, Lian Y, Fu H, et al. MitoRCA-seq reveals unbalanced cytocine to thymine transition in Polg mutant mice. Sci Rep. 2015;5:12049. PMID:26212336
17. Greaves LC, Reeve AK, Taylor RW, Turnbull DM. MtDNA and disease. J Pathol. 2012;226:274–86. PMID:21989606
18. Parfait B, Rustin P, Munnich A, Rotig A. Co-amplification of nuclear pseudogenes and assessment of heteroplasmy of mtDNA mutations. Biochem Biophys Res Commun. 1998;247:57–9. PMID:9636653
19. Samuels DC, Han L, Li J, Quanghu S, Clark TA, Shyr Y, et al. Finding the lost treasures in exome sequencing data. Trends Genet. 2013;29:593–9. PMID:23972387
20. Ameur A, Stewart JB, Freyer C, Hagstrom E, Ingman M, Larsson NG, et al. Ultra-deep sequencing of mouse mtDNA: mutational patterns and their origins. PLoS Genet. 2011;7:e1002028. PMID:21455489
21. Satoh M, Kuroiwa T. Organization of multiple nucleoids and DNA molecules in mitochondria of a human cell. Exp Cell Res. 1991;196:137–40. PMID:1715276
22. Sondheimer N, Glatz CE, Tirone JE, Deardorff MA, Krieger AM, Hakonarson H. Neutral mitochondrial heteroplasmy and the influence of aging. Hum Mol Genet. 2011;20:1653–9. PMID:21296868
23. Zhang C, Huang VH, Simon M, Sharma LK, Fan W, Haas R, et al. Heteroplasmic mutations of the mitochondrial genome cause paradoxical effects on mitochondrial functions. FASEB J. 2012;26:4914–24. PMID:22925728
24. Kloss-Brandstatter A, Weissensteiner H, Erhart G, Schafer G, Forer L, Schonherr S, et al. Validation of next-generation sequencing of entire mitochondrial genomes and the diversity of MtDNA mutations in oral squamous cell carcinoma. PloS One. 2015;10:e0135643. PMID:26262956
25. Seneca S, Vancampenhout K, Van Coster R, Smet J, Lissens W, Vanlander A, et al. Analysis of the whole mitochondrial genome: translation of the Ion Torrent Personal Genome Machine system to the diagnostic bench? Eur J Hum Genet. 2015;23:41–8. PMID:24667782

26. Weissensteiner H, Forer L, Fuchsberger C, Schopf B, Kloss-Brandstatter A, Specht G, et al. mtDNA-Server: next-generation sequencing data analysis of human mtDNA in the cloud. Nucleic Acids Res. 2016;44:W64–9. PMID:27084948
27. Tang S, Wang J, Zhang VW, Li FY, Landsverk M, Cui H, et al. Transition to next generation analysis of the whole mitochondrial genome: a summary of molecular defects. Hum Mutat. 2013;34:882–93. PMID:23463613
28. Griffin HR, Pyle A, Blakely EL, Alston CL, Duff J, Hudson G, et al. Accurate mtDNA sequencing using off-target reads provides a single test to identify pathogenic point mutations. Genetics Med. 2014;16:962–71. PMID:24901348
29. Guo Y, Li CI, Sheng Q, Winther JF, Cai Q, Boice JD, et al. Very low-level heteroplasmy mtDNA variations are inherited in humans. J Genet Genomics = Yi chuan xue bao. 2013;40:607–15. PMID:24377867
30. Gonzaga-Jauregui C, Lupski JR, Gibbs RA. Human genome sequencing in health and disease. Annu Rev Med. 2012;63:35–61. PMID:22248320
31. D'Erchia AM, Atlante A, Gadaleta G, Pavesi G, Chiara M, De Virgilio C, et al. Tissue-specific mtDNA abundance from exome data and its correlation with mitochondrial transcription, mass and respiratory activity. Mitochondrion. 2015;20:13–21. PMID:25446395
32. Rackham O, Shearwood AM, Mercer TR, Davies SM, Mattick JS, Filipovska A. Long noncoding RNAs are generated from the mitochondrial genome and regulated by nuclear-encoded proteins. RNA. 2011;17:2085–93. PMID:22028365
33. Tian Y, Smith DR. Recovering complete mitochondrial genome sequences from RNA-seq: a case study of Polytomella non-photosynthetic green algae. Mol Phylogenet Evol. 2016;98:57–62. PMID:26860338
34. Loher P, Telonis AG, Rigoutsos I. MINTmap: fast and exhaustive profiling of nuclear and mitochondrial tRNA fragments from short RNA-seq data. Sci Rep. 2017;7:41184. PMID:28220888
35. Yao YG, Kajigaya S, Young NS. MtDNA mutations in single human blood cells. Mutat Res. 2015;779:68–77. PMID:26149767
36. DiMauro S, Schon EA. Mitochondrial respiratory-chain diseases. N Engl J Med. 2003;348:2656–68. PMID:12826641
37. Medvedev R, Hildt E, Ploen D. Look who's talking-the crosstalk between oxidative stress and autophagy supports exosomal-dependent release of HCV particles. Cell Biol Toxicol. 2017;33(3):211–31. PMID:27987184
38. Vedi M, Sabina EP. Assessment of hepatoprotective and nephroprotective potential of withaferin A on bromobenzene-induced injury in Swiss albino mice: possible involvement of mitochondrial dysfunction and inflammation. Cell Biol Toxicol. 2016;32(5):373–90. PMID: 27250656
39. Tan AS, Baty JW, Dong LF, Bezawork-Geleta A, Endaya B, Goodwin J, et al. Mitochondrial genome acquisition restores respiratory function and tumorigenic potential of cancer cells without mtDNA. Cell Metab. 2015;21:81–94. PMID:25565207
40. Zhu LZ, Hou YJ, Zhao M, Yang MF, Fu XT, Sun JY, Fu XY, Shao LR, Zhang HF, Fan CD, Gao HL, Sun BL. Caudatin induces caspase-dependent apoptosis in human glioma cells with involvement of mitochondrial dysfunction and reactive oxygen species generation. Cell Biol Toxicol. 2016;32(4):333–45. PMID:27184666
41. Geurts-Giele WR, Gathier GH, Atmodimedjo PN, Dubbink HJ, Dinjens WN. Mitochondrial D310 mutation as clonal marker for solid tumors. Virchows Archiv: Int J Pathol. 2015;467:595–602. PMID:26276353
42. Carew JS, Huang P. Mitochondrial defects in cancer. Mol Cancer. 2002;1:9. PMID:12513701
43. Bai RK, Chang J, Yeh KT, Lou MA, Lu JF, Tan DJ, et al. MtDNA content varies with pathological characteristics of breast cancer. J Oncol. 2011;2011:496189. PMID:22028711
44. Aravamudan B, Thompson MA, Pabelick CM, Prakash YS. Mitochondria in lung diseases. Expert Rev Respir Med. 2013;7:631–46. PMID:23978003

45. Guerra F, Perrone AM, Kurelac I, Santini D, Ceccarelli C, Cricca M, et al. MtDNA mutation in serous ovarian cancer: implications for mitochondria-coded genes in chemoresistance. J Clin Oncol. 2012;30:e373–8. PMID:23150702
46. Taanman JW. The mitochondrial genome: structure, transcription, translation and replication. Biochim Biophys Acta. 1999;1410:103–23. PMID:10076021
47. Guerra F, Kurelac I, Cormio A, Zuntini R, Amato LB, Ceccarelli C, et al. Placing mtDNA mutations within the progression model of type I endometrial carcinoma. Hum Mol Genet. 2011;20:2394–405. PMID:21470976
48. Yu M, Zhou Y, Shi Y, Ning L, Yang Y, Wei X, et al. Reduced mtDNA copy number is correlated with tumor progression and prognosis in Chinese breast cancer patients. IUBMB Life. 2007;59:450–7. PMID:17654121
49. Fang Y, Huang J, Zhang J, Wang J, Qiao F, Chen HM, et al. Detecting the somatic mutations spectrum of Chinese lung cancer by analyzing the whole mtDNA genomes. MtDNA. 2015;26:56–60. PMID:24006865
50. Yang Ai SS, Hsu K, Herbert C, Cheng Z, Hunt J, Lewis CR, et al. MtDNA mutations in exhaled breath condensate of patients with lung cancer. Respir Med. 2013;107:911–8. PMID:23507584
51. Piantadosi CA, Suliman HB. Mitochondrial dysfunction in lung pathogenesis. Annu Rev Physiol. 2017;79:495–515. PMID:27959621
52. Gao YK, Xu XL, Miao F, Chen ZM. Whole mitochondrial genome sequencing and analysis for rat squamous cell carcinoma tissue cell. MtDNA Part A DNA Mapp Seq Anal. 2016;27:2143–4. PMID:25541317
53. Kazdal D, Harms A, Endris V, Penzel R, Kriegsmann M, Eichhorn F, et al. Prevalence of somatic mitochondrial mutations and spatial distribution of mitochondria in non-small cell lung cancer. Br J Cancer. 2017;117:220–6. PMID:28557978
54. Zhou S, Kachhap S, Sun W, Wu G, Chuang A, Poeta L, et al. Frequency and phenotypic implications of mtDNA mutations in human squamous cell cancers of the head and neck. Proc Natl Acad Sci U S A. 2007;104:7540–5. PMID:17456604
55. Morandi L, Tarsitano A, Gissi D, Leonardi E, Balbi T, Marchetti C, et al. Clonality analysis in primary oral squamous cell carcinoma and related lymph-node metastasis revealed by TP53 and mtDNA next generation sequencing analysis. J Craniomaxillofac Surg. 2015;43:208–13. PMID:25530302
56. Hearn JM, Romero-Canelon I, Munro AF, Fu Y, Pizarro AM, Garnett MJ, et al. Potent organo-osmium compound shifts metabolism in epithelial ovarian cancer cells. Proc Natl Acad Sci U S A. 2015;112:E3800–5. PMID:26162681
57. Pierini S, Fang C, Rafail S, Facciponte JG, Huang J, De Sanctis F, et al. A tumor mitochondria vaccine protects against experimental renal cell carcinoma. J Immunol. 2015;195:4020–7. PMID:26378078
58. Bai RK, Leal SM, Covarrubias D, Liu A, Wong LJ. Mitochondrial genetic background modifies breast cancer risk. Cancer Res. 2007;67:4687–94. PMID:17510395
59. Shen J, Platek M, Mahasneh A, Ambrosone CB, Zhao H. Mitochondrial copy number and risk of breast cancer: a pilot study. Mitochondrion. 2010;10:62–8. PMID:19788937
60. Kirches E. MtDNA As a cancer marker: a finally closed chapter? Curr Genomics. 2017;18:255–67. PMID:28659721
61. Schwartz S Jr, Alazzouzi H, Perucho M. Mutational dynamics in human tumors confirm the neutral intrinsic instability of the mitochondrial D-loop poly-cytidine repeat. Genes Chromosomes Cancer. 2006;45:770–80. PMID:16708351
62. Shakhssalim N, Houshmand M, Kamalidehghan B, Faraji A, Sarhangnejad R, Dadgar S, et al. The mitochondrial C16069T polymorphism, not mitochondrial D310 (D-loop) mononucleotide sequence variations, is associated with bladder cancer. Cancer Cell Int. 2013;13:120. PMID:24308421
63. Cui H, Huang P, Wang Z, Zhang Y, Zhang Z, Xu W, et al. Association of decreased mtDNA content with the progression of colorectal cancer. BMC Cancer. 2013;13:110. PMID:23497023

64. de Araujo LF, Fonseca AS, Muys BR, Placa JR, Bueno RB, Lorenzi JC, et al. Mitochondrial genome instability in colorectal adenoma and adenocarcinoma. Tumour Biol: J Int Soc Oncodev Biol Med. 2015;36:8869–79. PMID:26069104
65. Venesio T, Balsamo A, Errichiello E, Ranzani GN, Risio M. Oxidative DNA damage drives carcinogenesis in MUTYH-associated-polyposis by specific mutations of mitochondrial and MAPK genes. Modern Pathol. 2013;26:1371–81. PMID:23599153
66. Booker LM, Habermacher GM, Jessie BC, Sun QC, Baumann AK, Amin M, et al. North American white mitochondrial haplogroups in prostate and renal cancer. J Urol. 2006;175:468–72. Discussion 72-3. PMID:16406974
67. Datta S, Majumder M, Biswas NK, Sikdar N, Roy B. Increased risk of oral cancer in relation to common Indian mitochondrial polymorphisms and Autosomal GSTP1 locus. Cancer. 2007;110:1991–9. PMID:17886251
68. Amend K, Hicks D, Ambrosone CB. Breast cancer in African-American women: differences in tumor biology from European-American women. Cancer Res. 2006;66:8327–30. PMID:16951137
69. Mohammed F, Rezaee Khorasany AR, Mosaieby E, Houshmand M. Mitochondrial A12308G alteration in tRNA(Leu(CUN)) in colorectal cancer samples. Diagn Pathol. 2015;10:115. PMID:26189042
70. Lai CH, Huang SF, Liao CT, Chen IH, Wang HM, Hsieh LL. Clinical significance in oral cavity squamous cell carcinoma of pathogenic somatic mitochondrial mutations. PloS One. 2013;8:e65578. PMID:23799027
71. Cloonan SM, Choi AM. Mitochondria in lung disease. J Clin Invest. 2016;126:809–20. PMID:26928034
72. Claiborne A, English R, Kahn J, editors. Mitochondrial replacement techniques: ethical, social, and policy considerations. Washington, DC: National Academies Press; 2016.
73. Fogleman S, Santana C, Bishop C, Miller A, Capco DG. CRISPR/Cas9 and mitochondrial gene replacement therapy: promising techniques and ethical considerations. Am J Stem Cells. 2016;5:39–52. PMID:27725916
74. Gomez-Tatay L, Hernandez-Andreu JM, Aznar J. Mitochondrial modification techniques and ethical issues. J Clin Med. 2017;6:25. PMID:28245555
75. Devarakonda S, Govindan R, Hammerman PS. Cancer gene sequencing: ethical challenges and promises. Virtual Mentor. 2012;14:868–72. PMID:23351899
76. Richardson DS. Success of transdisciplinary science requires monodisciplinary support. Cell Biol Toxicol. 2016;32(1):5–6. PMID: 27002607
77. Gu J, Wang X. New future of cell biology and toxicology: thinking deeper. Cell Biol Toxicol. 2016;32(1):1–3. PMID: 26874518
78. Thomas PD, Kahn M. Kat3 coactivators in somatic stem cells and cancer stem cells: biological roles, evolution, and pharmacologic manipulation. Cell Biol Toxicol. 2016;32(1):61–81. PMID: 27008332
79. Tlotleng N, Vetten MA, Keter FK, Skepu A, Tshikhudo R, Gulumian M. Cytotoxicity, intracellular localization and exocytosis of citrate capped and PEG functionalized gold nanoparticles in human hepatocyte and kidney cells. Cell Biol Toxicol. 2016;32(4):305–21. PMID: 27184667
80. Giromini C, Rebucci R, Fusi E, Rossi L, Saccone F, Baldi A. Cytotoxicity, apoptosis, DNA damage and methylation in mammary and kidney epithelial cell lines exposed to ochratoxin A. Cell Biol Toxicol. 2016;32(3):249–58. PMID: 27154019
81. Fang T, Hou J, He M, Wang L, Zheng M, Wang X, Xia J. Actinidia chinensis planch root extract (acRoots) inhibits hepatocellular carcinoma progression by inhibiting EP3 expression. Cell Biol Toxicol. 2016;32(6):499–511. PMID: 27475644
82. Zerin T, Kim JS, Gil HW, Song HY, Hong SY. Effects of formaldehyde on mitochondrial dysfunction and apoptosis in SK-N-SH neuroblastoma cells. Cell Biol Toxicol. 2015;31(6):261–72. PMID: 26728267

83. Bao L, Diao H, Dong N, Su X, Wang B, Mo Q, Yu H, Wang X, Chen C. Histone deacetylase inhibitor induces cell apoptosis and cycle arrest in lung cancer cells via mitochondrial injury and p53 up-acetylation. Cell Biol Toxicol. 2016;32(6):469–82. PMID: 27423454
84. Ganta KK, Mandal A, Chaubey B. Depolarization of mitochondrial membrane potential is the initial event in non-nucleoside reverse transcriptase inhibitor efavirenz induced cytotoxicity. Cell Biol Toxicol. 2017;33(1):69–82. PMID: 27639578
85. Farombi EO. Genotoxicity of chloroquine in rat liver cells: protective role of free radical scavengers. Cell Biol Toxicol. 2006;22(3):159–67. PMID: 16532284
86. Lin J, Chen H, Luo L, Lai Y, Xie W, Kee K. Creating a monomeric endonuclease TALE-I-SceI with high specificity and low genotoxicity in human cells. Nucleic Acids Res. 2015;43(2):1112–22. PMID: 25541197
87. Fang H, Wang W. Could CRISPR be the solution for gene editing's Gordian knot? Cell Biol Toxicol. 2016;32(6):465–7. PMID: 27614448
88. Wang W, Gao D, Wang X. Can single-cell RNA sequencing crack the mystery of cells? Cell Biol Toxicol. 2017. https://doi.org/10.1007/s10565-017-9404-y. PMID: 28733864
89. Wang W, Zhu B, Wang X. Dynamic phenotypes: illustrating a single-cell odyssey. Cell Biol Toxicol. 2017;33:423–7. PMID: 28638956
90. Wang W, Wang X. Single-cell CRISPR screening in drug resistance. Cell Biol Toxicol. 2017;33(3):207–10. https://doi.org/10.1007/s10565-017-9396-7. PMID: 28474250

Mengjia Qian, BS, Researcher of Clinical Science Institute of Fudan University Zhongshan Hospital. Her main research is focused on next-generation sequencing technology and lung diseases. She is the author of more than 10 scientific publications with the impact factor about 60.

Claudio Spada, serves as President of the Scientific Society Healthcare Safety and Accountability. He is associated to the first hospital of Wenzhou Medical University, Zhejiang, P.R. China, and to the Healthcare Accountability Lab, Department of Biomedical Sciences for Health, University of Milan, Italy. His main fields of research include Biomedical Science and Technology application in Healthcare; Quality of Care and Patient's Safety; Continuing; Advanced Medical Simulation Systems, Bioethics; and Medical Education. He also serves as member of the Editorial Board and as Reviewer for International Scientific Journals.

Xiangdong Wang, MD, PhD, is a Distinguished Professor of Medicine, Director of Shanghai Institute of Clinical Bioinformatics, Executive Director of Clinical Science Institute of Fudan University Zhongshan Hospital, Director of Fudan University Center of Clinical Bioinformatics, Deputy Director of Shanghai Respiratory Research Institute, and visiting professor of King's College of London. His main research is focused on clinical bioinformatics, disease-specific biomarkers, lung chronic diseases, cancer immunology, and molecular and cellular therapies. He is the author of more than 200 scientific publications with the impact factor about 900, citation number about 6918, h-index 46, i10-index 181, and cited journal impact factor about 7000.

Chapter 4
Roles of Mitochondrial DNA Signaling in Immune Responses

Lingyan Wang, Michael N. Liebmen, and Xiangdong Wang

Abstract Mitochondrial DNA (mtDNA) plays an important role in immune responses during the evolution. The present chapter systemically describes its role on immune-related diseases and its interaction on immune responses. It is important to explore the main function and mechanisms of mtDNA in immune responses by which mtDNA regulates the signaling pathways of Toll-like receptor 9, autophagy, and STING. There are potentials to discover therapeutic targets of mtDNA in immune diseases and inflammation. It will be more exciting if the CRISPR-Cas9 method can be applied for mtDNA gene editing to cure diseases and provide a novel insight of mtDNA in immune responses as well as new therapies.

Keywords mtDNA • Immune responses • TLR9 • Autophagy • STING • CRISPR-Cas9

Abbreviation

ALI	Acute lung injury
CPB	Cardiopulmonary bypass
CRISPR-Cas9	Prokaryotic type II clustered regularly interspaced short palindromic repeats-CRISPR-associated 9
DAMPs	Damage-associated molecular patterns

L. Wang • M.N. Liebmen
Zhongshan Hospital Institute of Clinical Science, Fudan University, Shanghai, China

Shanghai Institute of Clinical Bioinformatics, Biomedical Research Center, Shanghai, China
e-mail: alick333@126.com; michael.liebman@ipqanalytics.com

X. Wang (✉)
Zhongshan Hospital Institute of Clinical Science, Fudan University,
Shanghai Medical College, Shanghai, China
e-mail: xiangdong.wang@clintransmed.org

H. Sun, X. Wang (eds.), *Mitochondrial DNA and Diseases*, Advances in Experimental Medicine and Biology 1038,
https://doi.org/10.1007/978-981-10-6674-0_4

HMGB1	High-mobility group protein B1
IRF7	Interferon regulatory factor 7
IRP	Immune-related pancytopenia
ISG	Interferon-stimulated genes
LC3B	Microtubule-associated protein1 light chain 3B
MAPKs	Mitogen-activated protein kinases
MELAS	Stroke-like episodes
MPT	Mitochondrial permeability transition
mtDNA	Mitochondrial DNA
MYD88	Response protein 88
NAFLD	Nonalcoholic fatty liver disease
nDNA	Nuclear DNA
NET	Neutrophil extracellular traps
NF-κB	Nuclear factor-κB
OXPHOS	Oxidative phosphorylation
PMN	Polymorphonuclear neutrophils
RA	Rheumatoid arthritis
ROS	Reactive oxygen species
SHR	Spontaneously hypertensive rats
TLR9	Toll-like receptor 9

1 Introduction

Mitochondria are a double-membrane organelle, to encode essential proteins of the oxidative phosphorylation system [1]. Mitochondria function not only in ATP production and regulation of reactive oxygen species (ROS), traditionally considered as energy generator, but also in metabolism, apoptosis, and calcium homeostasis [2–6]. The mitochondrial DNA (mtDNA) consists of the coding region and transcription and replication sites. The mtDNA is more susceptible to damage than nuclear DNA (nDNA), probably due to the lack of protection of genetic materials in the DNA repair system of mtDNA [7]. In the cytosolic step, the mtDNA release relies on the switch of mitochondrial permeability transition pores, which was caused by the concentrations of calcium and other signals at the inner membrane [8, 9]. In terms of extracellular release, cellular stress and necrosis are important factors in the liberation of mtDNA. Endosomes are small vesicles and contain substances such as protein and DNA and carry mtDNA as a mtDNA transporter to be involved in inflammatory response pathway (Fig. 4.1) [10, 11]. The present article overviews the close relationship, interaction, and communication between mtDNA and immune function and describes the importance of mtDNA in the maintenance of the immune system and in the pathogenesis of diseases. We specially focus on molecular mechanisms by which mtDNA contributes to the occurrence of diseases; regulates the transcriptional signaling and biological functions, e.g., metabolism, apoptosis, and autophagy; and is involved in cell sensitivity and susceptibility to therapies. mtDNA-associated therapies are also explored.

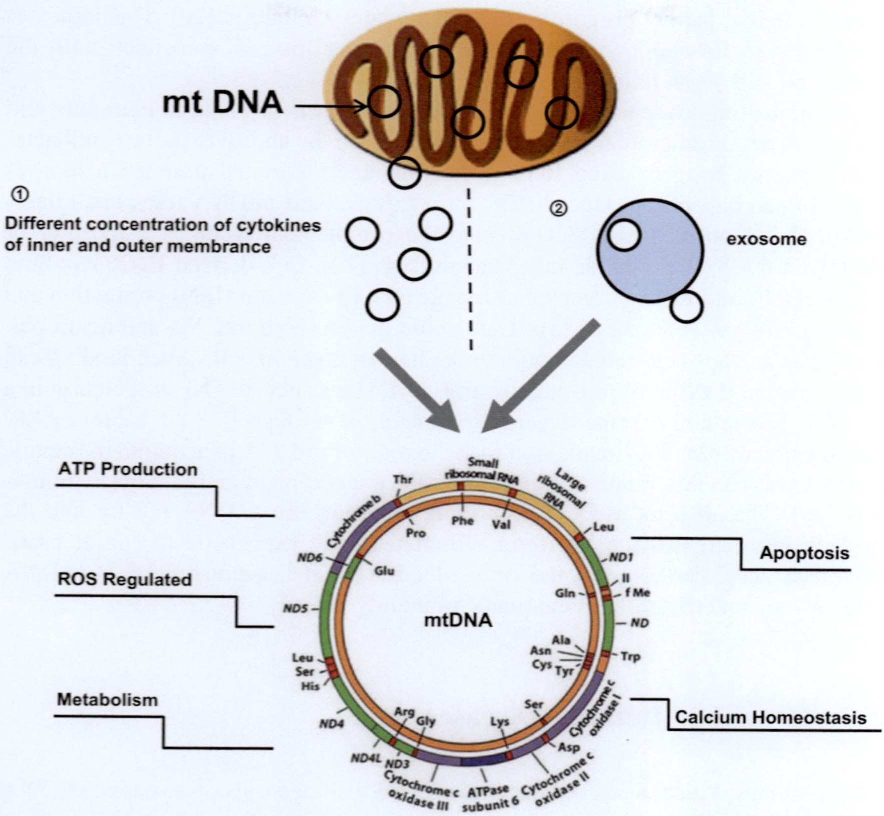

Fig. 4.1 Types of mtDNA release and its functions. The picture showed that mtDNA can be released by mitochondrial permeability transition pores and endosomes and has the different functions of ATP production, ROS regulation, metabolism, apoptosis, and calcium homeostasis due to different regions of coding, transcription, and replication sites

2 mtDNA and Immune Responses

Mitochondria have a number of biological functions, e.g., the synthesis of N-formylated proteins, amount of unmethylated DNA present, the system in triggering innate immune responses, and inducing formation of neutrophil extracellular traps [12–15]. mtDNA is widely involved in multi-processes of energy production, oxidative stress, metabolism, inflammation, and carcinogenesis [14–16]. mtDNA also plays an important role in immune responses when released into the circulation. mtDNA can activate human neutrophils through the immune signaling pathway, leading to leukocyte migration and degranulation [17]. Viable neutrophils per se release the mtDNA to form neutrophil extracellular traps after trauma [18, 19]. The decreased formation of neutrophil extracellular traps may become a critical factor in aging, patient prognosis, and inflammatory initiation [20]. This indicates that impaired formation of neutrophil extracellular traps is associated with the trauma-related organ failure in patients.

Accumulating evidences also suggest that mtDNA is involved in immunity and plays a direct antimicrobial role. Mitochondria have the ability to induce antibacterial immunity by generating ROS in macrophage-associated immune responses [21]. Human basophil-produced ROS form extracellular mtDNA traps upon interleukin (IL)-3 priming and then activate the complement factor 5, a receptor of FcεR1, in the NADPH oxide-independent form [22]. Cell-derived IL-26 can bind with mtDNA and form a complex to trigger type I interferon (IFN) production and achieve antivirus resistance [23]. IL-26 could bind to both mtDNA and neutrophil extracellular traps and transfer them to the human myeloid cells. Such binding can trigger the production of pro-inflammatory cytokines such as TNF-α secretion in a STING- and inflammasome-dependent manner, or through P2Y1 receptor-cAMP signal pathway [24, 25]. In addition to those roles of mtDNA in acquired immunity, mtDNA was recently found to contribute to the integration of antiviral innate immunity [26]. The aberrant mtDNA packaging promotes the mtDNA release into the cytosol and increases the interferon-stimulated gene expression to confer broad viral resistance. Furthermore, the virus infection could cause disruption of mtDNA homeostasis and trigger antiviral innate immunity [26].

3 mtDNA and Immune Diseases

mtDNA plays critical and necessary roles in the pathogenesis of diseases. mtDNA was detected in the synovial fluids of patients with rheumatoid arthritis [25] and in the plasma of patients with femur facture [17], rather than in healthy individuals. More free mtDNA was detected in the circulation in septic patients compared with non-septic patients [27]. It seems that mtDNA is more disease specific in the location of immune diseases, while more severity specific in the circulation of systemic severe inflammation. Furthermore, pressure-overload released mtDNA could cause inflammatory responses in cardiomyocytes and induce myocarditis and dilated cardiomyopathy [28], indicating that mtDNA plays a key role in the heart inflammation responses. mtDNA are also detected in the systemic inflammation in acute liver failure [29], atherosclerosis [30], and Parkinson's disease [31]. Cancer is a diseases related with microenvironment immunity with pro-inflammatory cells impaired and immunosuppressive cells increased [32]. mtDNA acts as a critical message to travel and communicate between tumor cells and neighbor non-tumor cells. Tumor cells can acquire functional mtDNA from healthy cells to restore respiratory function and metabolic activity which enabled them to proliferate [33]. In the absence of mtDNA, tumor cells were inhibited in the growth ability. The outcome of mtDNA horizontal transfer could induce chemoresistance in the treatment [34]. The mechanism by which mtDNA induced inflammation in immunology diseases seems different from the one by which mtDNA induces tumor progression and metastasis.

4 mtDNA-Dependent TLR9 Signaling Pathways

Toll-like receptor 9 (TLR9) is one of the first defenses in the innate immune system and can bridge the adaptive immunity with innate immunity during immune responses. TLR9 is an important sensor to damaged components and to unmethylated CpG dinucleotides [35]. TLR9 signaling pathway acts as an important signaling to link one of the many unmethylated CpGs present in mtDNA during immune responses [36]. The intercommunication between mtDNA and TLR9 is through the signaling pathways of protein 88, mitogen-activated protein kinases, and nuclear factor-κB (NF-κB) in inflammation and could enhance type I IFN responses through interferon regulatory factor 7 in immune responses [35]. mtDNA acted as damage-associated molecular patterns (DAMPs) and activated TLR in patients with nonalcoholic fatty liver disease. When blocking the TLR9, the disease progression could be reversed [37]. Those data indicate that mtDNA-activated TLR9 plays important roles in immune diseases.

On the other hand, mtDNA can activate the TLR9 signaling to provoke systemic inflammation, e.g., overproduction of inflammatory mediators, over-activation of leukocytes, and lung injury [38]. Such mtDNA-induced tissue injury can be prevented by pretreatment with the specific TLR9 inhibitor ODN2008. However, the extraordinary roles of nDNA differ from mtDNA in TLR-mediated immune responses, since nDNA hardly acts as mtDNA in the systemic inflammation. The mechanisms of such difference between nDNA and mtDNA in inflammation remain unclear, even though some differences, e.g., genetic code, inheritance, and structure, have been described. In addition, mtDNA could induce the overproduction of IL-12 and the acute renal injury through a TLR9-dependent pathway, evidenced by the inhibitory effects of TLR9KO [39]. These phenomena highlight that TLR9 is a potential therapy target in tissue injury. In septic patients, researchers also found that mtDNA activated TLR9 signaling possibly linked with initial inflammation and associated with increased mortality in critically ill patients [40]. mtDNA induced vascular dysfunction in spontaneously hypertensive rats through the TLR9 pathway [41].

The endogenous mtDNA was a direct dangerous signal effect on cardiomyocytes through TLR9-dependent NF-κB activation [42]. The mtDNA induced local inflammation and leukocyte recruitment through mtDNA-mediated TLR9-p38 MAPK signal pathway [43]. mtDNA released into the circulation could activate human neutrophils through TLR9 and promote neutrophil Ca(2+) flux as well as MAPK phosphorylation, increasing the migration and degranulation of neutrophils [17]. The surgery like cardiopulmonary bypass induced the lease of mtDNA and hyper-expression of neutrophil IL-6 gene [44]. Critical illness like shock can also induce the release of mtDNA accompanied with the overproduction of MMP-8/MMP-9 from neutrophils through TLR9 signaling pathway linked with p38 phosphorylation, but not p44/42 of MAPK [45]. It indicated that a specific phosphorylation site of MAPK is involved in mtDNA-induced neutrophil over-activation.

It is a challenge to identify and validate diagnostic biomarkers or therapeutic targets from mtDNA, mtDNA-TLR9 interaction, and TLR9-dependent signals, especially with disease specificity. Some epigenetic biomarkers were discovered when mtDNA activated neutrophil through TLR9 signaling in immune responses. For example, mtDNA released by hepatocytes and activated neutrophils can cause the occurrence of acute liver failure through TLR9 [37]. In such process, microRNA-223 could be upregulated as a negative feedback to attenuate TLR9/NF-κB-mediated inflammation and liver injury. It is questioned if microRNA-223 can be selected as the biomarker or target, since there are a large number of increased molecules during the induction. It is also questioned whether those molecules are stable and measurable in diseases; have the specificity of disease pathology, severity, and duration; as well as process biological specificity among the complex networks and interactions [46–48].

5 mtDNA and Autophagy Pathways

Autophagy as a lysosomal degradation pathway of cells is a process by which protein complexes are relocated into to the endolysosomal system for degradation. Autophagy plays critical roles in the protection against internal and external challenges and in the pathogenesis of diseases [49]. mtDNA can gather with cathelicidin antimicrobial peptide to form a complex and blind the autophagic recognition in TLR9-mediated inflammatory responses [50]. mtDNA can activate NLR family pyrin domain containing 3 (NLRP3) inflammasome complex, which is the activator of NF-kB signaling and enables the recruitment and activation of caspase-1, an important apoptosis protein in the innate immune system [21, 51, 52]. It is possible that mtDNA is involved in the process of autophagy through overproduction of inflammatory mediators and apoptotic signal pathways, since apoptosis can interact with autophagy [53]. Of those signal pathways, the activation of TLR/NF-kB pathway is a necessary priming step to upregulate NLRP3 and subsequent downstream signaling. mtDNA activated NLRP3 inflammasome in the autophagy pathways. The microtubule-associated protein1 light chain 3B/Beclin-1 can negatively regulate the NLRP3 activation [54]. On the other hand, autophagy may be a negative regulator of mtDNA-mediated inflammatory responses, since autophagy could inhibit mtDNA-induced TLR9 overexpression accompanying inflammatory state [55]. The release of mtDNA is accompanied by the secretion of protein that reside in autolysosomes such as LC3-II and cathepsin D in sepsis patients [56]. It indicated that mtDNA might be associated with LC3-II and cathepsin D in autophagy signaling in the function of immune responses.

6 mtDNA and cGAS-STING Signaling Pathway

Mitochondria are the major source of damage-associated molecular patterns (DAMPs). The cyclic GMP-AMP synthase (cGAS)/stimulator of interferon genes (STING) signaling plays a central role in mtDNA activity. The heterozygosity of the histone-like mitochondrial transcription factor triggers the disruption and release of mtDNA which can activate the STING-mediated pathway. The activation of the STING pathway is correlated with cyclic GMP-AMP (cGAMP) synthase (cGAS) [57]. The mechanism by which STING interacts with mtDNA is that STING as an adapter transmembrane protein located in the endoplasmic reticulum is changed and recruited into the Golgi after cGAMP is activated. After then, STING recruits tank-binding kinase 1 and phosphorylates the transcription factor interferon regulatory factor 3, to drive the transcription of type I IFNs such as IFNα and IFNβ and initiate the process of autophagy [58]. Through such mechanism, virus and toxicants may induce the maturation of autophagosomes and the process of autophagy with the involvement of Nrf2-ROS endosomal pathway [21, 59]. mtDNA promotes the anti-virus immunity through STING signaling pathway [26].

Palmitic acid inhibits endothelial cell proliferation and mtDNA release to cytosol through the activation of the cGAS-STING-IRF3 signaling [60]. After that, macrophage stimulating 1 (MST1) gene has been induced and Hippo-YAP pathway dysregulated, which caused angiogenesis suppression. Palmitic acid treatment can disconnect the mtDNA-activated STING signaling, leading to IRF3 phosphorylation and induced intercellular adhesion molecule 1 (ICAM-1) expression [61]. ICAM-1 plays important roles in leukocyte migration and infiltration into the inflamed tissue and development of organ dysfunction and injury, e.g., infection, organ failure, and immune hyperresponsiveness [62]. A deletion of the ICAM-1 could ameliorate mtDNA-mediated neutrophil infiltration and liver injury [37]. A large number of signal pathways are involved in the interaction between mtDNA and autophagy, e.g., STING, MST1, and ICAM-1, which may become potential therapeutic targets for the prevention and therapy for immune-regulated diseases. STING was also an important player in the RIG-1-IFN pathway [63]. West et al. found that [26] when deleting cGAS or STING in null mice, it exhibited significantly impaired IFN secretion, reflecting mtDNA stimulation [64].

The mtDNA involved cGAS-STING pathway also has some actions in cell apoptosis. mtDNA activates STING/IRF3 which could interact with pro-apoptotic molecule, BAX/BAK, and leads to hepatocyte apoptosis in an alcoholic liver disease model [15, 65]. mtDNA can stimulate cGAMP/STING signaling to upregulate the expression of IFN and apoptotic genes, inducing IFN-dependent responses and cell apoptosis [20, 66]. Furthermore, mtDNA released during BAX- and BAK-mediated apoptosis can activate cGAS-STING-IRF3-type I IFN signaling responses and caspase 9-associated responses. After then, caspases 3 and c7 and IFNs induce cGAS-STING pathway which were activated by mtDNA released from dying cells and could be silenced [67, 68]. Thus, mtDNA in IFN response and apoptosis reaction plays critical roles in switch off/on between the immunological silence and inflammatory reactions, to maintain the host immune homeostasis.

7 mtDNA-Targeted Potential Therapies

A number of factors and signal pathways, e.g., TLR9, autophagy, and STING signaling pathways, contribute to the immune responses and the mtDNA-caused immune activation (Fig. 4.2). Simultaneously, the corresponding molecular inhibitors are being discovered and developed for therapies. For example, ODN2088, a TLR9 inhibitor, could significantly ameliorate mtDNA-induced cardiac inflammation, spontaneous hypertension, and tissue injury [41, 69]. Another TLR9 inhibitor, chloroquine, could relieve the mtDNA-specific cell proliferative inhibition and TLR9-MAPK activation which are not caused by nDNA [70]. Chloroquine can benefit patients with heart failure by inhibiting the mtDNA-induced TLR9-NF-kB activation [11]. The inhibitor of autophagy, 3MA, could relieve the mtDNA-mediated liver cell injury [56].

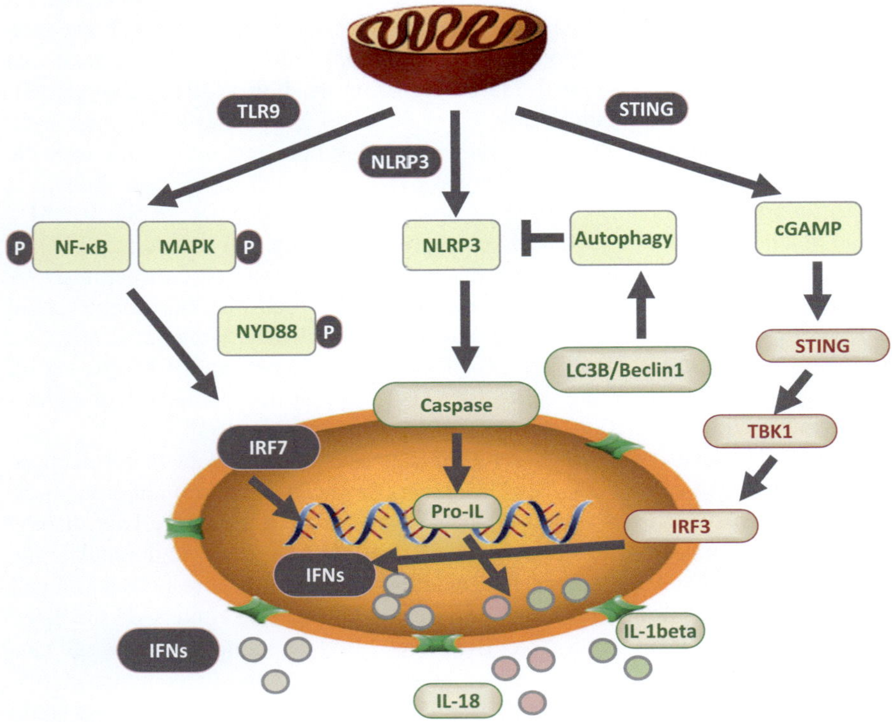

Fig. 4.2 Signaling pathways of mtDNA in immune responses. The picture elucidates the three main signaling pathways of mtDNA in immune responses. In the first pathway, mtDNA activates TLR9 and then phosphorylates MAPK and NF-κB and promotes the transcript of interferon regulatory factor 7, which stimulates the secretion of IFNs. In the second pathway, mtDNA activates the NLRP3 and increased the secretion of IL-1β and IL-18, and the autophagy proteins could reverse this effect. In the third pathway of STING, mtDNA activates cGAMP-STING; STING then recruits TBK1 and phosphorylates the IRF3 which drives the transcription of type I IFNs such as IFNα and IFNβ

The mutation of mtDNA is involved in a number of immune diseases. Normally, human immune surveillance recognizes mtDNA as self-antigen, while spontaneous mutations and sequence changes of mtDNA occur and are recognized as foreign subtracts by its own immune system for the immune responses, through the binding to MHC ligand [71]. Various mutations of mtDNA occurred in D-LOOP region of peripheral lymphocytes from patients with immune-related pancytopenia, correlated with cellular IgM and IgG levels [72]. The mutation of mtDNA A3243G was correlated with stroke-like episodes [73]. Between nDNA and mtDNA mutation, specific mtDNA mutation enhanced tumor progression with the involvement of ROS [74]. mtDNA mutation in complex I subunit 5 could gradually decrease oxidative phosphorylation while increasing glucose-dependent lactate production to enhance tumor growth [75].

Gene editing as a new strategy to cure mtDNA mutation can be an alternative to develop new therapies. Clustered regularly interspaced short palindromic repeats (CRISPR)-associated 9 (CRISPR-Cas9) system is rapidly developed and matured for gene editing of nDNA, although there are still a number of challenges to be faced [76–78]. mtDNA CRISPR-Cas9 is recently proposed as one of the novel therapies for mitochondrial disease [87]. Using CRISPR-Cas9 technology, specific sequences of mutated DNA can be deleted to treat patients with disease caused by mutant mtDNA; a number of ethical considerations have been raised. The mitochondrial 16S rRNA can contribute to cell proliferation and can regulate NGRN, WBSCR16, RPUSD3, RPISD4, TRUB2, and FASTKD2 after the oxidative phosphorylation (OXPHOS) gene was edited using CRISPR-Cas9 [79]. The mitochondrial gene PGC1α could influence the mitochondrial energetic metabolism and suppressed metastasis of melanoma after gene editing (Fig. 4.3) [80].

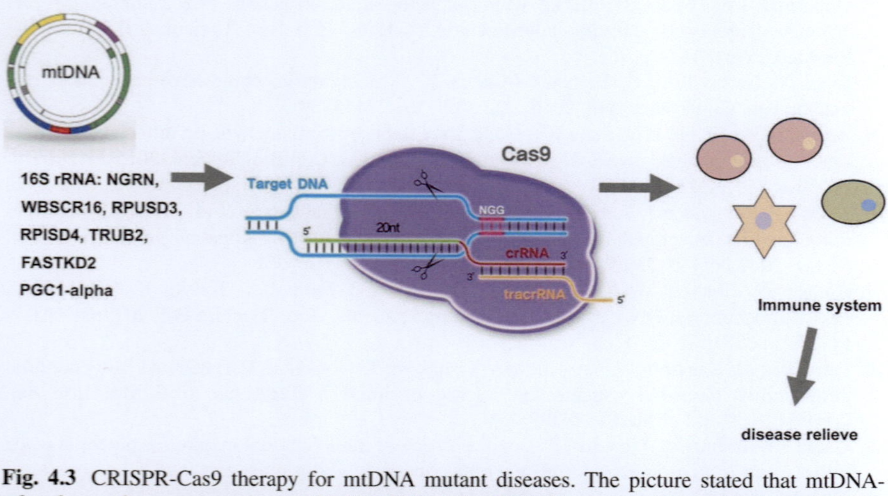

Fig. 4.3 CRISPR-Cas9 therapy for mtDNA mutant diseases. The picture stated that mtDNA-related mutation genes such as NGRN, PGC1-alpha, TRUB2, etc. could be edited by CRISPR-Cas9 systems. Then, the editing process resulted in the different responses to the immune system which relieve the mtDNA mutation-related immune diseases

8 Conclusion

mtDNA plays an important role in immune responses and the occurrence of immune diseases, through the TLR9, autophagy, and STING signaling pathways. The inhibitors of those signaling pathways can be strong candidates for new drug discovery and development. mtDNA gene editing can be an alternative to treat mtDNA mutation-dependent immune diseases, although it will need more efforts to be validated in the future.

Acknowledgments The work was supported by Zhongshan Distinguished Professor Grant (XDW), the National Nature Science Foundation of China (91230204, 81270099, 81320108001, 81270131, 81300010), the Shanghai Committee of Science and Technology (12JC1402200, 12431900207, 11410708600, 14431905100), Operation funding of Shanghai Institute of Clinical Bioinformatics, Ministry of Education for Academic Special Science and Research Foundation for PhD Education (20130071110043), and National Key Research and Development Program (2016YFC0902400, 2017YFSF090207).

Conflict of Interest The authors declare no conflict of interest.

References

1. Finsterer J, Zarrouk-Mahjoub S. Is chronic fatigue syndrome truly associated with haplogroups or mtDNA single nucleotide polymorphisms? J Transl Med. 2016;14:182. PubMed:27317438
2. Zhu LZ, Hou YJ, Zhao M, Yang MF, XT F, Sun JY, et al. Caudatin induces caspase-dependent apoptosis in human glioma cells with involvement of mitochondrial dysfunction and reactive oxygen species generation. Cell Biol Toxicol. 2016;32:333–45. PubMed:27184666
3. Opperman CM, Sishi BJ. Tumor necrosis factor alpha stimulates p62 accumulation and enhances proteasome activity independently of ROS. Cell Biol Toxicol. 2015;31:83–94. PubMed:25761618
4. Ribas V, Garcia-Ruiz C, Fernandez-Checa JC. Mitochondria, cholesterol and cancer cell metabolism. Clin Transl Med. 2016;5:22. PubMed:27455839
5. Zerin T, Kim JS, Gil HW, Song HY, Hong SY. Effects of formaldehyde on mitochondrial dysfunction and apoptosis in SK-N-SH neuroblastoma cells. Cell Biol Toxicol. 2015;31:261–72. PubMed:26728267
6. Khan NA, Nikkanen J, Yatsuga S, Jackson C, Wang L, Pradhan S, et al. mTORC1 regulates mitochondrial integrated stress response and mitochondrial myopathy progression. Cell Metab. 2017;26:419–28 e5. PubMed:28768179
7. Kodron A, Ghanim M, Krawczyk KK, Stelmaszczyk-Emmel A, Tonska K, Demkow U, et al. Mitochondrial DNA in pediatric leukemia patients. Acta Biochim Pol. 2017;64:183–7. PubMed:28284021
8. Patrushev M, Kasymov V, Patrusheva V, Ushakova T, Gogvadze V, Gaziev A. Mitochondrial permeability transition triggers the release of mtDNA fragments. Cell Mol Life Sci. 2004;61:3100–3. PubMed:15583871
9. Ganta KK, Mandal A, Chaubey B. Depolarization of mitochondrial membrane potential is the initial event in non-nucleoside reverse transcriptase inhibitor efavirenz induced cytotoxicity. Cell Biol Toxicol. 2017;33:69–82. PubMed:27639578

10. Hammerling BC, Shires SE, Leon LJ, Cortez MQ, Gustafsson AB. Isolation of Rab5-positive endosomes reveals a new mitochondrial degradation pathway utilized by BNIP3 and Parkin. Small GTPases. 2017:1–8. PubMed:28696827
11. Meng N, Han L, Pan X, Su L, Jiang Z, Lin Z, et al. Nano-Mg(OH)2-induced proliferation inhibition and dysfunction of human umbilical vein vascular endothelial cells through caveolin-1-mediated endocytosis. Cell Biol Toxicol. 2015;31:15–27. PubMed:25575676
12. Bao W, Xia H, Liang Y, Ye Y, Lu Y, Xu X, et al. Toll-like receptor 9 can be activated by endogenous mitochondrial DNA to induce podocyte apoptosis. Sci Rep. 2016;22579:6. PubMed:26934958
13. EP Y, Bennett MR. Mitochondrial DNA damage and atherosclerosis. Trends Endocrinol Metab. 2014;25:481–7. PubMed:25034130
14. Kikuchi S, Ninomiya T, Kohno T, Kojima T, Tatsumi H. Cobalt inhibits motility of axonal mitochondria and induces axonal degeneration in cultured dorsal root ganglion cells of rat. Cell Biol Toxicol. 2017. https://doi.org/10.1007/s10565-017-9402-0. PubMed:28656345
15. Giromini C, Rebucci R, Fusi E, Rossi L, Saccone F, Baldi A. Cytotoxicity, apoptosis, DNA damage and methylation in mammary and kidney epithelial cell lines exposed to ochratoxin A. Cell Biol Toxicol. 2016;32:249–58. PubMed:27154019
16. Vedi M, Sabina EP. Assessment of hepatoprotective and nephroprotective potential of withaferin A on bromobenzene-induced injury in Swiss albino mice: possible involvement of mitochondrial dysfunction and inflammation. Cell Biol Toxicol. 2016;32:373–90. PubMed:27250656
17. Zhang Q, Raoof M, Chen Y, Sumi Y, Sursal T, Junger W, et al. Circulating mitochondrial DAMPs cause inflammatory responses to injury. Nature. 2010;464:104–7. PubMed:20203610
18. Yousefi S, Mihalache C, Kozlowski E, Schmid I, Simon HU. Viable neutrophils release mitochondrial DNA to form neutrophil extracellular traps. Cell Death Differ. 2009;16:1438–44. PubMed:19609275
19. McIlroy DJ, Jarnicki AG, Au GG, Lott N, Smith DW, Hansbro PM, et al. Mitochondrial DNA neutrophil extracellular traps are formed after trauma and subsequent surgery. J Crit Care. 2014;29(1133):e1–5. PubMed:25128442
20. Venkatesan T, Choi YW, Mun SP, Kim YK. Pinus radiata bark extract induces caspase-independent apoptosis-like cell death in MCF-7 human breast cancer cells. Cell Biol Toxicol. 2016;32:451–64. PubMed:27400986
21. Medvedev R, Hildt E, Ploen D. Look who's talking-the crosstalk between oxidative stress and autophagy supports exosomal-dependent release of HCV particles. Cell Biol Toxicol. 2017;33:211–31. PubMed:27987184
22. Morshed M, Hlushchuk R, Simon D, Walls AF, Obata-Ninomiya K, Karasuyama H, et al. NADPH oxidase-independent formation of extracellular DNA traps by basophils. J Immunol. 2014;192:5314–23. PubMed:24771850
23. Meller S, Di Domizio J, Voo KS, Friedrich HC, Chamilos G, Ganguly D, et al. T(H)17 cells promote microbial killing and innate immune sensing of DNA via interleukin 26. Nat Immunol. 2015;16:970–9. PubMed:26168081
24. Collins LV, Hajizadeh S, Holme E, Jonsson IM, Tarkowski A. Endogenously oxidized mitochondrial DNA induces in vivo and in vitro inflammatory responses. J Leukoc Biol. 2004;75:995–1000. PubMed:14982943
25. Seo JB, Jung SR, Hille B, Koh DS, Extracellular ATP. protects pancreatic duct epithelial cells from alcohol-induced damage through P2Y1 receptor-cAMP signal pathway. Cell Biol Toxicol. 2016;32:229–47. PubMed:27197531
26. West AP, Khoury-Hanold W, Staron M, Tal MC, Pineda CM, Lang SM, et al. Mitochondrial DNA stress primes the antiviral innate immune response. Nature. 2015;520:553–7. PubMed:25642965
27. Di Caro V, Walko TD 3rd, Bola RA, Hong JD, Pang D, Hsue V, et al. Plasma mitochondrial DNA – a novel DAMP in pediatric sepsis. Shock. 2016;45:506–11. PubMed:26682947

28. Oka T, Hikoso S, Yamaguchi O, Taneike M, Takeda T, Tamai T, et al. Mitochondrial DNA that escapes from autophagy causes inflammation and heart failure. Nature. 2012;485:251–5. PubMed:22535248
29. Marques PE, Amaral SS, Pires DA, Nogueira LL, Soriani FM, Lima BH, et al. Chemokines and mitochondrial products activate neutrophils to amplify organ injury during mouse acute liver failure. Hepatology. 2012;56:1971–82. PubMed:22532075
30. Kaarniranta K, Tokarz P, Koskela A, Paterno J, Blasiak J. Autophagy regulates death of retinal pigment epithelium cells in age-related macular degeneration. Cell Biol Toxicol. 2017;33:113–28. PubMed:27900566
31. Celardo I, Martins LM, Gandhi S. Unravelling mitochondrial pathways to Parkinson's disease. Br J Pharmacol. 2014;171:1943–57. PubMed:24117181
32. Lin C, Chen S, Li Y. T cell modulation in immunotherapy for hematological malignancies. Cell Biol Toxicol. 2017;33:323–7. PubMed:28474249
33. Berridge MV, Dong L, Neuzil J. Mitochondrial DNA in tumor initiation, progression, and metastasis: role of horizontal mtDNA transfer. Cancer Res. 2015;75:3203–8. PubMed:26224121
34. Pasquier J, Guerrouahen BS, Al Thawadi H, Ghiabi P, Maleki M, Abu-Kaoud N, et al. Preferential transfer of mitochondria from endothelial to cancer cells through tunneling nanotubes modulates chemoresistance. J Transl Med. 2013;11:94. PubMed:23574623
35. Bai L, Chen W, Chen J, Li W, Zhou L, Niu C, et al. Heterogeneity of toll-like receptor 9 signaling in B cell malignancies and its potential therapeutic application. J Transl Med. 2017;51:15. PubMed:28241765
36. Barber GN. Cytoplasmic DNA innate immune pathways. Immunol Rev. 2011;243:99–108. PubMed:21884170
37. He Y, Feng D, Li M, Gao Y, Ramirez T, Cao H, et al. Hepatic mitochondrial DNA/Toll-like receptor 9/MicroRNA-223 forms a negative feedback loop to limit neutrophil overactivation and acetaminophen hepatotoxicity in mice. Hepatology. 2017;220-34:66. PubMed:28295449
38. Zhang L, Deng S, Zhao S, Ai Y, Pan P, Su X, et al. Intra-peritoneal administration of mitochondrial DNA provokes acute lung injury and systemic inflammation via toll-like receptor 9. Int J Mol Sci. 2016;17:1425. PubMed:27589725
39. Tsuji N, Tsuji T, Ohashi N, Kato A, Fujigaki Y, Yasuda H. Role of mitochondrial DNA in septic AKI via toll-like receptor 9. J Am Soc Nephrol. 2016;27:2009–20. PubMed:26574043
40. Schafer ST, Franken L, Adamzik M, Schumak B, Scherag A, Engler A, et al. Mitochondrial DNA: an endogenous trigger for immune paralysis. Anesthesiology. 2016;124:923–33. PubMed:26808636
41. McCarthy CG, Wenceslau CF, Goulopoulou S, Ogbi S, Baban B, Sullivan JC, et al. Circulating mitochondrial DNA and Toll-like receptor 9 are associated with vascular dysfunction in spontaneously hypertensive rats. Cardiovasc Res. 2015;119-30:107. PubMed:25910936
42. Bliksoen M, Mariero LH, Torp MK, Baysa A, Ytrehus K, Haugen F, et al. Extracellular mtDNA activates NF-kappaB via toll-like receptor 9 and induces cell death in cardiomyocytes. Basic Res Cardiol. 2016;42:111. PubMed:27164906
43. Gu X, Wu G, Yao Y, Zeng J, Shi D, Lv T, et al. Intratracheal administration of mitochondrial DNA directly provokes lung inflammation through the TLR9-p38 MAPK pathway. Free Radic Biol Med. 2015;83:149–58. PubMed:25772007
44. Sandler N, Kaczmarek E, Itagaki K, Zheng Y, Otterbein L, Khabbaz K, et al. Mitochondrial DAMPs are released during cardiopulmonary bypass surgery and are associated with postoperative atrial fibrillation. Heart Lung Circ. 2017. https://doi.org/10.1016/j.hlc.2017.02.014. PubMed:28487062
45. Zhang Q, Itagaki K, Hauser CJ. Mitochondrial DNA is released by shock and activates neutrophils via p38 map kinase. Shock. 2010;34:55–9. PubMed:19997055
46. Wang X. New biomarkers and therapeutics can be discovered during COPD-lung cancer transition. Cell Biol Toxicol. 2016;32:359–61. PubMed:27405768

47. Chen C, Shi L, Li Y, Wang X, Yang S. Disease-specific dynamic biomarkers selected by integrating inflammatory mediators with clinical informatics in ARDS patients with severe pneumonia. Cell Biol Toxicol. 2016;32:169–84. PubMed:27095254
48. Arumugam P, Samson A, Ki J, Song JM. Knockdown of clusterin alters mitochondrial dynamics, facilitates necrosis in camptothecin-induced cancer stem cells. Cell Biol Toxicol. 2017;33:307–21. PubMed:28064403
49. Lippai M, Szatmari Z. Autophagy-from molecular mechanisms to clinical relevance. Cell Biol Toxicol. 2017;33:145–68. PubMed:27957648
50. Zhang Z, Meng P, Han Y, Shen C, Li B, Hakim MA, et al. Mitochondrial DNA-LL-37 complex promotes atherosclerosis by escaping from autophagic recognition. Immunity. 2015;1137-47:43. PubMed:26680206
51. Carlos D, Costa FR, Pereira CA, Rocha FA, Yaochite JN, Oliveira GG, et al. Mitochondrial DNA activates the NLRP3 inflammasome and predisposes to type 1 diabetes in Murine model. Front Immunol. 2017;8:164. PubMed:28289409
52. Denardin CC, Martins LA, Parisi MM, Vieira MQ, Terra SR, Barbe-Tuana FM, et al. Autophagy induced by purple pitanga (Eugenia uniflora L.) extract triggered a cooperative effect on inducing the hepatic stellate cell death. Cell Biol Toxicol. 2017;33:197–206. PubMed:27744523
53. Cao C, Wang W, Lu L, Wang L, Chen X, Guo R, et al. Inactivation of Beclin-1-dependent autophagy promotes ursolic acid-induced apoptosis in hypertrophic scar fibroblasts. Exp Dermatol. 2017. https://doi.org/10.1111/exd.13410. PubMed:28767174
54. Nakahira K, Haspel JA, Rathinam VA, Lee SJ, Dolinay T, Lam HC, et al. Autophagy proteins regulate innate immune responses by inhibiting the release of mitochondrial DNA mediated by the NALP3 inflammasome. Nat Immunol. 2011;12:222–30. PubMed:21151103
55. Rodgers MA, Bowman JW, Liang Q, Jung JU. Regulation where autophagy intersects the inflammasome. Antioxid Redox Signal. 2014;20:495–506. PubMed:23642014
56. Unuma K, Aki T, Funakoshi T, Hashimoto K, Uemura K. Extrusion of mitochondrial contents from lipopolysaccharide-stimulated cells: Involvement of autophagy. Autophagy. 2015;11:1520–36. PubMed:26102061
57. Liu S, Zhang Y, Ren J, Li J. Microbial DNA recognition by cGAS-STING and other sensors in dendritic cells in inflammatory bowel diseases. Inflamm Bowel Dis. 2015;21:901–11. PubMed:25581829
58. Rongvaux A, Jackson R, Harman CC, Li T, West AP, de Zoete MR, et al. Apoptotic caspases prevent the induction of type I interferons by mitochondrial DNA. Cell. 2014;159:1563–77. PubMed:25525875
59. Zhang X, Yin H, Li Z, Zhang T, Yang Z. Nano-TiO2 induces autophagy to protect against cell death through antioxidative mechanism in podocytes. Cell Biol Toxicol. 2016;32:513–27. PubMed:27430495
60. Yuan L, Mao Y, Luo W, Wu W, Xu H, Wang XL, et al. Palmitic acid dysregulates the Hippo-YAP pathway and inhibits angiogenesis by inducing mitochondrial damage and activating the cytosolic DNA sensor cGAS-STING-IRF3 signaling. J Biol Chem. 2017. https://doi.org/10.1074/jbc.M117.804005. PubMed:28698384
61. Mao Y, Luo W, Zhang L, Wu W, Yuan L, Xu H, et al. STING-IRF3 triggers endothelial inflammation in response to free fatty acid-induced mitochondrial damage in diet-induced obesity. Arterioscler Thromb Vasc Biol. 2017;37:920–9. PubMed:28302626
62. Zhu D, Liu Z, Pan Z, Qian M, Wang L, Zhu T, et al. A new method for classifying different phenotypes of kidney transplantation. Cell Biol Toxicol. 2016;32:323–32. PubMed:27278387
63. Biacchesi S, Merour E, Lamoureux A, Bernard J, Bremont M. Both STING and MAVS fish orthologs contribute to the induction of interferon mediated by RIG-I. PLoS One. 2012;7:e47737. PubMed:23091644
64. White MJ, McArthur K, Metcalf D, Lane RM, Cambier JC, Herold MJ, et al. Apoptotic caspases suppress mtDNA-induced STING-mediated type I IFN production. Cell. 2014;159:1549–62. PubMed:25525874

65. Petrasek J, Iracheta-Vellve A, Csak T, Satishchandran A, Kodys K, Kurt-Jones EA, et al. STING-IRF3 pathway links endoplasmic reticulum stress with hepatocyte apoptosis in early alcoholic liver disease. Proc Natl Acad Sci U S A. 2013;110:16544–9. PubMed:24052526
66. Onuora S. Autoinflammation: a new STING-associated monogenic autoinflammatory disease. Nat Rev Rheumatol. 2014;10:512. PubMed:25090944
67. Li XD, Wu J, Gao D, Wang H, Sun L, Chen ZJ. Pivotal roles of cGAS-cGAMP signaling in antiviral defense and immune adjuvant effects. Science. 2013;341:1390–4. PubMed:23989956
68. Gao D, Wu J, YT W, Du F, Aroh C, Yan N, et al. Cyclic GMP-AMP synthase is an innate immune sensor of HIV and other retroviruses. Science. 2013;341:903–6. PubMed:23929945
69. Wu B, Ni H, Li J, Zhuang X, Zhang J, Qi Z, et al. The impact of circulating mitochondrial DNA on cardiomyocyte apoptosis and myocardial injury after TLR4 activation in experimental autoimmune myocarditis. Cell Physiol Biochem. 2017;42:713–28. PubMed:28618428
70. Xie L, Liu S, Cheng J, Wang L, Liu J, Gong J. Exogenous administration of mitochondrial DNA promotes ischemia reperfusion injury via TLR9-p38 MAPK pathway. Regul Toxicol Pharmacol. 2017;89:148–54. PubMed:28757323
71. Duvvuri B, Duvvuri VR, Wang C, Chen L, Wagar LE, Jamnik V, et al. The human immune system recognizes neopeptides derived from mitochondrial DNA deletions. J Immunol. 2014;192:4581–91. PubMed:24733843
72. Zhou QF, SM X, Wang HQ, Xing LM, Fu R, Shao ZH. Single nucleotide polymorphism of mitochondrial DNA D-LOOP region in peripheral blood lymphocytes of immuno-related pancytopenia patients. Zhongguo Shi Yan Xue Ye Xue Za Zhi. 2017;25:186–91. PubMed:28245399
73. Wang W, Zhuang Q, Ji K, Wen B, Lin P, Zhao Y, et al. Identification of miRNA, lncRNA and mRNA-associated ceRNA networks and potential biomarker for MELAS with mitochondrial DNA A3243G mutation. Sci Rep. 2017;7:41639. PubMed:28139706
74. Imanishi H, Takibuchi G, Kobayashi T, Ishikawa K, Nakada K, Mori M, et al. Specific mtDNA mutations in mouse carcinoma cells suppress their tumor formation via activation of the host innate immune system. PLoS One. 2013;8:e75981. PubMed:24098752
75. Feng S, Xiong L, Ji Z, Cheng W, Yang H. Correlation between increased ND2 expression and demethylated displacement loop of mtDNA in colorectal cancer. Mol Med Rep. 2012;6:125–30. PubMed:22505229
76. Fang H, Wang W. Could CRISPR be the solution for gene editing's Gordian knot? Cell Biol Toxicol. 2016;32:465–7. PubMed:27614448
77. Wang W, Wang X. Single-cell CRISPR screening in drug resistance. Cell Biol Toxicol. 2017;33:207–10. PubMed:28474250
78. Paes B, Moco PD, Pereira CG, Porto GS, de Sousa Russo EM, Reis LCJ, et al. Ten years of iPSC: clinical potential and advances in vitro hematopoietic differentiation. Cell Biol Toxicol. 2017;33:233–50. PubMed:28039590
79. Arroyo JD, Jourdain AA, Calvo SE, Ballarano CA, Doench JG, Root DE, et al. A genome-wide CRISPR death screen identifies genes essential for oxidative phosphorylation. Cell Metab. 2016;24:875–85. PubMed:27667664
80. Luo C, Lim JH, Lee Y, Granter SR, Thomas A, Vazquez F, et al. A PGC1alpha-mediated transcriptional axis suppresses melanoma metastasis. Nature. 2016;537:422–6. PubMed:27580028

Lingyan Wang, BS, is a project leader of high-throughput Cellomics, at Clinical Science Institute of Fudan University Zhongshan Hospital, Shanghai, China. She is the author of more than 30 scientific publications. She is interested in the field of immunology and oncology.

Michael N. Liebman, PhD (theoretical chemistry and protein crystallography), is the Managing Director of IPQ Analytics, LLC and Strategic Medicine, Inc. after serving as the Executive Director of the Windber Research Institute (now Chan Soon-Shiong Institute for Molecular Medicine) from 2003 to 2007. He is an Adjunct Professor of Pharmacology and Physiology at Drexel College of Medicine and Adjunct Professor of Drug Discovery, First Hospital of Wenzhou Medical University and also Fudan University. Previously, he was Director, Computational Biology and Biomedical Informatics, at the University of Pennsylvania Cancer Center 2000–2003. He served as Global Head of Computational Genomics, Roche Pharmaceuticals; Director, Bioinformatics and Pharmacogenomics, Wyeth Pharmaceuticals; and Director of Genomics for Vysis, Inc. He is a cofounder of Prosanos, Inc. (now United BioSource) (2000). He was an Associate Professor of Pharmacology and of Physiology/Biophysics at Mount Sinai School of Medicine. Michael is Chair of the Informatics Program and also Translational Medicine and Therapeutics for the PhRMA Foundation. He is a member of the IUPAC Division on Human Health's Medicinal Chemistry subcommittee. He has served on the External Advisory Board for the INBRE (NIH) program for the state of Delaware since 2000. He is an Invited Professor at the Shanghai Center for Bioinformatics Technology and of the Chinese Academy of Sciences. His research focuses on computational models of disease progression that stress risk detection, disease processes and clinical pathway modeling, and disease stratification from the clinical perspective. He utilizes systems-based approaches and design thinking to represent and analyze risk/benefit analysis in pharmaceutical development and healthcare.

Xiangdong Wang, MD, PhD, is a Distinguished Professor of Medicine, Director of Shanghai Institute of Clinical Bioinformatics, Executive Director of Clinical Science Institute of Fudan University Zhongshan Hospital, Director of Fudan University Center of Clinical Bioinformatics, Deputy Director of Shanghai Respiratory Research Institute, and visiting professor of King's College London. His main research is focused on clinical bioinformatics, disease-specific biomarkers, lung chronic diseases, cancer immunology, and molecular and cellular therapies. He is the author of more than 200 scientific publications with the impact factor of about 900, citation number of about 6918, h-index of 46, i10-index of 181, and cited journal impact factor of about 7000.

Chapter 5
Mitochondrial DNA in Telocytes

Dongli Song, Dragos Cretoiu, and Xiangdong Wang

Abstract Telocyte (TC) is a new identified interstitial cell type with a small nuclear and one or several long and thin prolongations with enlargements on them. They were found in many mammals including humans, mouse, rats, dogs, and monkeys and play vital roles in many physiological and pathological conditions. The ultrastructure of mitochondria was observed in TCs, and the alterations were found in TCs from inflammatory ureter tissue. MtDNA is associated with mitochondria normal functions and involved in physiological and pathological processes. However, mitochondria and mtDNA in TCs were not investigated deeply. This review will introduce the origin, distribution, morphology, and functions of TCs and the distribution and functions of TC mitochondria in order to improve a better understanding of the potential functions of mtDNA in TCs.

Keywords Mitochondria • DNA • Telocytes

D. Song (✉)
Zhongshan Hospital Institute of Clinical Science, Fudan University,
Shanghai Medical College, Shanghai, China

Shanghai Institute of Clinical Bioinformatics, Shanghai, China
e-mail: songdongli37@126.com

D. Cretoiu
Division of Cellular and Molecular Biology and Histology, Department of Morphological Sciences, Carol Davila University of Medicine and Pharmacy, Bucharest, Romania

Victor Babeş National Institute of Pathology, Bucharest, Romania
e-mail: dragos@cretoiu.ro

X. Wang (✉)
Zhongshan Hospital Institute of Clinical Science, Fudan University,
Shanghai Medical College, Shanghai, China
e-mail: xiangdong.wang@clintransmed.org

H. Sun, X. Wang (eds.), *Mitochondrial DNA and Diseases*, Advances in Experimental Medicine and Biology 1038,
https://doi.org/10.1007/978-981-10-6674-0_5

1 Introduction

Telocyte (TC), a new identified cell type, is a kind of interstitial cell. It is different from stem cells and fibroblast from the characteristic features; one or several long and thin prolongations with enlargements on them, named telopodes (Tps); and specific markers detected by immunocytochemistry. TCs are distributed widely in many organs and tissues in human bodies, and they were also found in many other mammals, i.e., mouse, rats, dogs, and monkeys. The roles of TCs include embryogenesis, homeostasis, regeneration, remodeling, repair, angiogenesis, and tumorigenesis.

Energy production is the main function of mitochondria; besides, ATP transport and stress responses are involved in mitochondria. MtDNA is the own genome that the mitochondria individually encapsulate. Mitochondria were observed in the cytoplasm of TCs under scanning electron microscope. The roles of mitochondria and mtDNA in TCs should not be ignored; however, there is no direct research on the roles of them under biological and pathological process [1]. Therefore, we will summarize the functions of mitochondria and mtDNA involved in the physiological and pathophysiological roles of TCs and supply perspectives of TCs associated with mitochondria and mtDNA.

2 Brief Introduction of TCs

2.1 Discovery of TCs

The beginning of TCs is from 100 years ago, when S. Ramon y Cajal discovered a cell type in the gut and named it "interstitial neuron" [2]. This type of cells has features looking like neurons and located in the interstitial. Although the existence of this cell is controversial in the scientific community after discovery, their ultrastructure was observed, and the existents were demonstrated by the application of electron microscope [3]. Then, they were demonstrated not neurons by L. Thuneberg and named interstitial cells of Cajal (ICC) later [4]. When Popescu and his team looked for Cajal cells in the pancreas, "interstitial neurons" were found, and these cells were named interstitial Cajal-like cells (ICLC) by a series of publications [5]. After Popescu distinguished ICLC from ICC in 2010, the name "telocytes" was proposed instead of Cajal-like cells. The brief history of TCs discovery was summarized in Fig. 5.1.

2.2 Cell Morphology and Ultrastructural Aspects

TCs were described as small cell body with one to five long and thin prolongations, named telopodes (Tps), containing dilatations termed podoms and podomeres. And the presence of Tps is the most characteristic feature of TCs. Mitochondria, small Golgi apparatus, endoplasmic reticulum, and lipid blobs were shown in the cytoplasm of TCs (Fig. 5.1). The morphology of TCs is similar with the phenotypes of fibroblasts, pericytes, and neurons, while neurons have more prolongations with

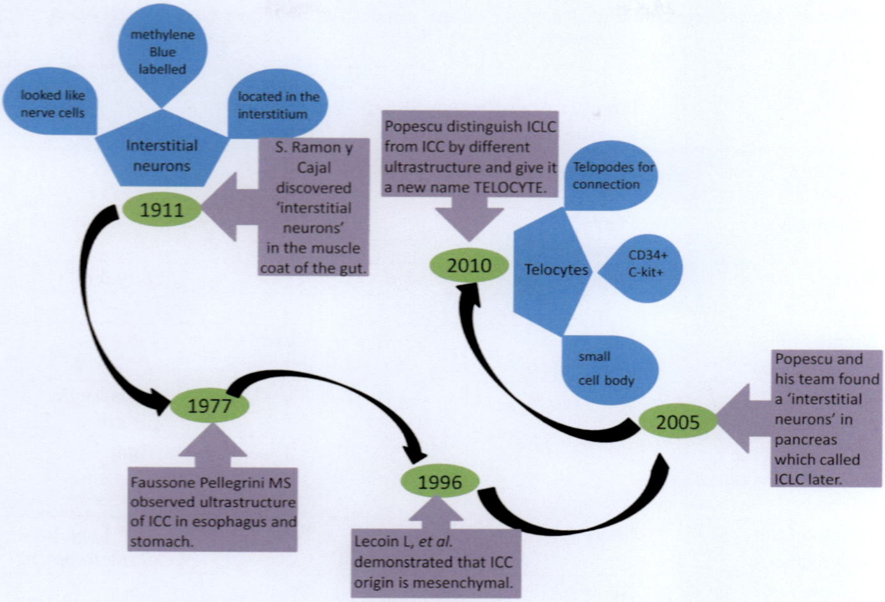

Fig. 5.1 The brief discovery history of telocytes. S. Ramon Cajal discovered "interstitial neurons" in the muscle coat of the gut in 1911. Faussone-Pellegrini MS observed ultrastructure of ICC in the esophagus and stomach in 1977. Popescu and his team found an "interstitial neuron" in the pancreas which is called ICLC in 2005. Popescu distinguished "ICLC" from ICC by different ultrastructure and gave it a new name, TELOCYTE, in 2010

complex branches and no expansions like Tps on the cell body. Synapses are different with Tps since they are located in the terminal of a neuron and can form synaptic junction. The cell body of pericytes is flattened with various shapes, and their nuclei are elongated and fail to have Tps [6]. Table 5.1 demonstrates the comparisons of ultrastructure of TCs with other interstitial cells including fibroblast-type interstitial cells, myoid interstitial cells, and branched interstitial cells [7].

The shape of cultured TC body depends on the number of Tps including piriform, spindle, triangular, and stellate shaped corresponding to one to four or more Tps, respectively [8]. The formation of the shapes of TCs mainly depends on movement, presented as cellular polymorphism [9]. Mitochondria, Golgi, actin, and endoplasmic reticulum were also observed by molecular probes using laser scanning confocal microscope (Fig. 5.2). TCs can release exosomes as biological mediators, and cultured TCs produce exosome as well (Fig. 5.3).

2.3 Immunophenotypes of TCs

TCs show a broad range of staining affinity positive to CD117/c-kit, caveolin-1, vimentin, CD34, desmin, S-100 protein, and progesterone and estrogen receptors alpha. However, the immunophenotypes of TCs are dependent on organs, tissues,

Table 5.1 Morphology and ultrastructural features of telocytes and three types of fibroblasts

Cell types	Fibroblast-type interstitial cells	Myoid interstitial cells	Branched interstitial cells	Telocytes
Cell body	Large	Sheet like	Thin (1–4 μm)	Flat
Cytoplasm	Plenty (prominent)			
Process	Slender	Slender		Very thin
Nucleus				
Number of prolongations			Variable number	1–5
Branching of the prolongations				Dichotomic pattern
Sprouting of the prolongations out of the cell body				Thin
Length of the prolongations	Short		Long	Very long, tens of micrometers
Podomeres	No	No	Yes	Yes
Podoms	No	No	Yes	Yes
Mitochondria	Abundant	Little and small		Yes
rER and sER	Prominent	Little		Yes
Basal lamina	No	No		
Microfilament bundles	No	Yes	Yes	
Fibronexus-type adhesion plaques	No	No	Numerous	
Intercellular junctions	No	Few	Yes	Yes

and embryonic periods. TCs are located in the gastrointestinal tract and do not express c-kit [10]. TCs in the skeletal muscle of human and rat interstitium express c-kit, caveolin-1, vimentin, and VEGF. TCs in lung tissue are positive reactions for c-kit, vimentin, and CD34. Double immunolabeling for c-kit/CD34 and for c-kit/vimentin is used to identify epicardial TCs and myocardial TCs in humans. The expression of c-kit of myocardial TCs was different during the embryonic life since c-kit-negative TCs were observed during the early stage of embryonic and in newborns [11]. No single immuno-marker can identify TCs alone, since there are $CD34^+$ stromal cells, fibroblasts, fibrocytes, TCs, and mesenchymal cells [12].

TCs belong to the endothelial lineage since they share the similar markers of CD34 and PDGFR-α with endothelial cells, CD34 with endothelial progenitor cells, and α-SMA with pericytes in larger vessels. However, the characteristic structure of Tps and double immunolabeling for CD34/c-kit c-kit/vimentin expression were ignored [13]. Moreover, TCs were demonstrated with stem cells and fibroblasts on morphology, ultrastructure, and profiling [14–18].

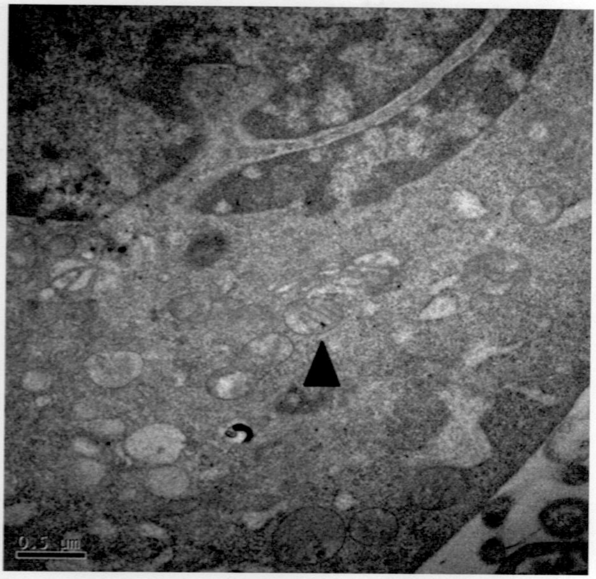

Fig. 5.2 The ultrastructure of telocytes under transmission electron microscopy. Cultured telocytes were observed under transmission electron microscopy, and the mitochondria were indicated with black arrowhead

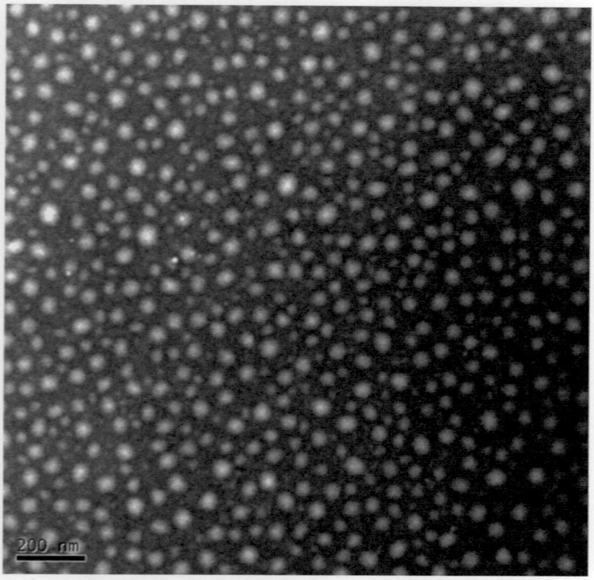

Fig. 5.3 Exosomes released by cultured telocytes (Under transmission electron microscopy). Exosomes were isolated from cultured telocytes and observed under transmission electron microscopy

2.4 Location and Roles

TCs are located in the various organs and tissues, including the heart, thoracic duct, trachea and lungs, pancreas, liver, gallbladder, spleen, cornea, skin, pleura, bone marrow, meninges, choroid plexus, dura mater, urinary system, uterus, oviduct, gland, placenta, testicles, prostate, parotid gland, pleura, minor salivary glands, gastrointestinal tract, mesentery, fascia lata, temporomandibular joint disc, skeletal muscles, blood vessels, neuromuscular spindles, and trigeminal ganglion. They were also discovered in cancerous tissues: squamous cell carcinoma and basal cell carcinoma, mammary gland carcinoma, as well as in non-small-cell lung cancer (Fig. 5.4). TCs are involved in physiological and pathophysiological processes. They were first considered to be communication cells, and the classical roles of TCs were regeneration and repair. Activated CD34+ SC/TCs were considered to be the substrate of tissue repair through granulation tissue [19].

TCs play an essential role in maintaining the integrity of the myocardium and in the regeneration of damaged myocardium. The pretransplantation of TCs and the combination with MSCs will relieve lung tissue inflammation and improve airway hyperresponsiveness and could be a new therapeutic role for asthma [20]. Pretreatment of TCs decreases the infarct size and improves myocardial functions during myocardial infarction. Translation of TCs increased angiogenesis at the infarct site, decreased myocardium fibrosis, and improved pathological reconstruction of ventricles [8, 12, 21]. Macromolecular assembly would be one type of TC-CM connection that is essential for cardiac development, renewal, repair, and cardiac physiology. Calcium signaling is another mechanism involved in TCs in

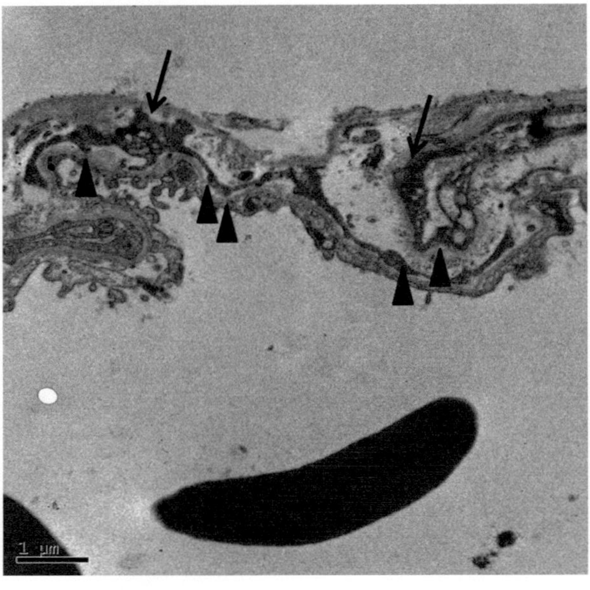

Fig. 5.4 Telocytes in tumor tissue of non-small-cell lung cancer. Tumor tissue of non-small-cell lung cancer was isolated and observed under transmission electron microscopy. Telocytes were indicated with black arrow, and telopodes were indicated with black arrowhead

cardiac pathologies [22]. TCs were involved in inflammatory/repair processes in intestine diseases, including appendicitis, diverticulitis of the large bowel, and Crohn's disease of the terminal ileum, and mediated inflammatory, proliferative, and remodeling [23].

Recently, the roles of TCs in tissue homeostasis maintenance, organ development regulation, and immune surveillance were found. TCs were demonstrated to mediate liver regeneration and could give therapeutic effects on liver fibrosis. TCs in the spleen might alleviate the progress of splenic malfunction via recruiting hematopoietic stem cells and macrophages in type C1 Niemann-Pick disease. TCs are potential functional players in local immunoregulatory and immunosurveillance in gynecologic diseases and infertility. TCs in the prostate were in normal metabolism and prostate development [9].

3 Mitochondria in TCs

3.1 Distribution of Mitochondria in TCs

Accommodate mitochondria marked with Janus Green and MitoTracker were shown in TCs of lung tissue and they located in TC cell body and as well as the level of podoms. Mitochondria were observed under transmission electron microscope together with vesicles accommodated in the thick bead-like portions of TCs which are named podoms. Although the cell body of TCs was small, their cytoplasm contained numerous mitochondria. Significant accumulations of mitochondria were present within some of the podoms as well. The vesicles together with mitochondria in podoms are including endoplasmic reticulum and caveolae and alternate with podomeres. Ribosomes were observed existing nearby mitochondria in Tps of TCs from lung tissues under transmission electron microscopy [20].

Mitochondria were swollen and had a clear matrix, and few cristae in TCs of different stages of systemic sclerosis of the skin indicate that mitochondria in TCs display ultrastructural abnormalities. And the ultrastructural abnormalities of numerous swollen mitochondria were shown in advanced diffuse cutaneous systemic sclerosis of the skin. Few mitochondria and cisternae of endoplasmic reticulum and a small Golgi apparatus were observed in advanced limited cutaneous systemic sclerosis of the skin indicating the content of mitochondria would change in diseases [24].

3.2 Mitochondrial Respiration and ATP Production in TCs

The central role of mitochondria is energy production. Mitochondria oxidate food molecules and convert energy in the form of ATP that fuels cell functions. Mitochondrial efficiency has been suggested to be a major physiological adaptation for regulating energy homeostasis in extreme environmental conditions of starvation [25]. The motility of TCs with rich mitochondria in Tps suggest that energy providing of mitochondria for movement of TCs [9].

Furthermore, the hypothesis of an ischemic injury is strongly supported by the presence in TCs of numerous swollen mitochondria and extensive cytoplasmic vacuolization [24]. Mitochondria are involved in human diseases, including acute lung injury, myocardial infarction, and many tumor types. Mitochondria mediate oxidative stress and DNA damage, trigger death signals inside the cells, and regulate cell apoptosis pathways. The release of cytochrome c and apoptosis-inducing factor and caspase activation by mitochondria are the mechanisms of apoptosis regulations. Variation in the amount of ATP has significant consequences at all levels of biological organization, including energy homeostasis, tissue function, disease pathologies, and normal aging [26–28]. Levels of small heat shock proteins in mitochondrial of the failing heart following myocardial infarction were decreased and the induction of small heat shock proteins mediated by geranylgeranylacetone improved mitochondrial function and augment cardiac contractile function in rats myocardial infarction model. Ultrastructural alterations including swollen mitochondria in TCs may indicate a cellular degenerative process in early diffuse cutaneous systemic sclerosis and be more marked in the advanced stage of both systemic sclerosis subsets [24].

The mechanisms involved in mitochondrial fragmentation in tissue injury had been studied. Ultraviolet B radiation damages mitochondria and reduces ATP production by increasing reactive oxygen species (ROS) formation [29]. AKT/PKB signaling pathway was involved in mancozeb induces a mild oxidative stress, affects mitochondrial activity and ATP production in mouse and human granulosa cells [30]. Glucocorticoid receptor-γ has the role in modulating mitochondrial function, including increased mitochondrial mass, basal respiration, and ATP generation [31]. E2F1/miR-421/Pink axis was found to be a promoter of mitochondrial fragmentation, cardiomyocyte apoptosis, and myocardial infarction [32]. Recent research suggested pyrimidinol antioxidant analogues as potential therapeutic agents for neurodegenerative and mitochondrial disorders since they suppressed lipid peroxidation and reactive oxygen species (ROS), to preserve mitochondrial membrane potential and support ATP production [33].

3.3 Endoplasmic Reticulum-Mitochondria Interactions in TCs

It is not difficult to find from the ultrastructure of TCs that mitochondria are usually nearby endoplasmic reticulum which creates the physical basis for the endoplasmic reticulum (ER)-mitochondria interactions in TCs. The endoplasmic reticulum-mitochondria encounter structure (ERMES) is a protein complex that is located in the ER-mitochondria interactions, associates the two organelles to each other physically, creates communication between them, and plays roles in the regulation of various cellular functions: Ca^{2+} homeostasis, mitochondrial dynamics, transport, autophagy, ER stress, apoptotic signaling, and inflammation [34].

Alterations of the ER-mitochondria axis could be responsible for the onset and progression of several diseases, including neurodegeneration, cancer, and obesity [35]. ERMES exchanges calcium (Ca^{2+}), regulates cellular homeostasis, controls hepatic metabolism, and plays a central role in metabolic inflexibility and insulin resistance during the procession of hepatic metabolic diseases [34]. Enriched ER-mitochondria interaction could increase mitochondrial Ca^{2+} which is associated with increased apoptosis and compromised cytoplasmic maturation in oocytes from obese mice models. Downregulation of ERMES-related protein IP3R1 and PACS-2 in oocytes from obese mice decreases mitochondrial Ca^{2+}, reduces the rate of apoptosis, and improved cytoplasmic maturation of oocytes from obese mice. This findings suggest ERMES should be a therapeutic target for obesity-induced oocyte defects [36]. The stimulation of ER-mitochondria cross talk could inhibit apoptotic pathways in STZ-induced type I diabetes [37]. Manipulation of ER-mitochondria interaction on Ca^{2+} flux is the key mechanism underlying the action of several tumor suppressor and oncogenic genes. Besides, providing asylum to a number of proteins with oncogenes and tumor suppressor properties represents one of the key functions of ER-mitochondria [38].

3.4 Mitochondria and Cancers

TCs showed promotion on typical breast structure assembly, induced breast cancer cell proliferation, and inhibited apoptosis, mediated or associated with mitochondria. Moreover, the ultrastructure of TCs was investigated in tumor stroma in basal cell carcinoma and squamous cell carcinoma using transmission electron microscopy [39]. Since Ca^{2+} signaling-mediated tumor-stroma interaction is necessary for tumor progression and ER-mitochondria interaction in TCs is involved in exchanges of Ca^{2+} [34], mitochondria of TCs are probably involved in tumor-stroma interaction. Myricetin-activated caspase cascades upregulated the release of apoptosis-inducing factor, induced the mitochondria dysfunction, and finally led to the death of human papillary thyroid cancer cells. Those results provided therapeutic agents for human thyroid cancers [40, 41]. Reactive oxygen species (ROS) induction and alterations in mitochondrial membrane potential induced by caffeic acid n-butyl ester were determined to be the mechanisms involved in A549 cell death induction [42].

4 Mitochondria DNA in TCs

Mitochondrial DNA (mtDNA) is located near the inner mitochondrial membrane and electron transport system, making it prone to oxidative stress from the electron transport system. Human mtDNA, similar to that of its bacterial ancestors, is a double-stranded circular molecule. On one hand, the mtDNA encodes the mitochondrial respiratory chain complex associated protein under physiological conditions. On the other hand, mtDNA is generally released and plays an important role in the development of different kinds of inflammatory diseases, including acute and chronic ischemic diseases, traumatic brain injury, and infectious diseases. Moreover, expression of mtDNA-encoded genes is largely regulated by mtDNA copy number [43].

4.1 mtDNA Injury in TCs

mtDNA oxidative damage resulted in acute lung injury and was demonstrated in rat cultured pulmonary artery endothelial cells, perfused lungs, and intact mice. ROS, one of second messengers, were involved in [44]. Human uterus TCs are considered to be mechanical sensors and involved in transferring a signal which leads to the activation of gene expression [45]. PI3K signaling pathways was found mediated oxidative mtDNA damage and accompanied molecular patterns in the development of trauma-related multiple organ dysfunction syndrome. Pretreatment with the PI3K inhibitor LY294002 resulted in damage suppression and indicates a new strategy for preventing mtDNA damage and associated inflammatory factor release and prevents patients from trauma-related multiple organ dysfunction syndrome [46].

Alterations of mtDNA sequence caused by heteroplasmic mtDNA variants have been detected in types of tumors. Urinary bladder carcinoma would release tumor cells with high copy number of tumor-derived mtDNA into bodily fluids. Recently researchers studied mtDNA copy number in patients with head and neck cancer and compared with that of malignancy-free controls. The results showed significantly higher mtDNA copy number in head and neck cancer patients, and the mtDNA copy number was associated with cancer stage and survival. A study on 109 gastric cancer samples showed that mitochondrial DNA copy number was significantly higher in cancer tissue compared with that in the adjacent normal mucosa and might be involved in gastric cancer initiation and progression. Bacteria of the gastrointestinal tract including *Helicobacter pylori* and *Enterococcus faecalis* were demonstrated to induce genomic instability and mitochondrial dysfunction in cancer [47].

4.2 mtDNA Copy Number Disorders in TCs

TCs are involved in inflammation and regeneration. mtDNA release can mediate acute and chronic inflammatory diseases, including acute lung injury, rheumatoid arthritis, and acute myocardial infarction, by activating neutrophils and inducing inflammation and regeneration. Higher plasma leukocyte mtDNA copy number would be a selectable prognostic biomarker in myocardial infarction patients since they were associated with improvement of left ventricular shape and myocardium remodeling [48].

Mitochondria damage resulted in the release of mtDNA into blood circulation which activates inflammatory signaling pathways and tissue injury. The accumulation of mtDNA and inflammatory factors, such as TNF-α and IL-6, was detected in the plasma of thermal injury rat models and caused severe ARDS. Moreover, inflammatory factors would recruit neutrophils to the alveolar capillary membrane, release various proteases, and cause endothelium damage. Injection with mtDNA significantly augments the concentrations of inflammatory factors. Limiting mtDNA release by epigallocatechin gallate would relieve acute lung injury and have a protective role on acute respiratory distress syndrome (ARDS) in a rat model [49].

5 Conclusion and Perspectives

Although the mitochondria and mtDNA of TCs and the signaling mechanisms are incompletely understood, great therapeutic opportunities were opened for predominantly organs and diseases, on which mitochondria and mtDNA have vital roles. Targeting mitochondria and mtDNA of TCs may provide a novel prognostic biomarker for diseases. The single-cell biology is an alternative to explore human telocyte biology and heterogeneity, although we should consider the irregular shapes of telocytes which make the single-cell isolation more difficult [50–54]. Telocyte mtDNA should be furthermore investigated and compared with mtDNA of other interstitial cells as well as own nDNA by rapid development of clustered regularly interspaced short palindromic repeats [55–57]. One of the most difficult challenges is to discover and develop telocyte-specific biomarkers to monitor telocyte evolution, differentiation, and development, which are reliable, repeatable, measurable, and targetable, as described for criteria of other biomarkers [58–65].

Acknowledgments The work was supported by Zhongshan Distinguished Professor Grant (XDW), the National Nature Science Foundation of China (91230204, 81270099, 81320108001, 81270131, 81300010), the Shanghai Committee of Science and Technology (12JC1402200, 12431900207, 11410708600, 14431905100), Operation funding of Shanghai Institute of Clinical Bioinformatics, Ministry of Education for Academic Special Science and Research Foundation for PhD Education (20130071110043), and National Key Research and Development Program (2016YFC0902400, 2017YFSF090207).

References

1. Wang J, Jin M, Ma WH, Zhu Z, Wang X. The history of Telocyte discovery and understanding. Adv Exp Med Biol. 2016;913:1–21. [Pubmed:27796877]
2. Popescu LM, Faussone-Pellegrini MS. TELOCYTES - a case of serendipity: the winding way from interstitial cells of Cajal (ICC), via interstitial Cajal-like cells (ICLC) to TELOCYTES. J Cell Mol Med. 2010;14:729–40. [Pubmed:20367664]

3. Faussone Pellegrini MS. C Cortesini and P Romagnoli. [Ultrastructure of the tunica muscularis of the cardial portion of the human esophagus and stomach, with special reference to the so-called Cajal's interstitial cells]. Arch Ital Anat Embriol. 1977;82:157–77. [Pubmed:613989]
4. Thuneberg L. Interstitial cells of Cajal: intestinal pacemaker cells? Adv Anat Embryol Cell Biol. 1982;71:1–130. [Pubmed:7090872]
5. Popescu LM, Ciontea SM, Cretoiu D, Hinescu ME, Radu E, Ionescu N, Ceausu M, Gherghiceanu M, Braga RI, Vasilescu F, Zagrean L, Ardeleanu C. Novel type of interstitial cell (Cajal-like) in human fallopian tube. J Cell Mol Med. 2005;9:479–523. [Pubmed:15963270]
6. van der Scheer HT, Doelman A. Synapse fits neuron: joint reduction by model inversion. Biol Cybern. 2017;111:309–34. [Pubmed:28689352]
7. Vandecasteele T, Cornillie P, Vandevelde K, Logothetidou A, Couck L, van Loon G, Van den Broeck W. Presence of ganglia and telocytes in proximity to myocardial sleeve tissue in the porcine pulmonary Veins Wall. Anat Histol Embryol. 2017;46:325–33. [Pubmed:28421621]
8. Kucybala I, Janas P, Ciuk S, Cholopiak W, Klimek-Piotrowska W, Holda MK. A comprehensive guide to telocytes and their great potential in cardiovascular system. Bratisl Lek Listy. 2017;118:302–9. [Pubmed:28516795]
9. Song D, Cretoiu D, Cretoiu SM, Wang X. Telocytes and lung disease. Histol Histopathol. 2016;31:1303–14. [Pubmed:27463150]
10. Ibba-Manneschi L, Rosa I, Manetti M. Telocyte implications in human pathology: an overview. Semin Cell Dev Biol. 2016;55:62–9. [Pubmed:26805444]
11. Faussone-Pellegrini MS, Bani D. Relationships between telocytes and cardiomyocytes during pre- and post-natal life. J Cell Mol Med. 2010;14:1061–3. [Pubmed:20455994]
12. Zheng Y, Wang X. Roles of Telocytes in the development of angiogenesis. Adv Exp Med Biol. 2016;913:253–61. [Pubmed:27796893]
13. Popescu LM, Manole CG, Gherghiceanu M, Ardelean A, Nicolescu MI, Hinescu ME, Kostin S. Telocytes in human epicardium. J Cell Mol Med. 2010;14:2085–93. [Pubmed:20629996]
14. Popescu LM, Gherghiceanu M, Suciu LC, Manole CG, Hinescu ME. Telocytes and putative stem cells in the lungs: electron microscopy, electron tomography and laser scanning microscopy. Cell Tissue Res. 2011;345:391–403. [Pubmed:21858462]
15. Song D, Cretoiu D, Zheng M, Qian M, Zhang M, Cretoiu SM, Chen L, Fang H, Popescu LM, Wang X. Comparison of chromosome 4 gene expression profile between lung telocytes and other local cell types. J Cell Mol Med. 2016;20:71–80. [Pubmed:26678350]
16. Zheng Y, Cretoiu D, Yan G, Cretoiu SM, Popescu LM, Fang H, Wang X. Protein profiling of human lung telocytes and microvascular endothelial cells using iTRAQ quantitative proteomics. J Cell Mol Med. 2014;18:1035–59. [Pubmed:25059386]
17. Zheng Y, Cretoiu D, Yan G, Cretoiu SM, Popescu LM, Wang X. Comparative proteomic analysis of human lung telocytes with fibroblasts. J Cell Mol Med. 2014;18:568–89. [Pubmed:24674459]
18. Zheng Y, Zhang M, Qian M, Wang L, Cismasiu VB, Bai C, Popescu LM, Wang X. Genetic comparison of mouse lung telocytes with mesenchymal stem cells and fibroblasts. J Cell Mol Med. 2013;17:567–77. [Pubmed:23621815]
19. Wang DC, Wang X. Systems heterogeneity: an integrative way to understand cancer heterogeneity. Semin Cell Dev Biol. 2017;64:1–4. [Pubmed:27552921]
20. Ye L, D Song, M Jin, X Wang. Therapeutic roles of telocytes in OVA-induced acute asthma in mice. J Cell Mol Med. 2017. [Pubmed:28524369]
21. Zhaofu L, Dongqing C. Cardiac telocytes in regeneration of myocardium after myocardial infarction. Adv Exp Med Biol. 2016;913:229–39. [Pubmed:27796891]
22. Radu BM, Banciu A, Banciu DD, Radu M, Cretoiu D, Cretoiu SM. Calcium signaling in interstitial cells: focus on telocytes. Int J Mol Sci. 2017;18(2):E397. [Pubmed:28208829]
23. Diaz-Flores L, Gutierrez R, Garcia MP, Gonzalez M, Saez FJ, Aparicio F, Diaz-Flores L Jr, Madrid JF. Human resident CD34+ stromal cells/telocytes have progenitor capacity and are a source of alphaSMA+ cells during repair. Histol Histopathol. 2015;30:615–27. [Pubmed:25500909]

24. Manetti M, Guiducci S, Ruffo M, Rosa I, Faussone-Pellegrini MS, Matucci-Cerinic M, Ibba-Manneschi L. Evidence for progressive reduction and loss of telocytes in the dermal cellular network of systemic sclerosis. J Cell Mol Med. 2013;17:482–96. [Pubmed:23444845]
25. Salin K, Villasevil EM, Auer SK, Anderson GJ, Selman C, Metcalfe NB, Chinopoulos C. Simultaneous measurement of mitochondrial respiration and ATP production in tissue homogenates and calculation of effective P/O ratios. Physiol Rep. 2016;4(20):pii: e13007. [Pubmed:27798358]
26. Layec G, Bringard A, Le Fur Y, Micallef JP, Vilmen C, Perrey S, Cozzone PJ, Bendahan D. Opposite effects of hyperoxia on mitochondrial and contractile efficiency in human quadriceps muscles. Am J Physiol Regul Integr Comp Physiol. 2015;308:R724–33. [Pubmed:25695290]
27. Pham T, Loiselle D, Power A, Hickey AJ. Mitochondrial inefficiencies and anoxic ATP hydrolysis capacities in diabetic rat heart. Am J Physiol Cell Physiol. 2014;307:C499–507. [Pubmed:24920675]
28. Conley KE. Mitochondria to motion: optimizing oxidative phosphorylation to improve exercise performance. J Exp Biol. 2016;219:243–9. [Pubmed:26792336]
29. Ghanizadeh Kazerouni E, Franklin CE, Seebacher F. UV-B exposure reduces locomotor performance by impairing muscle function but not mitochondrial ATP production. J Exp Biol. 2016;219:96–102. [Pubmed:26567351]
30. Iorio R, Castellucci A, Rossi G, Cinque B, Cifone MG, Macchiarelli G, Cecconi S. Mancozeb affects mitochondrial activity, redox status and ATP production in mouse granulosa cells. Toxicol In Vitro. 2015;30:438–45. [Pubmed:26407525]
31. Morgan DJ, Poolman TM, Williamson AJ, Wang Z, Clark NR, Ma'ayan A, Whetton AD, Brass A, Matthews LC, Ray DW. Glucocorticoid receptor isoforms direct distinct mitochondrial programs to regulate ATP production. Sci Rep. 2016;6:26419. [Pubmed:27226058]
32. Wang K, Zhou LY, Wang JX, Wang Y, Sun T, Zhao B, Yang YJ, An T, Long B, Li N, Liu CY, Gong Y, Gao JN, Dong YH, Zhang J, Li PF. E2F1-dependent miR-421 regulates mitochondrial fragmentation and myocardial infarction by targeting Pink1. Nat Commun. 2015;6:7619. [Pubmed:26184432]
33. Chevalier A, Alam MP, Khdour OM, Schmierer M, Arce PM, Cripe CD, Hecht SM. Optimization of pyrimidinol antioxidants as mitochondrial protective agents: ATP production and metabolic stability. Bioorg Med Chem. 2016;24:5206–20. [Pubmed:27624526]
34. Xue Y, Schmollinger S, Attar N, Campos OA, Vogelauer M, Carey MF, Merchant SS, Kurdistani SK. Endoplasmic reticulum-mitochondria junction is required for iron homeostasis. J Biol Chem. 2017;292:13197–204. [Pubmed:28637866]
35. Filadi R, Theurey P, Pizzo P. The endoplasmic reticulum-mitochondria coupling in health and disease: molecules, functions and significance. Cell Calcium. 2017;62:1–15. [Pubmed:28108029]
36. Zhao L, Lu T, L Gao XF, Zhu S, Hou Y. Enriched endoplasmic reticulum-mitochondria interactions result in mitochondrial dysfunction and apoptosis in oocytes from obese mice. J Anim Sci Biotechnol. 2017;8:62. [Pubmed:28781772]
37. Yang F, Yu X, Li T, Wu J, Zhao Y, Liu J, Sun A, Dong S, Wu J, Zhong X, Xu C, Lu F, Zhang W. Exogenous H2S regulates endoplasmic reticulum-mitochondria cross-talk to inhibit apoptotic pathways in STZ-induced type I diabetes. Am J Physiol Endocrinol Metab. 2017;312:E190–203. [Pubmed:27998959]
38. Missiroli S, Danese A, Iannitti T, Patergnani S, Perrone M, Previati M, Giorgi C, Pinton P. Endoplasmic reticulum-mitochondria Ca2+ crosstalk in the control of the tumor cell fate. Biochim Biophys Acta. 2017;1864:858–64. [Pubmed:28064002]
39. Tao L, Wang H, Wang X, Kong X, Li X. Cardiac Telocytes. Curr Stem Cell Res Ther. 2016;11:404–9. [Pubmed:25584905]
40. Ha TK, Jung I, Kim ME, Bae SK, Lee JS. Anti-cancer activity of myricetin against human papillary thyroid cancer cells involves mitochondrial dysfunction-mediated apoptosis. Biomed Pharmacother. 2017;91:378–84. [Pubmed:28463801]

41. Jo S, Ha TK, Han SH, Kim ME, Jung I, Lee HW, Bae SK, Lee JS. Myricetin induces apoptosis of human anaplastic thyroid cancer cells via mitochondria dysfunction. Anticancer Res. 2017;37:1705–10. [Pubmed:28373432
42. Zhang YX, Yu PF, Gao ZM, Yuan J, Zhang Z. Caffeic acid n-butyl ester-triggered necrosis-like cell death in lung cancer cell line A549 is prompted by ROS mediated alterations in mitochondrial membrane potential. Eur Rev Med Pharmacol Sci. 2017;21:1665–71. [Pubmed:28429338]
43. Zekri A, Mesbahi Y, Ghanizadeh-Vesali S, Alimoghaddam K, Ghavamzadeh A, Ghaffari SH. Reactive oxygen species generation and increase in mitochondrial copy number: new insight into the potential mechanism of cytotoxicity induced by aurora kinase inhibitor, AZD1152-HQPA. Anticancer Drugs. 2017;28(8):841–51. [Pubmed:28639950]
44. Matsuura TR, Bartos JA, Tsangaris A, Shekar KC, Olson MD, Riess ML, Bienengraeber M, Aufderheide TP, Neumar RW, Rees JN, McKnite SH, Dikalova AE, Dikalov SI, Douglas HF, Yannopoulos D. Early effects of prolonged cardiac arrest and ischemic postconditioning during cardiopulmonary resuscitation on cardiac and brain mitochondrial function in pigs. Resuscitation. 2017;116:8–15. [Pubmed:28408349]
45. Cretoiu SM. Immunohistochemistry of telocytes in the uterus and fallopian tubes. Adv Exp Med Biol. 2016;913:335–57. [Pubmed:27796898]
46. Black GE, Sokol KK, Moe DM, Simmons J, Muscat D, Pastukh V, Capley G, Gorodnya O, Ruchko M, Roth MB, Gillespie M, Martin MJ. Impact of a novel PI3-KINASE inhibitor in preventing mitochondrial DNA damage and damage associated molecular pattern accumulation: results from the biochronicity project. J Trauma Acute Care Surg. 2017. [Pubmed:28538623]
47. Kirches E. MtDNA as a cancer marker: a finally closed chapter? Curr Genomics. 2017;18:255–67. [Pubmed:28659721].
48. Zheng Y, Bai C, Wang X. Potential significance of telocytes in the pathogenesis of lung diseases. Expert Rev Respir Med. 2012;6:45–9. [Pubmed:22283578]
49. Liu R, Xu F, Si S, Zhao X, Bi S, Cen Y. Mitochondrial DNA-induced inflammatory responses and lung injury in thermal injury rat model: protective effect of epigallocatechin gallate. J Burn Care Res. 2017. [Pubmed:28198717]
50. Chu MP, Kriangkum J, Venner CP, Sandhu I, Hewitt J, Belch AR, Pilarski LM. Addressing heterogeneity of individual blood cancers: the need for single cell analysis. Cell Biol Toxicol. 2017;33(2):83–97. PMID: 27761761
51. Wang W, Gao D, Wang X. Can single-cell RNA sequencing crack the mystery of cells? Cell Biol Toxicol. 2017. https://doi.org/10.1007/s10565-017-9404-y. PMID: 28733864
52. Wang W, Zhu B, Wang X. Dynamic phenotypes: illustrating a single-cell odyssey. Cell Biol Toxicol. 2017;33(5):423–7. PMID: 28638956
53. Wang W, Wang X. Single-cell CRISPR screening in drug resistance. Cell Biol Toxicol. 2017;33(3):207–10. https://doi.org/10.1007/s10565-017-9396-7. PMID: 28474250
54. Gu J, Wang X. New future of cell biology and toxicology: thinking deeper. Cell Biol Toxicol. 2016;32(1):1–3. PMID: 26874518
55. Fang H, Wang W. Could CRISPR be the solution for gene editing's Gordian knot? Cell Biol Toxicol. 2016;32(6):465–7.
56. Paes BCMF, Moço PD, Pereira CG, Porto GS, de Sousa Russo EM, Reis LCJ, Covas DT, Picanço-Castro V. Ten years of iPSC: clinical potential and advances in vitro hematopoietic differentiation. Cell Biol Toxicol. 2017;33(3):233–50. https://doi.org/10.1007/s10565-016-9377-2. PMID: 28039590
57. Kumar D, Anand T, Kues WA. Clinical potential of human-induced pluripotent stem cells: perspectives of induced pluripotent stem cells. Cell Biol Toxicol. 2017;33(2):99–112. PMID: 27900567
58. Wang X. New biomarkers and therapeutics can be discovered during COPD-lung cancer transition. Cell Biol Toxicol. 2016;32(5):359–61. PMID: 27405768
59. Chen C, Shi L, Li Y, Wang X, Yang S. Disease-specific dynamic biomarkers selected by integrating inflammatory mediators with clinical informatics in ARDS patients with severe pneumonia. Cell Biol Toxicol. 2016;32(3):169–84. PMID: 27095254

60. Zhu D, Liu Z, Pan Z, Qian M, Wang L, Zhu T, Xue Y, Wu DA. New method for classifying different phenotypes of kidney transplantation. Cell Biol Toxicol. 2016;32(4):323–32. PMID: 27278387
61. Bao L, Zhang Y, Wang J, Wang H, Dong N, Su X, Xu M, Wang X. Variations of chromosome 2 gene expressions among patients with lung cancer or non-cancer. Cell Biol Toxicol. 2016;32(5):419–35. PMID: 27301951
62. Bao L, Diao H, Dong N, Su X, Wang B, Mo Q, Yu H, Wang X, Chen C. Histone deacetylase inhibitor induces cell apoptosis and cycle arrest in lung cancer cells via mitochondrial injury and p53 up-acetylation. Cell Biol Toxicol. 2016;32(6):469–82. PMID: 27423454
63. Wang XCBT. Profiles of cabozantinib approved for advanced renal cell carcinomas. Cell Biol Toxicol. 2016;32(4):259–61. PMID: 27383755
64. Giromini C, Rebucci R, Fusi E, Rossi L, Saccone F, Baldi A. Cytotoxicity, apoptosis, DNA damage and methylation in mammary and kidney epithelial cell lines exposed to ochratoxin a. Cell Biol Toxicol. 2016;32(3):249–58. PMID: 27154019
65. Kang SJ, Lee HM, Park YI, Yi H, Lee H, So B, Song JY, Kang HG. Chemically induced hepatotoxicity in human stem cell-induced hepatocytes compared with primary hepatocytes and HepG2. Cell Biol Toxicol. 2016;32(5):403–17. PMID: 27287938

Dongli Song, BM, is a project leader of flow cytometry and responsible for immune research in the Clinical Science Institute of Fudan University Zhongshan Hospital, Shanghai, China. She has already studied on telocytes for 2 years.

Dragos Cretoiu, MD, PhD. Current position: Division of Cellular and Molecular Biology and Histology, Department of Morphological Sciences, Carol Davila University of Medicine and Pharmacy, Bucharest, Romania, and Department of Molecular Medicine, Victor Babes National Institute of Pathology, Bucharest, Romania. Education: MD from Carol Davila University, Bucharest (2003); PhD from Carol Davila University, Bucharest (2012); and Postdoctoral Fellowship at Victor Babes National Institute of Pathology, Bucharest, Romania (2014–2016). Research interests: I am particularly interested in telocytes, a newly discovered type of cell, and I have spent the past years working on computer 3D reconstructions and medical image processing and genomic/proteomic data analysis. My primary interest is in understanding telocytes' 3D configuration and the interrelations between these cells and the surrounding structures. Moreover, I have a great interest in studying telocytes in cell culture by various microscopy techniques, such as fluorescence and scanning confocal microscopy, time-lapse videomicroscopy, and electron microscopy.

Xiangdong Wang, MD, PhD, is a Distinguished Professor of Medicine, Director of Shanghai Institute of Clinical Bioinformatics, Executive Director of Clinical Science Institute of Fudan University Zhongshan Hospital, Director of Fudan University Center of Clinical Bioinformatics, Deputy Director of Shanghai Respiratory Research Institute, and Visiting Professor of King's College London. His main research is focused on clinical bioinformatics, disease-specific biomarkers, lung chronic diseases, cancer immunology, and molecular and cellular therapies. He is the author of more than 200 scientific publications with the impact factor of about 900, citation number of about 6918, h-index of 46, i10-index of 181, and cited journal impact factor of about 7000.

Chapter 6
Roles of Mitochondrial DNA in Energy Metabolism

Jiapei Lv, Madhav Bhatia, and Xiangdong Wang

Abstract Mitochondria are independent double-membrane organelles responsible for energy production, specifically by completing oxidative phosphorylation. Mitochondria are essential to regulate energy metabolism, signaling pathways, and cell death. Mitochondrial DNA (mtDNA) can be altered by metabolic disorders, oxidative stress, or inflammation in the progression and development of various diseases. In this chapter, we overview the role of mtDNA in energy metabolism and the diseases that are associated with mtDNA abnormality, with a special focus on the major factors which regulate the mechanism of mtDNA in metabolism.

Keywords Mitochondria • Metabolism • Energy • mtDNA • Disease

Abbreviations

ETC	Electron transport chain
HSP1	Heavy-strand promoter 1
LSP	Light-strand promoter
mtDNA	Mitochondrial DNA
OXPHOS	Oxidative phosphorylation
POLRMT	Single-subunit RNA polymerase
ROS	Reactive oxygen species
TFAM	Mitochondrial transcription factor A
TFB2M	Mitochondrial transcription factor B2

J. Lv
Zhongshan Hospital Institute of Fudan University, Shanghai Medical School, Shanghai, China

M. Bhatia
Department of Pathology, University of Otago, Wellington, New Zealand

X. Wang (✉)
Zhongshan Hospital Institute of Clinical Science, Fudan University,
Shanghai Medical College, Shanghai, China
e-mail: xiangdong.wang@clintransmed.org

H. Sun, X. Wang (eds.), *Mitochondrial DNA and Diseases*, Advances in Experimental Medicine and Biology 1038,
https://doi.org/10.1007/978-981-10-6674-0_6

1 Introduction

Mitochondria are the powerhouse in cells where most of the metabolic energy is generated for cellular use by many metabolic pathways, including citric acid cycle, oxidative phosphorylation (OXPHOS), oxidation of fatty acids, ketogenesis, portions of the pathways related to urea cycle, heme synthesis, or gluconeogenesis [1–3]. The functions and contents of mitochondrial DNA (mtDNA) vary according to a large number of factors and physiopathological changes. Recent study covered 310 human subjects and found the correlation between increased peripheral blood mtDNA content with low metabolic score, high neutrophil-to-lymphocyte ratio, and overproduction of inflammatory mediators [4]. It seems that an imbalance of systemic metabolisms and immune responses can change mtDNA vulnerability to oxidative stress, energy homeostasis, senescence, and aging [5].Single species of mtDNA deletion develops to high heteroplasmy during clonal expansion. Primary mitochondrial disorders can change mtDNA molecules and metabolic rate by signing transcriptional elements to affect the rate of clonal expansion in single cells [6].

Transcriptional elements, e.g., mitochondrial transcription factor A (TFAM) and mitochondrial transcription factor B2 (TFB2M), play the important roles in the regulation of nuclear-encoded single-subunit RNA polymerases in transcription of the mitochondrial genome [7, 8] TFAM as a DNA-binding protein maintains transcriptional activation and mtDNA organization [8]. The primary part of TFB2M stabilizes the open promoter complex and thaws the promoter by coinstantaneous binding of the priming substrate and the templating DNA base [9]. Transcription factors play critical roles in the regulation and bridging of signaling and interactions between mtDNA and mitochondria-dominated metabolism, mitochondrial dysfunction and drug metabolism, or mitochondrial metabolism and cell apoptosis/autophagy [10–13]. In this chapter, we will focus on the mechanisms of mtDNA in energy metabolism and the pivotal factors in mtDNA transcription.

2 Function of mtDNA in Energy Metabolism

Mitochondria execute multiple essential functions, such as OXPHOS, which generates energy, for ATP formation, by the electron transport chain (ETC) complexes in the inner mitochondrial membrane. Complexes I, II, III, and IV constitute the ETC, of which complex V is the ATP synthase [14]. Hydrogen atoms produced from different catabolic pathways bind to flavin adenine dinucleotide and nicotinamide adenine dinucleotide (NAD+) to yield FADH2 and NADH, respectively. NADH is oxidized by NADH dehydrogenase (complex I), whereas the electrons are transferred through multiple iron–sulfur (Fe–S) and flavin mononucleotide centers in complex I until they are transferred to coenzyme Q10 (CoQ10). CoQ10 receives hydrogen atoms from FADH2 produced by β-oxidation and the tricarboxylic acid enzyme succinate dehydrogenase (complex II). Electrons are subsequently

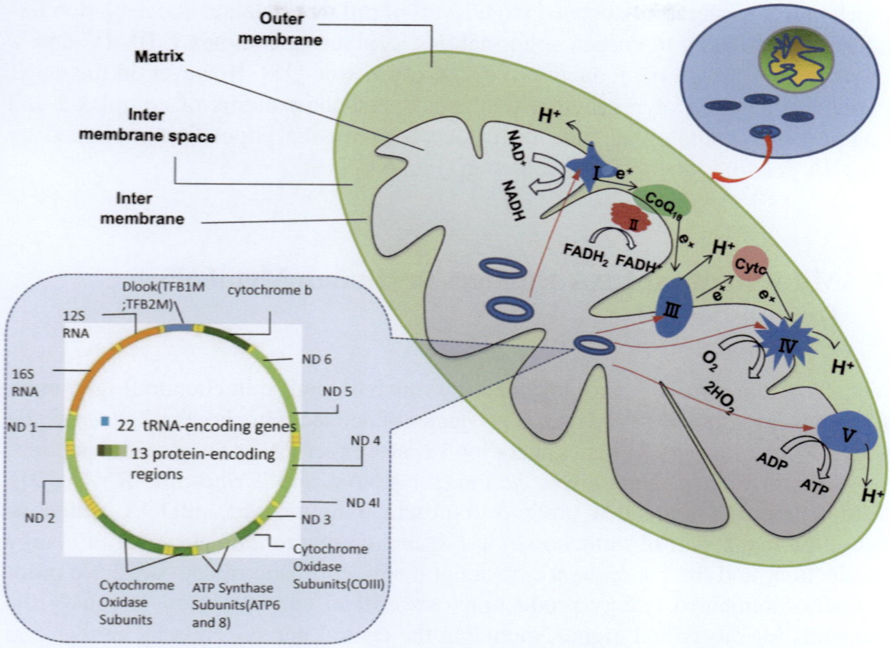

Fig. 6.1 Mitochondria are the place where oxidative phosphorylation proceeds, complexes I, II, III, and IV constitute the electron transport chain, or complex V acts as the ATP synthase. Diverse mitochondrial genome regions regulate the activities of complexes I, III, IV, and V, respectively, whereas complex II is controlled by nuclear genome

transported from CoQ10 to bc1 complex (complex III) within which the electrons move through the Fe–S components, cytochrome b, and cytochrome c1 [15]. The electrons which transfer the electrons to cytochrome c oxidase (complex IV) are then transferred from complex III to cytochrome c. Ultimately, the electrons are transferred through copper centers and cytochromes a and a3 and combine with O_2 to generate H_2O within this complex. The energy is released during electron transfer and accustomed to pump protons from inside the mitochondrial matrix across the inner mitochondrial membrane into the intermembrane space through complexes I, III, and IV. The resulting electrochemical gradient impels protons to move back through a proton channel in ATP synthase (complex V), which use this energy in compounding ATP. Both mtDNA and nuclear DNA (nDNA) encoded the ETC complexes, of which multipolypeptides are encoded entirely by nDNA except for complex II (Fig. 6.1) [16].

The mtDNA encodes 13 most significant OXPHOS polypeptides, including 7 of the 45 polypeptides of OXPHOS complex I (ND1-3, ND4L, ND4-6), 1 of the 11 polypeptides of complex III (cytochrome b, cytb), 3 of the 13 polypeptides of complex IV (COI-III), and 2 of the 15 polypeptides of complex V (ATP6 and 8) [17]. mtDNA deletions are formed, critical mutation threshold level reduces, and then the biochemical vulnerability occurs. High levels of mtDNA deletion and depletion frequently contributed to certain abnormalities, containing complex I, III, IV, and V dysfunction through the reduction of COX expression [18]. However, in the unexplained population of respiratory-deficient fibers, the patterns of complex I and complex IV subunits had little difference between the population explained by mtDNA deletions and mtDNA depletion (Fig. 6.1) [19].

3 Mitochondrial DNA Changes in Human Metabolism Diseases

Mitochondrial proteins are encoded by nuclear and mitochondrial genomes. Chromosomes include more than 1000 nuclear-encoded mitochondrial genes, while mtDNA contains only 37 genes encoding 13 messenger RNAs for essential peptides of respiratory chain complexes, 22 transfer RNAs, and 2 ribosome RNAs [20]. Although mtDNA-encoded proteins represent a minor part, mtDNA mutations cause the major part of mitochondrial dysfunction due to the deficiency of energy production, and the consequence of organ dysfunction and injury, since the mitochondria-dominated energy productions are critical and necessary to satisfy the demands for diversified organs, including the central nervous system, cardiac and skeletal muscles, liver, pancreas, and endocrine system (Fig. 6.2) [21].

The cardiac muscle, similar to the skeletal muscle, has a high demand for ATP and is governed by mitochondrial high-energy phosphate production. Compared with cardiomyopathy children without mitochondrial diseases, the mortality of cardiomyopathy patients with mitochondrial diseases is significantly higher [22]. mtDNA can cause the deficiency of the OXPHOS in the cardiac muscle. mtDNA mutations, e.g., MTND1 and MTND5, in cardiomyopathy might encode complex I subunits [23], and the MTCYB gene encoding cytochrome b leads to complex III deficiency and mutations in individuals with histiocytoid, dilated, and hypertrophic cardiomyopathies [24]. Such cardiomyopathies occurred in complex IV deficiencies related to mutations in complex IV subunit genes (MTCO2, MTCO3, and COX6B1) and complex IV assembly factor genes (SCO2 and SURF1) [25].

mtDNA also serves as a key regulator to impel and maintain inflammation in the cardiac muscles and escapes from autophagy cell autonomously. mtDNA regulates Toll-like receptor 9-mediated inflammatory responses in cardiomyocytes and induces dilated cardiomyopathy and myocarditis [26]. Any genetic manipulation leading to prominently decreased mtDNA could accelerate the aging process and cause the adverse myocardial remodeling and dysfunction. In aortic wall, decreased mtDNA content preceded the development of perceptible atherosclerotic lesions [27].

Mitochondrial dysfunction leads to the fat imbalance and an overproduction of reactive oxygen species (ROS) in hepatocytes. The ROS induce fatal hepatocyte injury related to nonalcoholic fatty liver disease. Oxidative stress may break the homeostasis at cellular level, such as cell death by apoptosis, cell degeneration and

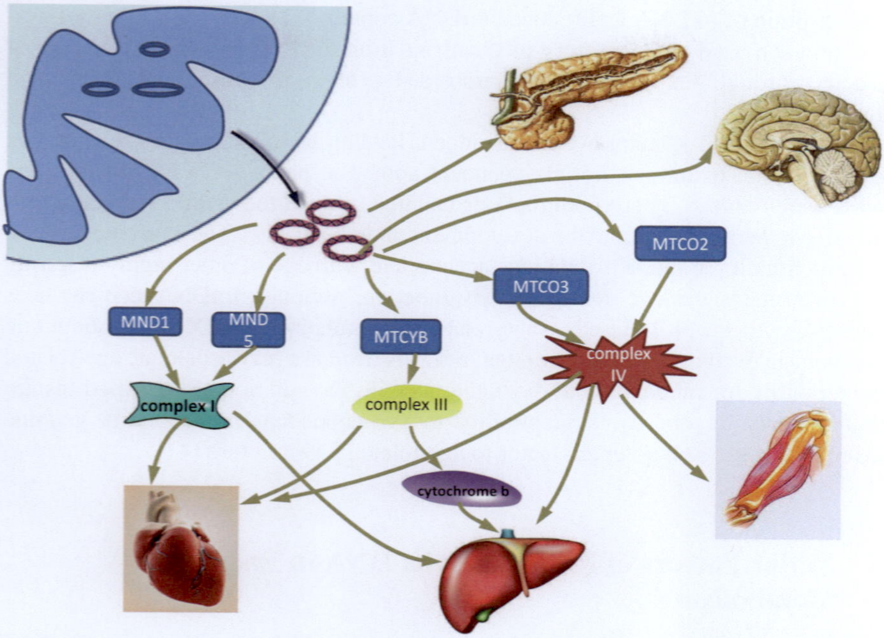

Fig. 6.2 Mutations of mtDNA lead to vulnerabilities of electron transport chain complexes to challenges and cause mitochondrial dysfunction due to the deficiency of the energy production to meet the demands for various organs

necrosis, membrane lipid peroxidation, overproduction of proinflammatory cytokines, hepatic stellate cell activation, or fibrogenesis [28–31]. Different types of mtDNA damage can cause point mutations and deletions, increase 8-hydroxydeoxyguanosine level, and the development of nonalcoholic fatty liver disease [29, 32]. Levels of mutations in the mitochondria-encoded subunit I and subunit II genes of complex I and complex IV are higher in patients with nonalcoholic fatty liver disease than those with fatty liver alone [33]. On the other hand, the expression of enzymes in lipid metabolism is downregulated, including CPT1 and acyl-CoA synthetase genes in the translocation of long-chain fatty acids or other enzymes involved in β-oxidation, leading to the accumulation of long-chain fatty acids in plasma. Defective β-oxidation is probably responsible for the dNTP depletion and gradual loss of liver mtDNA [34].

Insulin release from pancreatic β-cells is governed by plasma glucose levels via mitochondrial fuel metabolism. The insulin secretion is critically dependent on mtDNA and the proteins mtDNA encodes. In diabetes, mitochondrial dysfunction or decreased mtDNA copy number can cause insulin resistance and dysregulation of lipid metabolism [35]. Mitochondrial regulation of β-cell function occurs through many various pathways, including maintenance of mitochondrial mass, generation of reactive oxygen species, metabolic coupling, or interaction with other cellular organelles [36, 37]. For example, loss of TFB2M led to diabetes due to disrupted transcription of mtDNA and reduced mtDNA content [38]. The mitochondrial dysfunction activated compensatory mechanisms limit the dysfunction and damage of β-cells, through the mitochondrial unfolded protein response, mitophagy, and autophagy.

Cell-protective systems were overridden, leading to mitochondrial dysfunction and activation of mitochondrial-dependent apoptotic pathways. Cell function and mass were reduced. These perturbations resulted in impaired insulin secretion, progressive hyperglycemia, and the development of diabetes [39]. The exact mechanism by which multiple mtDNA deletions accumulate with age in other highly respiring tissues remains unclear during cell dysfunction, although mtDNA content does decrease with age [40]. Based on age-related accumulation of mtDNA mutations and deletions in decreased mtDNA content, mtDNA decrease may accelerate age-related vulnerability for mitochondrial dysfunction. With the addition of increased insulin requirements by obesity, susceptibility to age-dependent mitochondria genome instability could be a potential factor to age-related β-cell failure [41].

4 Major Factors of Mitochondrial DNA in Energy Metabolism

mtDNA transcription is regulated by mitochondrial RNA polymerase (POLRMT), TFB1M, TFB2M, and TFAM [42, 43]. The mitochondrial methyltransferase TFB1M depends on its primary sequence, similar to the yeast transcription factor mtTFB [44]. By methylating 12SrRNA, TFB1M regulates the expression of mitochondrial genes and ribosome subunit assembly, while the loss of TFB1M leads to the origin of the small ribosomal subunit damage and the unaffected transcription [45]. Either TFB1M mutation or overexpression in the nearby ribosomal sequence is associated with hypermethylation, to activate the proapoptotic transcription factor E2F1, increase mitochondrial reactive ROS, alter mitochondrial OXPHOS defects and ribosome function, or culminate in the progressive loss of critical cells [46]. For example, a common single-nucleotide polymorphism of TFB1M locus has a strong correlation with a decreased complex I activity and content in human pancreatic islets and increased risk D for diabetes in females [44].

There is another mtTFB homolog, the mitochondrial methyltransferase TFB2M, in mammalian cells [39]. Both TFB1M and TFB2M in sequence are similar to a large family of rRNA methyltransferases in archaea, bacteria, and eukaryotes [39]. TFB1M usually stands for the original methylation, while TFB2M without the methyltransferase activity represents a gene replication for a specialized mitochondrial transcription factor [47–49]. TFB2M can facilitate the biogenesis of transcription from light-strand promoter and the heavy-strand promoter 1 in the appearance of single-subunit POLRMT and TFAM (Fig. 6.3) [39]. TFB2M is essential for the structural changes that occur around the transcription initiation site and for the formation of the first phosphodiester bond [9]. Mitochondrial TFB2M controls the transcription of mitochondrial-encoded genes. Loss of TFB2M leads to diabetes

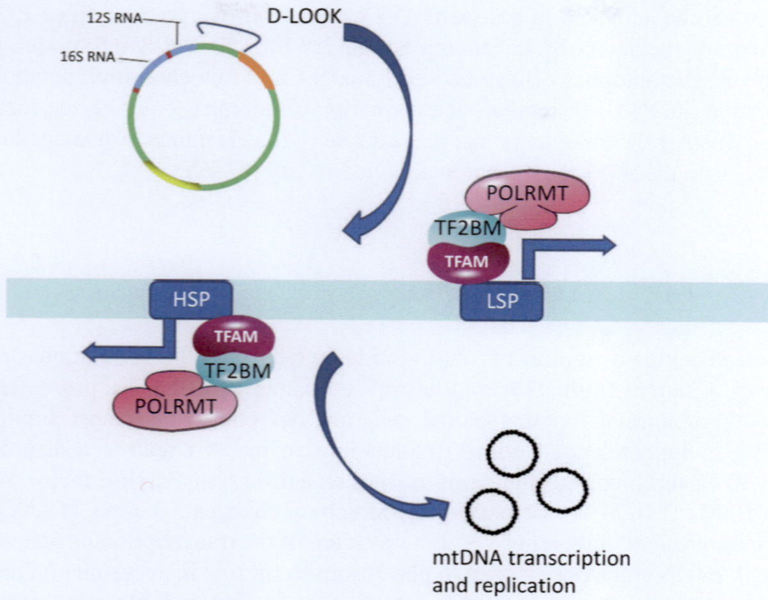

Fig. 6.3 A model of transcription initiation on light-strand promoter and heavy-strand promoter which is located in D-LOOK region in human mitochondria DNA. TFAM, TFB2M, and POLRMT are bound to the upstream site. The resulting POLRMT + TFAM + TFB2M promoter DNA complex plays an important role in maintaining mtDNA transcription and replication (Abbreviation: TFAM, mitochondrial transcription factor A; TFB2M, mitochondrial transcription factor B2; POLRMT, single-subunit RNA polymerase)

due to disrupted transcription of mtDNA and reduced mtDNA content [50]. TFAM and TFB2M were the only two essential initiation factors to activate transcription of the mitochondrial genome, as well as light-strand promoter and heavy-strand promoter 1 [51].

TFAM is a high-mobility group class protein, binds to mitochondrial DNA, and protects a region 14–35 bp upstream of the light-strand promoter transcription initiation site [52]. TFAM induces the assembly of the initiation complex by causing initial melting of the promoter and inducing POLRMT-TFB1M and/or POLRMT-TFB2M [53]. As a sequence-specific transcription factor, TFAM serve as a key regulator to maintain normal levels of transcripts from the light-strand promoter and the heavy-strand promoter 1 of the mtDNA [54], since truncated transcripts from light-strand promoter are used to prime DNA synthesis during mtDNA replication. Thus, lacking TFAM can impair mtDNA transcription and fail to maintain mtDNA stability, resulting in bioenergetic failure and embryonic lethality [55–57]. In addition, TFAM is necessary for energy production from oxidative phosphorylation because of its function of encoding for 13 essential components of the respiratory chain. mtDNA can decide the expression and activity of a large number of metabolism-involved proteins indirectly through the signal of transcriptional factors or ROS or directly through the release of mtDNA. For example, Nrf2 signaling pathway can be activated to reduce mtDNA release or overproduction of ROS or inflammatory mediators by enhancing autophagy [30, 58–60] is questioned what kind of role the autophagy plays between mtDNA and mitochondrial metabolism, whether the mtDNA can regulate the formation of autophagy during the metabolism, or whether the autophagy just plays a role in the clearance of damaged mitochondria or in the prevention from metabolic toxicity [61–65].

5 Conclusion

Mitochondria play a significant role in oxidative phosphorylation, autophagy, and apoptosis. Changes in mtDNA may disturb a number of metabolic processes and lead to the abnormal function of the cell. mtDNA encodes 13 most significant OXPHOS polypeptides. Deletions or mutations of mtDNA lead to a disorder in OXPHOS in the involved regulation of mitochondrial transcription factor TFAM and TFB2M. TFB2M is essential for the structural changes, whereas TFAM plays an important role in protecting a region upstream of the transcription initiation site. Although mtDNA has been shown to play a significant role in experimental models of various diseases, the values in clinic still need to be explored. More exact mechanism by which each mtDNA gene regulates the specific metabolism and control the metabolic process can be investigated by using clustered regularly interspaced short palindromic repeats in single cell [66–68]. We hope that active research in this direction will help us better understand disease processes and lead to the development of novel therapeutic approaches.

Acknowledgments The work was supported by the Zhongshan Distinguished Professor Grant (XDW), the National Natural Science Foundation of China (91230204, 81270099, 81320108001, 81270131, 81300010), the Shanghai Committee of Science and Technology (12JC1402200, 12431900207, 11410708600, 14431905100), Operation funding of Shanghai Institute of Clinical Bioinformatics, the Ministry of Education for Academic Special Science and Research Foundation for PhD Education (20130071110043), and the National Key Research and Development Program (2016YFC0902400, 2017YFSF090207).

References

1. Bailey CM, Anderson KS. A mechanistic view of human mitochondrial DNA polymerase gamma: providing insight into drug toxicity and mitochondrial disease. Biochim Biophys Acta. 2010;1213–22:1804. [PMID: 20083238]
2. Martinez-Reyes I, Diebold LP, Kong H, Schieber M, Huang H, Hensley CT, et al. TCA cycle and mitochondrial membrane potential are necessary for diverse biological functions. Mol Cell. 2016;199–209:61. [PMID: 26725009]
3. Zhu LZ, Hou YJ, Zhao M, Yang MF, XT F, Sun JY, et al. Caudatin induces caspase-dependent apoptosis in human glioma cells with involvement of mitochondrial dysfunction and reactive oxygen species generation. Cell Biol Toxicol. 2016;32:333–45. [PMID: 27184666]

4. Knez J, Marrachelli VG, Cauwenberghs N, Winckelmans E, Zhang Z, Thijs L, et al. Peripheral blood mitochondrial DNA content in relation to circulating metabolites and inflammatory markers: a population study. PLoS One. 2017;12:e0181036. [PMID: 28704533]
5. Tranah GJ, Manini TM, Lohman KK, Nalls MA, Kritchevsky S, Newman AB, et al. Mitochondrial DNA variation in human metabolic rate and energy expenditure. Mitochondrion. 2011;11:855–61. [PMID: 21586348]
6. Payne BA, Cree L, Chinnery PF. Single-cell analysis of mitochondrial DNA. Methods Mol Biol. 2015;1264:67–76. [PMID: 25631004]
7. Hsu HC, Li SJ, Chen CY, Chen MF. Eicosapentaenoic acid protects cardiomyoblasts from lipotoxicity in an autophagy-dependent manner. Cell Biol Toxicol. 2017, July 24. doi:https://doi.org/10.1007/s10565-017-9406-9. [PMID: 28741157]
8. Devine H, Patani R. The translational potential of human induced pluripotent stem cells for clinical neurology: the translational potential of hiPSCs in neurology. Cell Biol Toxicol. 2017;33:129–44. [PMID: 27915387]
9. Sologub M, Litonin D, Anikin M, Mustaev A, Temiakov D. TFB2 is a transient component of the catalytic site of the human mitochondrial RNA polymerase. Cell. 2009;139:934–44. [PMID: 19945377]
10. Fathi H, Ebrahimzadeh MA, Ziar A, Mohammadi H. Oxidative damage induced by retching; antiemetic and neuroprotective role of Sambucus Ebulus L. Cell Biol Toxicol. 2015;31:231–9. [PMID: 26493312]
11. Kaarniranta K, Tokarz P, Koskela A, Paterno J, Blasiak J. Autophagy regulates death of retinal pigment epithelium cells in age-related macular degeneration. Cell Biol Toxicol. 2017;33:113–28. [PMID: 27900566]
12. Vedi M, Sabina EP. Assessment of hepatoprotective and nephroprotective potential of withaferin a on bromobenzene-induced injury in Swiss albino mice: possible involvement of mitochondrial dysfunction and inflammation. Cell Biol Toxicol. 2016;32:373–90. [PMID: 27250656]
13. Zerin T, Kim JS, Gil HW, Song HY, Hong SY. Effects of formaldehyde on mitochondrial dysfunction and apoptosis in SK-N-SH neuroblastoma cells. Cell Biol Toxicol. 2015;31:261–72. [PMID: 26728267]
14. Rotig A. Genetics of mitochondrial respiratory chain deficiencies. Rev Neurol. 2014;170:309–22. [PMID: 24798924]
15. Chance B, Williams GR. Respiratory enzymes in oxidative phosphorylation. III. The steady state. J Biol Chem. 1955;217:409–27. [PMID: 13271404]
16. Thorburn DR, Sugiana C, Salemi R, Kirby DM, Worgan L, Ohtake A, et al. Biochemical and molecular diagnosis of mitochondrial respiratory chain disorders. Biochim Biophys Acta. 2004;1659:121–8. [PMID: 15576043]
17. Wallace DC, Chalkia D. Mitochondrial DNA genetics and the heteroplasmy conundrum in evolution and disease. Cold Spring Harb Perspect Biol. 2013;a021220:5. [PMID: 24186072]
18. Larsson NG. Somatic mitochondrial DNA mutations in mammalian aging. Annu Rev Biochem. 2010;683–706:79. [PMID: 20350166]
19. Campbell GR, Reeve A, Ziabreva I, Polvikoski TM, Taylor RW, Reynolds R, et al. Mitochondrial DNA deletions and depletion within paraspinal muscles. Neuropathol Appl Neurobiol. 2013;39:377–89. [PMID: 22762368]
20. Goffart S, Kleist-Retzow v. JC, Wiesner RJ. Regulation of mitochondrial proliferation in the heart: power-plant failure contributes to cardiac failure in hypertrophy. Cardiovasc Res. 2004;64:198–207. [PMID: 15485678]
21. Chinnery PF. Mitochondrial Disorders Overview. In: Pagon RA, Adam MP, Ardinger HH, Wallace SE, Amemiya A, Bean LJH, et al., editors. GeneReviews(R). Seattle;1993.
22. Holmgren D, Wahlander H, Eriksson BO, Oldfors A, Holme E, Tulinius M. Cardiomyopathy in children with mitochondrial disease; clinical course and cardiological findings. Eur Heart J. 2003;24:280–8. [PMID: 12590906]
23. Fassone E, Rahman S. Complex I Deficiency: clinical features, biochemistry and molecular genetics. J Med Genet. 2012;49:578–90. [PMID: 22972949]

24. Hagen CM, Aidt FH, Havndrup O, Hedley PL, Jespersgaard C, Jensen M, et al. MT-CYB mutations in hypertrophic cardiomyopathy. Mol Genet Genomic Med. 2013;1:54–65. [PMID: 24498601]
25. Marin-Garcia J, Goldenthal MJ, Ananthakrishnan R, Pierpont ME. The complete sequence of mtDNA genes in idiopathic dilated cardiomyopathy shows novel missense and tRNA mutations. J Card Fail. 2000;6:321–9. [PMID: 11145757]
26. Oka T, Hikoso S, Yamaguchi O, Taneike M, Takeda T, Tamai T, et al. Mitochondrial DNA that escapes from autophagy causes inflammation and heart failure. Nature. 2012;485:251–5. [PMID: 22535248]
27. Villarroya J, Dorado B, Vila MR, Garcia-Arumi E, Domingo P, Giralt M, et al. Thymidine kinase 2 deficiency-induced mitochondrial DNA depletion causes abnormal development of adipose tissues and adipokine levels in mice. PLoS One. 2011;6:e29691. [PMID: 22216345]
28. Wu Y, Geng XC, Wang JF, Miao YF, YL L, Li B. The HepaRG cell line, a superior in vitro model to L-02, HepG2 and hiHeps cell lines for assessing drug-induced liver injury. Cell Biol Toxicol. 2016;32:37–59. [PMID: 27027780]
29. Fierbinteanu-Braticevici C, Sinescu C, Moldoveanu A, Petrisor A, Diaconu S, Cretoiu D, et al. Nonalcoholic fatty liver disease: one entity, multiple impacts on liver health. Cell Biol Toxicol. 2017;33:5–14. [PMID: 27680752]
30. Ganta KK, Mandal A, Chaubey B. Depolarization of mitochondrial membrane potential is the initial event in non-nucleoside reverse transcriptase inhibitor efavirenz induced cytotoxicity. Cell Biol Toxicol. 2017;33:69–82. [PMID: 27639578]
31. Liu T, Liu WH, Zhao JS, Meng FZ, Wang H. Lutein protects against beta-amyloid peptide-induced oxidative stress in cerebrovascular endothelial cells through modulation of Nrf-2 and NF-kappab. Cell Biol Toxicol. 2017;33:57–67. [PMID: 27878403]
32. Nomoto K, Tsuneyama K, Takahashi H, Murai Y, Takano Y. Cytoplasmic fine granular expression of 8-hydroxydeoxyguanosine reflects early mitochondrial oxidative DNA damage in nonalcoholic fatty liver disease. AIMM. 2008;71–5:16. [PMID: 18091316]
33. Kawahara H, Fukura M, Tsuchishima M, Takase S. Mutation of mitochondrial DNA in livers from patients with alcoholic hepatitis and nonalcoholic steatohepatitis. Alcohol Clin Exp Res. 2007;31:S54–60. [PMID: 17331167]
34. Zhou X, Kannisto K, Curbo S, von Dobeln U, Hultenby K, Isetun S, et al. Thymidine kinase 2 deficiency-induced mtDNA depletion in mouse liver leads to defect beta-oxidation. PLoS One. 2013;8:e58843. [PMID: 23505564]
35. Wada J, Nakatsuka A. Mitochondrial dynamics and mitochondrial dysfunction in diabetes. Acta Med Okayama. 2016;70:151–8. [PMID: 27339203]
36. Wang L. Mitochondrial purine and pyrimidine metabolism and beyond. Nucleosides Nucleotides Nucleic Acids. 2016;35:578–94. [PMID: 27906631]
37. Kaufman BA, Li C, Soleimanpour SA. Mitochondrial regulation of beta-cell function: maintaining the momentum for insulin release. Mol Asp Med. 2015;42:91–104. [PMID: 25659350]
38. Malik AN, Czajka AI. Mitochondrial DNA content a potential biomarker of mitochondrial dysfunction? Mitochondrion. 2013;13:481–92. [PMID: 23085537]
39. Falkenberg M, Gaspari M, Rantanen A, Trifunovic A, Larsson NG, Gustafsson CM. Mitochondrial transcription factors B1 and B2 activate transcription of human mtDNA. Nat Genet. 2002;31:289–94. [PMID: 12068295]
40. Cree LM, Patel SK, Pyle A, Lynn S, Turnbull DM, Chinnery PF, et al. Age-related decline in mitochondrial DNA copy number in isolated human pancreatic islets. Diabetologia. 2008;51:1440–3. [PMID: 18528676]
41. Damas J, Samuels DC, Carneiro J, Amorim A, Pereira F. Mitochondrial DNA rearrangements in health and disease--a comprehensive study. Hum Mutat. 2014;35:1–14. [PMID: 24115352]
42. Fernandez-Silva P, Enriquez JA, Montoya J. Replication and transcription of mammalian mitochondrial DNA. Exp Physiol. 2003;88:41–56. [PMID: 12525854]
43. Asin-Cayuela J, Gustafsson CM. Mitochondrial transcription and its regulation in mammalian cells. Trends Biochem Sci. 2007;32:111–7. [PMID: 17291767]

44. Koeck T, Olsson AH, Nitert MD, Sharoyko VV, Ladenvall C, Kotova O, et al. A common variant in TFB1M is associated with reduced insulin secretion and increased future risk of type 2 diabetes. Cell Metab. 2011;13:80–91. [PMID: 21195351]
45. Metodiev MD, Lesko N, Park CB, Camara Y, Shi Y, Wibom R, et al. Methylation of 12S rRNA is necessary for in vivo stability of the small subunit of the mammalian mitochondrial ribosome. Cell Metab. 2009;9:386–97. [PMID: 19356719]
46. Raimundo N, Song L, Shutt TE, McKay SE, Cotney J, Guan MX, et al. Mitochondrial stress engages E2F1 apoptotic signaling to cause deafness. Cell. 2012;148:716–26. [PMID: 22341444]
47. Moustafa IM, Uchida A, Wang Y, Yennawar N, Cameron CE. Structural models of mammalian mitochondrial transcription factor B2. Biochim Biophys Acta. 2015;987–1002:1849. [PMID: 26066983]
48. Bao L, Diao H, Dong N, Su X, Wang B, Mo Q, et al. Histone deacetylase inhibitor induces cell apoptosis and cycle arrest in lung cancer cells via mitochondrial injury and p53 up-acetylation. Cell Biol Toxicol. 2016;32:469–82. [PMID: 27423454]
49. Guja KE, Venkataraman K, Yakubovskaya E, Shi H, Mejia E, Hambardjieva E, et al. Structural basis for S-adenosylmethionine binding and methyltransferase activity by mitochondrial transcription factor B1. Nucleic Acids Res. 2013;41:7947–59. [PMID: 23804760]
50. Mangus DA, Jang SH, Jaehning JA. Release of the yeast mitochondrial RNA polymerase specificity factor from transcription complexes. J Biol Chem. 1994;269:26568–74. [PMID: 7929382]
51. Litonin D, Sologub M, Shi Y, Savkina M, Anikin M, Falkenberg M, et al. Human mitochondrial transcription revisited: only TFAM and TFB2M are required for transcription of the mitochondrial genes in vitro. J Biol Chem. 2010;285:18129–33. [PMID: 20410300]
52. Dairaghi DJ, Shadel GS, Clayton DA. Human mitochondrial transcription factor a and promoter spacing integrity are required for transcription initiation. Biochim Biophys Acta. 1995;1271:127–34. [PMID: 7599198]
53. Nicholas LM, Valtat B, Medina A, Andersson L, Abels M, Mollet IG, et al. Mitochondrial transcription factor B2 is essential for mitochondrial and cellular function in pancreatic beta-cells. Mol Metabol. 2017;6:651–63. [PMID: 28702322]
54. Shi Y, Dierckx A, Wanrooij PH, Wanrooij S, Larsson NG, Wilhelmsson LM, et al. Mammalian transcription factor a is a core component of the mitochondrial transcription machinery. Proc Natl Acad Sci U S A. 2012;109:16510–5. [PMID: 23012404]
55. Ramachandran A, Basu U, Sultana S, Nandakumar D, Patel SS. Human mitochondrial transcription factors TFAM and TFB2M work synergistically in promoter melting during transcription initiation. Nucleic Acids Res. 2017;45:861–74. [PMID: 27903899]
56. Rantanen A, Jansson M, Oldfors A, Larsson NG. Downregulation of Tfam and mtDNA copy number during mammalian spermatogenesis. Mamm Genome Off J Int Mamm Genome Soc. 2001;12:787–92. [PMID: 11668394]
57. McCulloch V, Seidel-Rogol BL, Shadel GSA. Human mitochondrial transcription factor is related to RNA adenine methyltransferases and binds S-adenosylmethionine. Mol Cell Biol. 2002;22:1116–25. [PMID: 11809803]
58. Soltani B, Ghaemi N, Sadeghizadeh M, Najafi F. Curcumin confers protection to irradiated THP-1 cells while its nanoformulation sensitizes these cells via apoptosis induction. Cell Biol Toxicol. 2016;32:543–61. [PMID: 27473378]
59. Steinberg P, Behnisch PA, Besselink H, Brouwer AA. Screening of molecular cell targets for carcinogenic heterocyclic aromatic amines by using CALUX(R) reporter gene assays. Cell Biol Toxicol. 2017;33:283–93. [PMID: 27942899]
60. Leclerc E, Hamon J, Claude I, Jellali R, Naudot M, Bois F. Investigation of acetaminophen toxicity in HepG2/C3a microscale cultures using a system biology model of glutathione depletion. Cell Biol Toxicol. 2015;31:173–85. [PMID: 25956491]

61. Denardin CC, Martins LA, Parisi MM, Vieira MQ, Terra SR, Barbe-Tuana FM, et al. Autophagy induced by purple pitanga (Eugenia Uniflora L.) extract triggered a cooperative effect on inducing the hepatic stellate cell death. Cell Biol Toxicol. 2017;33:197–206. [PMID: 27744523]
62. Lippai M, Szatmari Z. Autophagy-from molecular mechanisms to clinical relevance. Cell Biol Toxicol. 2017;33:145–68. [PMID: 27957648]
63. Imhoff BR, Hansen JM. Tert-butylhydroquinone induces mitochondrial oxidative stress causing Nrf2 activation. Cell Biol Toxicol. 2010;26:541–51. [PMID: 20429028]
64. Cristofori P, Sauer AV, Trevisan A. Three common pathways of nephrotoxicity induced by halogenated alkenes. Cell Biol Toxicol. 2015;31:1–13. [PMID: 25665826]
65. Zhang X, Yin H, Li Z, Zhang T, Yang Z. Nano-TiO2 induces autophagy to protect against cell death through antioxidative mechanism in podocytes. Cell Biol Toxicol. 2016;32:513–27. [PMID: 27430495]
66. Wang W, Gao D, Wang X. Can single-cell RNA sequencing crack the mystery of cells? Cell Biol Toxicol. 2017, July 21. doi:https://doi.org/10.1007/s10565-017-9404-y. [PMID: 28733864]
67. Wang W, Wang X, Single-cell CRISPR. Screening in drug resistance. Cell Biol Toxicol. 2017;33:207–10. [PMID: 28474250]
68. Wang X. New biomarkers and therapeutics can be discovered during COPD-lung cancer transition. Cell Biol Toxicol. 2016;32:359–61. [PMID: 27405768]

Jiapei Lv, MD, Wenzhou Medical University. The main study of the relationship between traditional Chinese medicine and lung cancer bioinformatics. A project leader of lung lipidomic research and studies the identification and validation of disease markers. Dr. Lv currently has published three articles, with the total impact factor of about 20 points.

Madhav Bhatia, PhD, heads the Inflammation Research Group in the Department of Pathology at the University of Otago, Christchurch. Research in his laboratory has shown hydrogen sulfide and substance P as mediators of inflammation and potential therapeutic targets for inflammatory diseases such as acute pancreatitis, sepsis, burn injuries, and joint inflammation. He has received numerous grants and given several invited presentations in different countries and is on editorial boards of 34 journals.

He has mentored several postgraduate students who are research staff and who are now independent investigators in their own right, leading research groups in different parts of the world. He is the author of more than 170 scientific publications with the citation number of about 9123 and h-index 49.

Xiangdong Wang, MD, PhD, is a Distinguished Professor of Medicine, Director of Shanghai Institute of Clinical Bioinformatics, Executive Director of Clinical Science Institute of Fudan University Zhongshan Hospital, Director of Fudan University Center of Clinical Bioinformatics, Deputy Director of Shanghai Respiratory Research Institute, and a visiting professor of King's College London. His main research is focused on clinical bioinformatics, disease-specific biomarkers, lung chronic diseases, cancer immunology, and molecular and cellular therapies. He is the author of more than 200 scientific publications with the impact factor of about 900, citation number of about 6918, h-index of 46, i10-index of 181, and cited journal impact factor of about 7000.

Chapter 7
The Role of DNA Repair in Maintaining Mitochondrial DNA Stability

Linlin Zhang, Aurelio Reyes, and Xiangdong Wang

Abstract Mitochondria are vital double-membrane organelles that act as a "powerhouse" inside the cell and have essential roles to maintain cellular functions, e.g., ATP production, iron-sulfur synthesis metabolism, and steroid synthesis. An important difference with other organelles is that they contain their own mitochondrial DNA (mtDNA). Such powerful organelles are also sensitive to both endogenous and exogenous factors that can cause lesions to their structural components and their mtDNA, resulting in gene mutations and eventually leading to diseases. In this review, we will mainly focus on mammalian mitochondrial DNA repair pathways that safeguard mitochondrial DNA integrity and several important factors involved in the repair process, especially on an essential pathway, base excision repair. We eagerly anticipate to explore more methods to treat related diseases by constantly groping for these complexes and precise repair mechanisms.

Keywords Mitochondrial DNA • Lesions • DNA repair • BER • Major pathway

1 Introduction

The nucleus contains most of the genetic material, while only a small fraction is present in mitochondria. Nuclear DNA (nDNA) is inherited from both parents unlike mitochondrial DNA (mtDNA) that is essentially maternally inherited [1].

L. Zhang (✉) • X. Wang (✉)
Zhongshan Hospital Institute of Clinical Science, Fudan University, Shanghai Medical College, Shanghai, China
e-mail: 865957231@qq.com; xiangdong.wang@clintransmed.org

A. Reyes (✉)
MRC Mitochondrial Biology Unit, University of Cambridge, Cambridge, UK
e-mail: art@mrc-mbu.cam.ac.uk

H. Sun, X. Wang (eds.), *Mitochondrial DNA and Diseases*, Advances in Experimental Medicine and Biology 1038,
https://doi.org/10.1007/978-981-10-6674-0_7

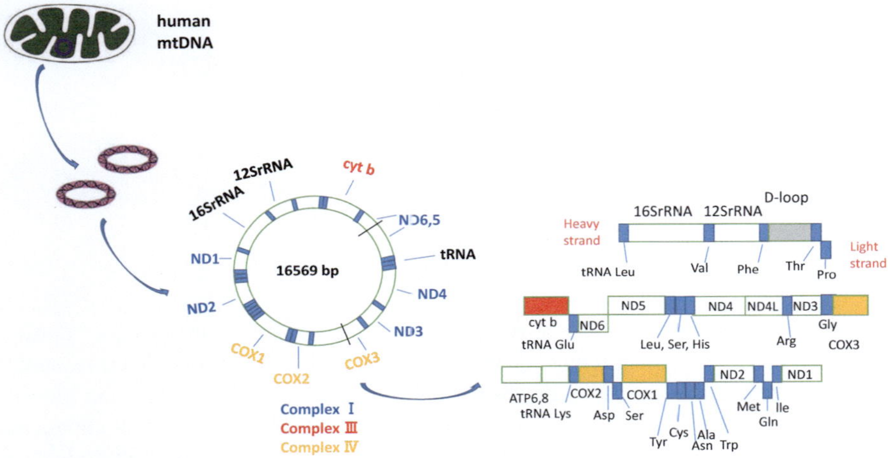

Fig. 7.1 The human mtDNA
The human mtDNA. It is a circular genome about 16.5 kb long which contains the heavy and the light strand, respectively. This mtDNA contains 37 genes: 22 transfer RNAs (tRNAs), 2 ribosomal RNAs (rRNAs), and 13 coding for proteins which are all connected with the oxidative phosphorylation system. ND1–6, NADH dehydrogenase, forming complex I; CYT b, cytochrome b of coenzyme Q, forming complex III; COX1–3, cytochrome c oxidase, forming complex IV; ATP8, ATP6, ATP synthase 8 and 6, forming complex V

Human mtDNA is a circular genome about 16.5 kb long, containing 37 genes: 13 coding for proteins which are structural components of the oxidative phosphorylation system, 22 transfer RNAs (tRNAs), and 2 ribosomal RNAs (rRNAs) essential for mitochondrial translation (Fig. 7.1) [2]. There are >100 mitochondria, of which each is comprised of 10 DNAs in eukaryotic cells [3]. Of genetic materials, mtDNA molecules with high mitochondrial densities are small in contrast to the size of the nuclear genome, accounting for only about 1%. nDNA contains 6 billion base pairs, coding with 25,000 protein-coding genes.

Despite representing a small amount as compared to nDNA, mtDNA plays an important role in maintaining the normal functioning of the organism. When mtDNA is damaged, repair is impossible or damage cannot be repaired in time, it can accumulate genetic mutations that eventually lead to various diseases involving diabetes, cancer, neurodegenerative disorders, and aging [4–9]. The mutation rate of mtDNA is estimated to be 10–100 times higher than nDNA [10]. mtDNA is more susceptible to damage (e.g., oxidative damage) as it is linked to the electron transport system, while nDNA is less vulnerable to such damage. mtDNA lacks certain protective proteins like histone [11]. Because of the biological characteristics of mitochondria, there are many channels to produce reactive oxygen species (ROS). For instance, nicotinamide adenine dinucleotide phosphate (NADPH) oxidase enzymes produce ROS in phagocytic cells, to fight against microbial infections [12]. Moreover, epithelial and endothelial cells can also produce ROS, e.g., Nox1 and Nox4, respectively, by NADPH oxidase enzymes [13, 14].

Mitochondria are the major site of ROS production. Superfluous ROS may lead to DNA damage in the form of oxidized bases and DNA strand breaks [12]. However, a variety of DNA repair systems are in place in mitochondria in order to maintain the integrity of mtDNA and the normal function of mitochondria. Base excision repair (BER) contains two ways of repairing DNA, the short-path BER (SP-BER) and long-path BER (LP-BER), existing in the nucleus and in the mitochondrion. Other repair systems like nonhomologous end joining, which is crucially important for the repair of DNA double-strand break [15], have been less studied, and related mechanisms remain unclear. Our purpose of this review is to sketch several types of mtDNA damage and the differentiation between nDNA and mtDNA as well as summarize several repair pathways and major enzymes.

2 Characteristics of Mitochondria

Mitochondria are involved in the synthesis of heme, cholesterol, and phospholipids, in immune responses [16], and in aging and apoptosis [17]. Despite mitochondria containing less genetic materials, their genetic information is at a higher density than the nDNA, since there are no introns, reduced or absent UTRs, no or very small intergenic regions, small regulatory regions, and some genes that overlap [1]. As a consequence, damage to mtDNA is likely to happen in a coding region. mtDNA damage often occurs in individual cells, due either to intracellular processes (e.g., DNA replication mismatch, base deamination, oxidative damage, depurination, or depyrimidination) [18] or to external factors (e.g., ionizing radiation, ultraviolet radiation, cisplatin) [19]. mtDNA damage can result in base substitutions, deletions, and missense mutations, which lead to mtDNA malfunction [20]. This does not mean that the damage of a single base definitely causes a disease. Only when mutation load in mtDNA reaches a high level, about 75% for most cases, will it have an impact in mitochondrial function [21]. Although DNA repair was present in the nucleus, there are a number of repair systems in mitochondria after the discovery of the presence of uracil-DNA glycosylase (UNG) within mitochondria [22].

3 Types of Damage

Mitochondrial DNA damage can be caused by endogenous or exogenous factors and either biological or chemical agents. Additionally, the proximity of mtDNA to damaging agents presents within mitochondria, and the high replication rate of mtDNA may also contribute to higher levels of DNA damage compared to the nucleus.

The alkylation is a type of mtDNA methylation by internal agents [18] like S-adenosylmethionine [23] and by external factors like chemotherapy for tumors [1]. Alkylation damage can lead to modified DNA bases, genetic instability, the normal function of cells, and development of cancer [24]. The alkylation-associated mtDNA damage is repaired by BER [25]. Hydrolytic damage includes the formation of abasic sites produced by hydrolysis of glycosidic bonds [26] and hydrolytic deamination of bases, mostly cytosine [27]. The formation of adducts can be generated by endogenous ROS (14) or estrogens [28] or by exogenous chemical agents [1], and these adducts are removed by the nucleotide excision repair (NER) system [29, 30]. However, there is no evidence of nucleotide excision repair in mitochondria. Mismatched bases of mtDNA damage originate from DNA replication errors, by incorporation of either wrong nucleotides or nucleotides mainly in oxidized bases [31]. Single-base mismatches are identified by Y-box-binding protein 1 (YB1) [32] and repaired by the mismatch repair (MMR) system [33]. DNA strand breaks include single-strand (SSBs) [34] and double-strand breaks (DSBs) [35]. DNA strand breaks may originate from ROS-induced DNA lesions [17] or during the BER repair if the repair intermediates cannot accomplish the proper steps [12]. Oxidative damage in mtDNA may occur as base modifications, abasic sites, and a variety of lesions [36]. Of those, repair of 8-oxoguanine is the most comprehensively studied one [4], and oxidative damage is the most prevalent damage in mitochondria [37]. The increased oxidative damage can lead to mitochondrial dysfunction and repaired BER or NER [38]. nDNA possesses unique types of injury with altered 3D chromatin architecture, rather than mtDNA [39]. Such a high-order chromatin structure is damaged by radical and prevented by antioxidant treatment [40].

4 Major Enzymes Involved in mtDNA

BER is one of the most studied DNA repair pathways both in the nucleus and mitochondria involved with a large number of proteinases, e.g., DNA glycosylases, AP endonuclease, DNA polymerase, DNA ligase, and other synergistic effect enzymes. Roles of those proteinases in the development of mtDNA damage and repair pathways remain unclear.

4.1 DNA Glycosylases

DNA glycosylase can damage and remove bases from the DNA and can be divided into four superfamilies [41], e.g., uracil-DNA glycosylase (UDG) family, alkyladenine DNA glycosylase (AAG) family, helix-hairpin-helix (HhH) family, and formamidopyrimidine DNA glycosylase (Fpg)/Nei (endonuclease VIII) or helix-two-turn-helix (H2TH) family.

The UDG Family UDG family can be further subdivided into five subfamilies of monofunctional glycosylases, e.g., uracil-DNA glycosylase (UNG), thymine DNA glycosylase (TDG), MUG (mismatch uracil-DNA glycosylase), single-strand-specific monofunctional uracil-DNA glycosylase (SMUG), and DNA glycosylases. Of those, only UNG was found in human mitochondria [41], of which UNG1 has a

30-amino acid leader peptide that targets the protein to mitochondria, while UNG2 lacks this domain and is targeted to the nucleus. The main role of UNG is to dislodge uracil from mtDNA [2] and remove other substances, e.g., isodialuric acid, 5-hydroxyuracil, and alloxan, albeit with low efficiency [42]. UNG has a single domain made of a β-sheet with four parallel β-strands flanked by two sets of α-helices and contributes to narrow and shallow DNA binding groove [30].

The AAG Family AAG, also known as MPG/MDG, is a monofunctional glycosylase able to recognize alkylated and deaminated bases. There are three isoforms, AAG-A, AAG-B, and AAG-C, originated by posttranscriptional processing of a single transcript [43]. Targeting to mitochondria happens thanks to an N-terminal mitochondrial targeting signal in isoforms A and B [43]. Similar to UDG, AAG also has a single domain, but in this case, it has a mixed α/β topology comprising a positive DNA binding groove [30].

The HhH Family HhH family comprises a diverse group of glycosylase that is further split into six subfamilies, e.g., endonuclease III (EndoIII or Nth) in several species, 8-oxo-7,8-dihydroguanine DNA glycosylase 1 (Ogg1) in eukaryotes, A/G mismatch-specific adenine glycosylase (MutY/Mig), alkyladenine DNA glycosylase (AlkA) in bacteria and eukaryotes but not in mammals, 8-oxoG DNA glycosylase 2 (Ogg2) in archaea, and N-methylpurine-DNA glycosylase II (MpgII) in bacteria and archaea. In the human genome, there are four representatives of this HhH family: NTHL1, OGG1, MUTYH, and MBD4 (methyl-CpG-binding domain protein 4). Human NTHL1 or NTH1 (*E. coli* endonuclease III-like 1) is present in the nucleus and mitochondria, while mouse Nth1 is primarily in mitochondria [44]. Human NTH1 is a bifunctional glycosylase to excise oxidized DNA bases such as 5-hydroxycytosine, 5-hydroxyuracil, as well as the ring-opened 2,6-diamino-4-hydroxy-5-formamidopyrimidine [45]. NTH1 contains an iron-sulfur (4Fe-4S) cluster formed by the N- and C-terminal ends for the protein involved in DNA binding [46]. Unlike NTH1, OGG1 can remove damaged DNA bases induced by ROS, 7,8-dihydro-8-oxoguanine (8-oxoG) [47]. DNA glycosylase removes basesl and has AP lyase activity [46]. Up to eight different isoforms containing an MTS are generated by alternative splicing, but their precise function is still to be established. MUTYH is a monofunctional glycosylase and plays a role in removing adenine opposite 8-oxoG, guanine, or cytosine. Unlike other members of this family, MUTYH has a unique mechanism to cleave the undamaged bases from the DNA opposite the damaged ones, such as a mismatch A:8-oxo-G [48], and can prevent mutations in DNA. Alternative splicing generates three primary transcripts that can further be spliced into an over 15 different transcripts in total. However, there seems to be a main mitochondrial isoform that contains an N-terminal MTS. The glycosylase also contains an iron-sulfur cluster within the catalytic region [47]. MBD4 is a monofunctional glycosylase with the peculiarity of having two functional domains: a methyl-binding domain at the N-terminus (MBD) and a glycosylase domain at the C-terminus [49]. Another feature is that this glycosylase was not found in mitochondria [30]. All these subfamilies have a common HhH motif with two α-helices that are important for DNA contact. The iron-sulfur clusters are not present in every subfamily and may play a role in lesion detection since the redox potential shift was detected upon DNA binding [47].

The Fpg/Nei Family Included under the Fpg/Nei family are three human Neil (Nei-like) DNA glycosylases in the nucleus, NEIL1, NEIL2, and NEIL3, of which NEIL1 and NEIL2 are present in human mitochondria and NEIL1 has a long N-terminal MTS [50]. Double-strand DNA (dsDNA) lesions are mainly cleaved by NEIL1, while lesions in single-strand DNA (ssDNA) are mostly excised by NEIL2. Both are able to recognize the same type of structures, e.g., bubbles, forks, and bulges [30]. Fpg has no homologous enzyme in humans. All family members present an N- and a C-terminal domain linked by a flexible region with the long axis of the protein placed orthogonal to DNA binding groove.

DNA glycosylases play a role in the identification of damaged bases and BER processes via creating intermediates or transferring the lesion to other repair methods [46]. X-ray repair cross-complementing protein 1 (XRCC1) [51], for example, acts as a BER scaffolding protein, but it can also team up with AAG, NTHL1, or NEIL1 to enhance the repair activity of BER.

4.2 AP Endonuclease

AP endonuclease exists both in the nucleus and in mitochondria and is encoded by a nuclear gene. In addition to excising abasic sites created during the repair process, it can regulate different transcription factors and inhibit ROS production as a way of controlling redox status in the cell [52]. The main hydrolytic AP endonuclease in mammalian cells is APE/Ref-1 (APE1) [26] and functions throughout the BER pathway [53]. The AP sites are left after the excision of the damaged bases by monofunctional glycosylases (e.g., UDG1) and excised by APE1 through a hydrolytic mechanism. By contrast, bifunctional glycosylases (e.g., OGG1, NTHL1) have both the capabilities of incising the damaged bases and cleaving the AP site from 3′ side in DNA backbone by the phosphodiesterase activity of APE1. AP site disrepair may result in blocking DNA transcription and replication [54]. APE1 can be subject to different posttranslational modifications and phosphorylation in repair activity; S-nitrosation targets the protein to the cytoplasm and binds to certain regulatory elements promoted by acetylation [55].

4.3 DNA Polymerase γ

The DNA polymerase γ (polγ) is encoded in the nucleus and consists of the catalytic and the accessory subunit [56] and plays a role in BER in mitochondria to fill the gaps created during repair [57]. The dRP lyase activity is essential to end the process of a combination of monofunctional glycosylases and APE1 [3]. Polγ is able to perform translesion synthesis (TLS), i.e., nucleotide incorporation opposite a damaged template and past the damage via extension of a DNA primer [56]. Polymerase β is present in the BER process in the nucleus and enhances mtDNA BER activity to maintain mitochondrial function [57].

4.4 DNA Ligase

Nuclear mammalian DNA ligases include ligases I, IV, and III, of which ligase III (Lig3) is only confined to vertebrates [58]. Alternative splicing of the human DNA ligase III gene originates both the nuclear and mitochondrial forms of the protein [59]. It plays a role in cell replication, recombination, and DNA repair (single-strand DNA repair and BER) [12]. Mitochondrial Lig3 also plays a role in maintaining cell survival under long-term oxidative stress and removing dysfunctional mitochondria via autophagy, rather than nuclear Lig3. The overexpression of Lig3 increases the rate of base excision repair pathway in mitochondria and is associated with XRCC1 in various DNA repair activities [58]. XRCC1 can protect mammalian cells from ionizing radiation [60] and can participate in the DNA repair process of alkylation or ultraviolet light [61] to maintain the integrity of DNA [62]. DNA Lig3 is closely bound to XRCC1, dependent on the expression of XRCC1 in the nucleus [63], but not in mitochondria [64].

5 mtDNA Repair Pathways

Lose of mtDNA integrity leads to mitochondrial dysfunction and ultimately to mitochondrial diseases. Reduced activity for DNA glycosylases, OGG1, NTH1, or UNG1, is accompanied by increased APE1 activity in aged brain mitochondria [65]. In the following section, we will discuss several significant repair pathways in mitochondria (Fig. 7.2).

5.1 SP-BER

BER repairs spontaneous base damages at a rate of more than 20,000 events per cell daily [66], associated with the repair of oxidation, alkylation, deamination, and SSBs. Uracil-DNA glycosylase could catalyze and cleave the mutagenic uracil-DNA base [67]. SP-BER and LP-BER are present in mitochondria. Although the BER pathways present in the nucleus and mitochondria contain similar steps [30], there are still significant differences. Mitochondrial BER has lower efficiency to repair oxidative DNA lesions than nuclear BER. The major type of DNA damage caused by ROS, 8-oxoguanine (8-oxoG), or thymine glycol (TG) [68], is more

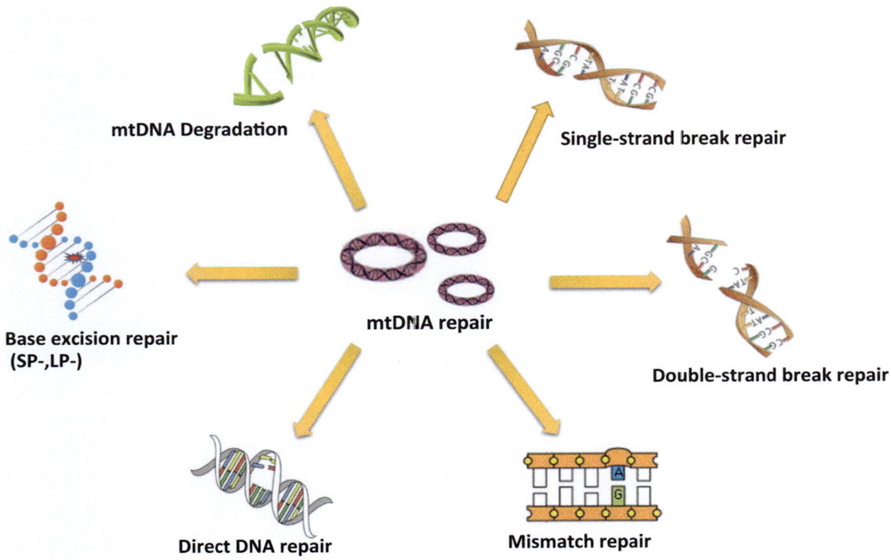

Fig. 7.2 MtDNA repair pathways
MtDNA repair pathways. In mitochondrial DNA, there are many pathways of repair, including BER (SP, LP), mtDNA degradation, direct DNA repair, mismatch repair, as well as DNA strand break repair (single, double)

efficient in mitochondrial BER than nuclear BER [12]. In the nucleus, there are multiple polymerases involved, including polβ, polδ, polε, polκ, and polλ, that connect with various sub-pathways, while human mitochondrial BER contains mainly Polγ [1], polβ, and PrimPol. BER comprises the recognition and incision of the damaged base by DNA glycosylases generating an abasic or AP site, removal of the AP site by AP endonuclease 1(APE1), gap filling by DNA polymerase, and finally ligation by DNA ligase [12].

Damaged bases are recognized by either mono- or bifunctional DNA glycosylases. Monofunctional DNA glycosylases are uracil-DNA glycosylase 1 (UDG1/UNG1) [22] and human MutY homolog glycosylase MUTYH (MYH) [69] in mitochondrial BER. After then, monofunctional DNA glycosylases remove the damaged base by hydrolysis of the N-glycosidic bond, leaving in the DNA a baseless or AP site (Fig. 7.3a). AP site is created and subsequently cleaved by APE1 from the 5′ end after the action of a monofunctional DNA glycosylase. A 5′-dRP is generated at one side and a 3′-OH at the other side [70]. Types of damages are instead recognized by bifunctional DNA glycosylases in mitochondria. There are four types of mitochondrial bifunctional glycosylases, e.g., 8-oxoguanine DNA glycosylase (OGG1) [71], NTH1/NTHL1 [72], NEIL1, or NEIL2 [50, 73], and have an inherent AP lyase activity. Bifunctional DNA glycosylases can recognize, remove the damaged base, and cleave the AP site in patterns of β- or β, δ-eliminations [74]. The specific process is shown in Fig. 7.1. Bifunctional glycosylases can be further subdivided into OGG1 and NTHL1 to possess β-lyase activity, while NEIL1 and NEIL2 are β, δ-lyases. Those different activities produce different cleaved products at the 3′ end. β-Lyases produce 5′P and 3′-phospho-α, β-unsaturated aldehyde (3′PUA) ends, while β, δ-lyases generate 5′P and 3′P at each end, respectively. The specific process is shown in Fig. 7.3b.

A 3′-OH and a 5′dRP are generated after the action of UNG1 or MUTYH1 and APE1 and accomplished by the mitochondrial polγ, thanks to its dRP lyase activity. Either a 3′PUA or a 3′P is generated after the action of bifunctional DNA glycosylases and completed by the 3′phosphodiesterase activity of APE1 [75] and the phosphatase activity of the polynucleotide kinase 3′-phosphatase (PNKP) [3], respectively. The single-nucleotide gaps with 5′P and 3′-OH groups are generated and can be filled by the mitochondrial polγ [70]. After gap filling by Polγ, a nick left in mtDNA is carried out by Lig3 to complete the DNA repair process, where Lig3 plays an essential role in SP-BER of mtDNA repair.

5.2 LP-BER

LP-BER is initiated when mtDNA is damaged beyond the capacity of SP-BER or the type of damage is not suitable for this type of repair system. 2-Deoxyribonolactone-induced damage is repaired via LP-BER pathway in mitochondria [76]. This repair process requires APE1, Polγ, and Lig3 [21], as well as other factors (Fig. 7.3c). In the LP-BER pathway, the 5′ blocking group becomes a single-stranded DNA flap with 6–9 nucleotides after incision by APE1 and displacement by Polγ as a consequence of the extension from the 3′-OH [17]. Then, the flap endonuclease 1(FEN1) [77] acts as 5′-endonuclease cooperating with DNA2 [78] or 5′ EXO/endonuclease (EXOG) [79] to incise the flap. The Lig3-dominant repair pathway and the machinery can be also caused by other oxidative lesions and could be involved in mtDNA replication [76]. The involvement in mtDNA replication is similar to the nucleus [78]. LP-BER exists both in the nucleus and in mitochondria, although there are differences regarding the enzymes or specific mechanisms. LP-BER occurs in the nucleus mainly in cell proliferation with the involvement of PCNA, FEN-1, polδ (ε), polβ, and Lig1 [75]. Polδ/ε is the preferred polymerase in proliferating cells, while Polβ is the main polymerase for repair in nonproliferating cells. Due to the weak dRP lyase of Polγ, it would be possible that LP-BER can actually be the predominant mode of repair for damaged bases in mitochondria.

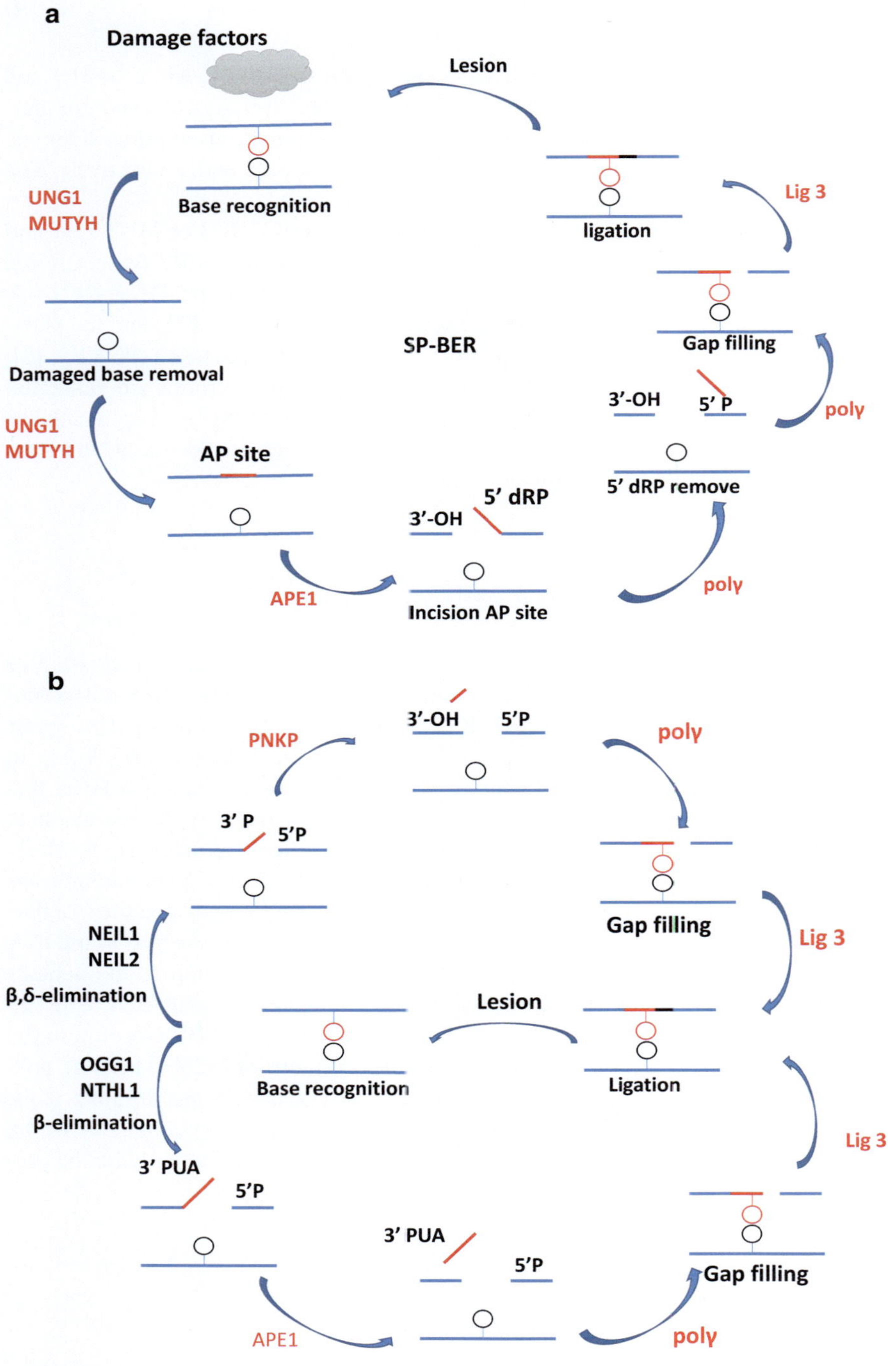

Fig. 7.3 BER
(a) SP-BER (Monofunctional DNA glycosylases) **(b)** SP-BER (Bifunctional DNA glycosylases)
(c) LP-BER

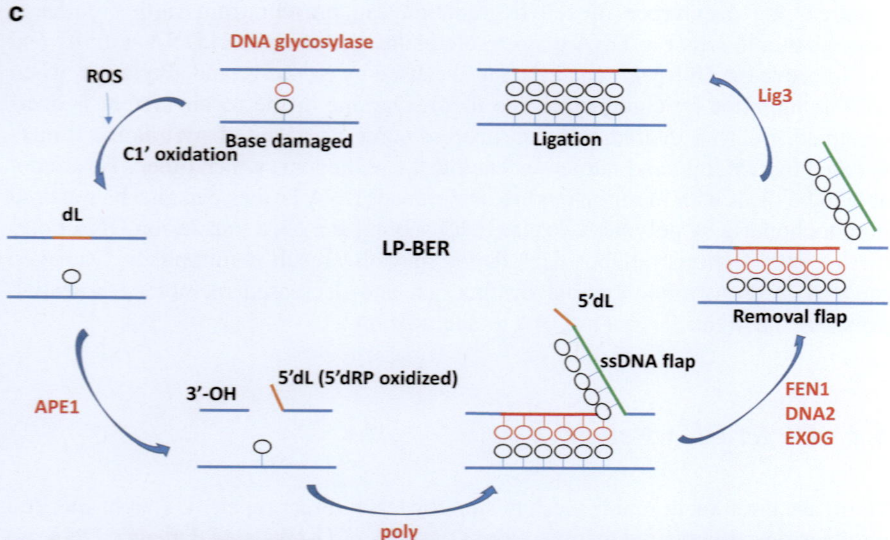

Fig. 7.3 (continued) The BER pathways. In SP-BER, the first step is the damaged base recognition and removal by monofunctional DNA glycosylase, remaining an AP site. This site is cleaved by APE1 from 5′ to the site. This step generates 5′-dRP at one side and a 3′-OH at another side. Since 5′-dRP cannot be ligated by DNA ligase, it must be removed. This step is accomplished by mitochondrial polγ, which possesses dRP lyase activity. But for bifunctional DNA glycosylases, other than that, they also have another characteristic which is inherent AP lyase activity. The bifunctional DNA glycosylases cleave the AP site through two methods: a β-elimination or β, δ-elimination reaction. Incision via the β-elimination reaction results in a strand break at the AP site with 3′PUA and 5′P, whereas β, δ-elimination leads to a single-nucleotide gap with 3′P and 5′P termini. In both cases, the ends cannot be ligated and have to be further processed. This procedure is completed by the 3′phosphodiesterase activity of APE1 (71) and phosphatase activity of the PNKP, respectively. As a result, this step generates a single-nucleotide gap with 5′P and 3′-OH groups. After gap filling by Polγ, a nick left in mtDNA is carried out by Lig3 to complete the DNA repair process. This procedure via DNA Lig3 closes the remaining nick. If the 5′ends produced by base excision and strand cleavage cannot be used to generate a sequent DNA strand, repair this damage via LP-BER. For instance, formation of the 2-deoxyribonolactone damage is not associated with glycosylase activity because the base is lost as a result of hydrogen abstraction and oxidation of the C1′ of deoxyribose and cannot be removed by dRP lyase activity of polγ. During LP-BER, the 5′ blocking group is displaced by DNA polymerase extending from the 3′-OH ending and forming a so-called flap. And the 5′ blocking group becomes part of a single-strand DNA flap. This flap in mitochondria has a length of up to 6–9 nucleotides. FEN1 acts as 5′-endonuclease cooperating with DNA2 or 5′ EXO/endonuclease (EXOG) to incise the flap. Finally, ligase III is used to fill the rest of the nick

5.3 *mtDNA Degradation*

mtDNA has many copies within mitochondria and is degraded without compromising organelle and cell function, different from nDNA. When mtDNA suffered from serious excessive injuries and repair is no longer possible or damage cannot be repaired, the occurrence of cell degradation can prevent from further damage. Degradation to repair mtDNA plays a role in the maintenance of DNA stability and can improve the ability to repair damages caused by oxidative and alkylation, when BER is inhibited by compounds like methoxyamine or the repair system is overwhelmed. mtDNA degradation can process some damages of pyrimidine dimers induced by UV light and numerous chemical carcinogens, which otherwise are not able to be dealt with in mitochondria. UV-related DNA lesions can also be repaired in mitochondria by polymerase zeta (Polζ) subunit REV3, a translesion DNA polymerase [80]. Indirect mtDNA degradation may also result in mitophagy. Damaged mtDNA leads to mitochondrial dysfunction and decreased membrane potential, directing the organelle and mtDNA to degradation.

5.4 Direct DNA Repair

There are three main repair mechanisms for DNA direct repair: UV light-induced photolesions are repaired by photolyases, a series of O-alkylated damaged DNA are reversed via O_6-alkylguanine-DNA alkyltransferases, and N-alkylated base adducts are reversed by AlkB family dioxygenases [81]. The photoreactivation of cyclobutene pyrimidine dimers by photolyase was described in yeast [82] or plant mitochondria [82]. Other direct reversal enzymes such as DNA methyltransferase and ABH2 dioxygenase are recently identified [83], although their mitochondrial localization is not described. However, the repair pathways in human mitochondria remain unclear.

5.5 Nucleotide Excision Repair

NER exists in the nucleus to repair damages induced by UV and various chemicals, while this seems not to be the case in human mitochondria [84]. Despite this, two related proteins, Cockayne syndrome A and B, form complexes with DNA glycosylase Ogg1 and ssDNA binding protein SSBP1 during oxidative stress and accumulate in mitochondria [70]. In addition, another NER-related protein, RAD23A, is present in human cardiomyocyte mitochondria co-localizing with mtDNA [85].

5.6 Mismatch Repair

MMR exists in the nucleus and in mitochondria to repair mismatches caused by alkylation, oxidation, deamination of bases, erroneous insertions, or slippage errors caused by dysfunction of the DNA polymerase in the DNA replication. Mitochondrial MMR is strand unbiased and not nick-directed unlike bacterial or nuclear counterparts, while nuclear MMR proteins are not found in mitochondria, suggesting a completely independent machinery. mtDNA instability is rarely associated with alterations in nuclear MMR genes, due to the variation between nuclear and mitochondria repair systems [70]. MMR begins with the recognition of mismatches and insertion/deletion loops in the nucleus by heterodimeric complexes MutSα and MutSβ, while those mismatches are recognized in mitochondria by the multifunctional Y-box-binding protein 1 [86, 87], although the mechanism by which other mitochondrial MMR proteins are recruited has been poorly understood. Y-box-binding protein 1 is also present in the nucleus and functions as a transcription factor.

5.7 *DNA Strand Break Repair*

Single-Strand Break Repair The single-stranded DNA is more susceptible to damage as compared to double-stranded DNA. SSB repair can be induced by ROS-induced disintegration of oxidized deoxyribose, genotoxic elements, aberrant hydrolytic processes, or ineffective enzymatic activities. The latter group includes abortive DNA topoisomerase 1 (TOP1) activity during replication and transcription [89], abortive DNA ligase activity at existing SSBs, and inefficient BER repair [70]. SSB repair pathway consists of break detection, DNA end processing, gap filling, and finally DNA ligation, similar to base excision repair pathway, of which the latter two are the same as BER [89]. The enzymes involved in SSB repair are different to distinct enzymes such as aprataxin [90] and tyrosyl-DNA phosphodiesterase [69] in DNA ligase and TOP1-related SSBs, respectively, rather than DNA glycosylases.

Double-Strand Break Repair In the nucleus, double-strand breaks can be repaired in nonhomologous end joining (NHEJ), homologous recombination [91], or microhomology-mediated end joining (MMEJ). In mitochondria, NHEJ and homologous recombination are initially described [4], and MMEJ is recently repaired as a very efficient repair system in mitochondria [92].

Homologous Recombination
Compared to the NHEJ and MMEJ of DSB repair, homologous recombination maintains genetic information and requires an intact copy of the DNA as template, of which the activity is restricted to S and G2 phases. The DSB is recognized by a complex of MRE11, RAD50, and NBS1. Then exonuclease 1 or STR-DNA2 produces an ssDNA which invades a homolog molecule of DNA with RAD51 to create a crossing the break site from DNA polymerase to synthesize DNA. Homologous recombination exists in yeast, *Chlamydomonas reinhardtii*, and plant mitochondria [93, 94]. Although homologous recombination occurs not widely in mammalian mitochondria, the large amounts of recombination junctions and catenation are often seen in human heart and associated to the initiation of replication in adult human heart, but not in infants and rodents [95]. Proteins involved in homologous recombination, e.g., RAD51, DNA2, and EXOG, are equivalent to the

mitochondrial exonuclease 1. We believe that homologous recombination plays a role in the maintenance of human mtDNA integrity.

Nonhomologous End Joining
NHEJ is the main form of DNA strand break repair in the nucleus of mammalian cells and is performed by direct rapid ligation of chromosome broken ends with no template requirement and at stages of the cell cycle except for M phase [96]. The repair process of NHEJ includes Ku heterodimer to prevent DSBs from degradation and promote the recruitment of other mediator elements. Of those, the DNA-dependent protein kinase catalytic subunit forms a bridge between two blunt DNA ends, XRCC4 and XLF (XRCC4-like factor), then binds to the complex, and facilitates sealing by DNA ligase IV (Lig4). The meticulous process of NHEJ seems to be involved in other repair processes, such as BER and NER [15]. Mitochondrial extracts also have blunt-ended DNA binding activities, and Ku proteins and MRE11 were identified in mitochondria [97, 98]. It is evidenced that NHEJ is an important repair pathway in mitochondria, albeit not the main one for resolving DSB.

Microhomology-Mediated End Joining
In addition to the classical NHEJ repair pathway, MMEJ functions in the cell with or without the deficiency of factors required for NHEJ. The junctions are often associated with larger deletions and the use of 5–25-nucleotide microhomology. The microhomology in regions is associated with mtDNA deletions in patients [99]. Proteins involved in mitochondrial MMEJ (MRE11, FEN1, PARP1, LIG3) may also contribute to this repair, instead of NHEJ [92].

6 Conclusion

The understanding of mtDNA repair experiences from the initial thought of no DNA repair pathway in mitochondria to the discovery of the BER repair pathway and to multiple mtDNA repair pathways. NER proteins are present in mitochondria, while their activities and functions are not clear [45]. The integrity of mtDNA is a prerequisite for maintaining mitochondrial function, dependent upon a variety of DNA repair pathways. Mitochondrial gene mutations cause mitochondrial diseases and other diseases, including cancer [100], diabetes [101], cardiovascular diseases, and neurodegenerative disorders, as well as aging [68]. There are clear differences of mtDNA function, regulation, and repair from nDNA, even though the repair machinery and some proteins can be shared between the nucleus and mitochondria. Such variations can reflect different environments and factors to DNA damage.

7 Future Outlooks

There are urgent and great needs to explore more precise mechanisms by which DNA is involved in the repair of various pathways, and mtDNA components interact between them owns and between mitochondria and nuclear. It is questioned whether actually NER exists in mitochondria and how mitochondrial DNA repair pathways function in pathological states. We should pay more attention to factors to control and regulate the balance between mtDNA injury and repair. In addition, new technical developments will benefit the understanding and potential application of gene editing and gene therapy as well as molecular mechanisms of mtDNA damage and repair. More efforts should be offered to investigate mtDNA damage dependent on cross talk, signaling, and interaction between mitochondria and the nucleus [102, 103].

Acknowledgments The work was supported by the Zhongshan Distinguished Professor Grant (XDW), the National Natural Science Foundation of China (91230204, 81270099, 81320108001, 81270131, 81300010), the Shanghai Committee of Science and Technology (12JC1402200, 12431900207, 11410708600, 14431905100), the operation funding of the Shanghai Institute of Clinical Bioinformatics, the Ministry of Education for Academic Special Science and Research Foundation for PhD Education (20130071110043), the National Key Research and Development Program (2016YFC0902400, 2017YFSF090207), and the Medical Research Council core funding (2015–2020).

References

1. Alexeyev M, Shokolenko I, Wilson G, LeDoux S. The maintenance of mitochondrial DNA integrity – critical analysis and update. Cold Spring Harb Perspect Biol. 2013;5:a012641. [PMID: 23637283]
2. Kaarniranta K, Tokarz P, Koskela A, Paterno J, Blasiak J. Autophagy regulates death of retinal pigment epithelium cells in age-related macular degeneration. Cell Biol Toxicol. 2017;33:113–28. [PMID: 27900566]
3. Tahbaz N, Subedi S, Weinfeld M. Role of polynucleotide kinase/phosphatase in mitochondrial DNA repair. Nucleic Acids Res. 2012;40:3484–95. [PMID: 22210862]
4. Zhu LZ, Hou YJ, Zhao M, Yang MF, XT F, Sun JY, XY F, Shao LR, Zhang HF, Fan CD, et al. Caudatin induces caspase-dependent apoptosis in human glioma cells with involvement of mitochondrial dysfunction and reactive oxygen species generation. Cell Biol Toxicol. 2016;32:333–45. [PMID: 27184666]
5. Medvedev R, Hildt E, Ploen D. Look who's talking-the crosstalk between oxidative stress and autophagy supports exosomal-dependent release of HCV particles. Cell Biol Toxicol. 2017;33(3):211–31.
6. van den Ouweland JM, Lemkes HH, Ruitenbeek W, Sandkuijl LA, de Vijlder MF, Struyvenberg PA, van de Kamp JJ, Maassen JA. Mutation in mitochondrial tRNA(Leu)(UUR) gene in a large pedigree with maternally transmitted type II diabetes mellitus and deafness. Nat Genet. 1992;1:368–71. [PMID: 1284550]
7. Bao L, Diao H, Dong N, Su X, Wang B, Mo Q, Yu H, Wang X, Chen C. Histone deacetylase inhibitor induces cell apoptosis and cycle arrest in lung cancer cells via mitochondrial injury and p53 up-acetylation. Cell Biol Toxicol. 2016;32:469–82. [PMID: 27423454]

8. Cormier M, Ghouili F, Roumaud P, Bauer W, Touaibia M, Martin LJ. Influences of flavones on cell viability and cAMP-dependent steroidogenic gene regulation in MA-10 Leydig cells. Cell Biol Toxicol. 2017; https://doi.org/10.1007/s10565-017-9395-8 . [PMID: 28455626]
9. Trifunovic A, Wredenberg A, Falkenberg M, Spelbrink JN, Rovio AT, Bruder CE, Bohlooly YM, Gidlof S, Oldfors A, Wibom R, et al. Premature ageing in mice expressing defective mitochondrial DNA polymerase. Nature. 2004;429:417–23. [PMID: 15164064]
10. Krokan HE, Bjoras M. Base excision repair. Cold Spring Harb Perspect Biol. 2013;5:a012583. [PMID: 23545420]
11. Thomas PD, Kahn M. Kat3 coactivators in somatic stem cells and cancer stem cells: biological roles, evolution, and pharmacologic manipulation. Cell Biol Toxicol. 2016;32:61–81. [PMID: 27008332]
12. Akbari M, Keijzers G, Maynard S, Scheibye-Knudsen M, Desler C, Hickson ID, Bohr VA. Overexpression of DNA ligase III in mitochondria protects cells against oxidative stress and improves mitochondrial DNA base excision repair. DNA Repair. 2014;16:44–53. [PMID: 24674627]
13. Liu T, Liu WH, Zhao JS, Meng FZ, Wang H. Lutein protects against beta-amyloid peptide-induced oxidative stress in cerebrovascular endothelial cells through modulation of Nrf-2 and NF-kappab. Cell Biol Toxicol. 2017;33:57–67. [PMID: 27878403]
14. Soltani B, Ghaemi N, Sadeghizadeh M, Najafi F. Curcumin confers protection to irradiated THP-1 cells while its nanoformulation sensitizes these cells via apoptosis induction. Cell Biol Toxicol. 2016;32:543–61. [PMID: 27473378]
15. Lobrich M, Jeggo P. A process of resection-dependent nonhomologous end joining involving the goddess artemis. Trends Biochem Sci. 2017;42:690–701. [PMID: 28739276]
16. Zerin T, Kim JS, Gil HW, Song HY, Hong SY. Effects of formaldehyde on mitochondrial dysfunction and apoptosis in SK-N-SH neuroblastoma cells. Cell Biol Toxicol. 2015;31:261–72. [PMID: 26728267]
17. Szczesny B, Tann AW, Longley MJ, Copeland WC, Mitra S. Long patch base excision repair in mammalian mitochondrial genomes. J Biol Chem. 2008;283:26349–56. [PMID: 18635552]
18. Corra S, Salvadori R, Bee L, Barbieri V, Mognato M. Analysis of DNA-damage response to ionizing radiation in serum-shock synchronized human fibroblasts. Cell Biol Toxicol. 2017;33:373–88. [PMID: 28466226]
19. Peng WX, Han X, Zhang CL, Ge L, Du FY, Jin J, Gong AH. FoxM1-mediated RFC5 expression promotes temozolomide resistance. Cell Biol Toxicol. 2017. [PMID: 28185110]
20. Rydberg B, Lindahl T. Nonenzymatic methylation of DNA by the intracellular methyl group donor S-adenosyl-L-methionine is a potentially mutagenic reaction. EMBO J. 1982;1:211–6. [PMID: 7188181]
21. Keta O, Bulat T, Golic I, Incerti S, Korac A, Petrovic I, Ristic-Fira A. The impact of autophagy on cell death modalities in CRL-5876 lung adenocarcinoma cells after their exposure to gamma-rays and/or erlotinib. Cell Biol Toxicol. 2016;32:83–101. [PMID: 27026538]
22. Akbari M, Otterlei M, Pena-Diaz J, Krokan HE. Different organization of base excision repair of uracil in DNA in nuclei and mitochondria and selective upregulation of mitochondrial uracil-DNA glycosylase after oxidative stress. Neuroscience. 2007;145:1201–12. [PMID: 17101234]
23. Giromini C, Rebucci R, Fusi E, Rossi L, Saccone F, Baldi A. Cytotoxicity, apoptosis, DNA damage and methylation in mammary and kidney epithelial cell lines exposed to ochratoxin A. Cell Biol Toxicol. 2016;32:249–58. [PMID: 27154019]
24. Fang H, Wang W. Could CRISPR be the solution for gene editing's Gordian knot? Cell Biol Toxicol. 2016;32:465–7. [PMID: 27614448]
25. Stepchenkova EI, Tarakhovskaya ER, Siebler HM, Pavlov YI. Defect of Fe-S cluster binding by DNA polymerase delta in yeast suppresses UV-induced mutagenesis, but enhances DNA polymerase zeta – dependent spontaneous mutagenesis. DNA Repair (Amst). 2017;49:60–9. [PMID: 28034630]

26. Tell G, Crivellato E, Pines A, Paron I, Pucillo C, Manzini G, Bandiera A, Kelley MR, Di Loreto C, Damante G. Mitochondrial localization of APE/Ref-1 in thyroid cells. Mutat Res. 2001;485:143–52. [PMID: 11182545]
27. Tlotleng N, Vetten MA, Keter FK, Skepu A, Tshikhudo R, Gulumian M. Cytotoxicity, intracellular localization and exocytosis of citrate capped and PEG functionalized gold nanoparticles in human hepatocyte and kidney cells. Cell Biol Toxicol. 2016;32:305–21. [PMID: 27184667]
28. Arumugam P, Samson A, Ki J, Song JM. Knockdown of clusterin alters mitochondrial dynamics, facilitates necrosis in camptothecin-induced cancer stem cells. Cell Biol Toxicol. 2017;33:307–21. [PMID: 28064403]
29. Gruz P, Shimizu M, Sugiyama KI, Honma M. Mutagenicity of omega-3 fatty acid peroxidation products in the Ames test. Mutat Res. 2017;819:14–9. [PMID: 28622825]
30. Prakash A, Doublie S. Base excision repair in the mitochondria. J Cell Biochem. 2015;116:1490–9. [PMID: 25754732]
31. Boesch P, Weber-Lotfi F, Ibrahim N, Tarasenko V, Cosset A, Paulus F, Lightowlers RN, Dietrich A. DNA repair in organelles: pathways, organization, regulation, relevance in disease and aging. Biochim Biophys Acta. 2011;1813:186–200. [PMID: 20950654]
32. Manlove AH, McKibbin PL, Doyle EL, Majumdar C, Hamm ML, David SS. Structure-activity relationships reveal key features of 8-oxoguanine: a mismatch detection by the MutY glycosylase. ACS Chem Biol. 2017;12:2335–44. [PMID: 28723094]
33. Pournara A, Kippler M, Holmlund T, Ceder R, Grafstrom R, Vahter M, Broberg K, Wallberg AE. Arsenic alters global histone modifications in lymphocytes in vitro and in vivo. Cell Biol Toxicol. 2016;32:275–84. [PMID: 27165195]
34. McNally JR, O'Brien PJ. Kinetic analyses of single-strand break repair by human DNA Ligase III isoforms reveals biochemical differences from DNA Ligase I. J Biol Chem. 2017. [PMID: 28751376]
35. Kasparek TR, Humphrey TC. DNA double-strand break repair pathways, chromosomal rearrangements and cancer. Semin Cell Dev Biol. 2011;22:886–97. [PMID: 22027614]
36. Fathi H, Ebrahimzadeh MA, Ziar A, Mohammadi H. Oxidative damage induced by retching; antiemetic and neuroprotective role of Sambucus ebulus L. Cell Biol Toxicol. 2015;31:231–9. [PMID: 26493312]
37. Ralph SJ, Rodriguez-Enriquez S, Neuzil J, Saavedra E, Moreno-Sanchez R. The causes of cancer revisited: "mitochondrial malignancy" and ROS-induced oncogenic transformation – why mitochondria are targets for cancer therapy. Mol Asp Med. 2010;31:145–70. [PMID: 20206201]
38. Patenge N. Quantification of DNA damage and repair in mitochondrial, nuclear, and bacterial genomes by real-time PCR. Methods Mol Biol. 2017;1644:159–66. [PMID: 28710762]
39. Ke Y, Xu Y, Chen X, Feng S, Liu Z, Sun Y, Yao X, Li F, Zhu W, Gao L, et al. 3D chromatin structures of mature gametes and structural reprogramming during Mammalian Embryogenesis. Cell. 2017;170:367–381 e20. [PMID: 28709003]
40. Styllou P, Styllou M, Hickel R, Hogg C, Reichl FX, Scherthan H. NAC ameliorates dental composite-induced DNA double-strand breaks and chromatin condensation. Dent Mater J 2017. [PMID: 28747595]
41. Coey CT, Drohat AC. Kinetic methods for studying DNA glycosylases functioning in base excision repair. Methods Enzymol. 2017;592:357–76. [PMID: 28668127]
42. Baehr CA, Huntoon CJ, Hoang SM, Jerde CR, Karnitz LM. Glycogen synthase kinase 3 (GSK-3)-mediated phosphorylation of uracil N-glycosylase 2 (UNG2) facilitates the repair of floxuridine-induced DNA lesions and promotes cell survival. J Biol Chem. 2016;291:26875–85. [PMID: 27875297]
43. van Loon B, Samson LD. Alkyladenine DNA glycosylase (AAG) localizes to mitochondria and interacts with mitochondrial single-stranded binding protein (mtSSB). DNA Repair (Amst). 2013;12:177–87. [PMID: 23290262]

44. Sampath H. Oxidative DNA damage in disease – insights gained from base excision repair glycosylase-deficient mouse models. Environ Mol Mutagen. 2014;55:689–703. [PMID: 25044514]
45. Hazra TK, Izumi T, Boldogh I, Imhoff B, Kow YW, Jaruga P, Dizdaroglu M, Mitra S. Identification and characterization of a human DNA glycosylase for repair of modified bases in oxidatively damaged DNA. Proc Natl Acad Sci U S A. 2002;99:3523–8. [PMID: 11904416]
46. Brooks SC, Adhikary S, Rubinson EH, Eichman BF. Recent advances in the structural mechanisms of DNA glycosylases. Biochim Biophys Acta. 1834;2013:247–71. [PMID: 23076011]
47. Grodick MA, Segal HM, Zwang TJ, Barton JK. DNA-mediated signaling by proteins with 4Fe-4S clusters is necessary for genomic integrity. J Am Chem Soc. 2014;136:6470–8. [PMID: 24738733]
48. Markkanen E, Dorn J, Hubscher UMUTYHDNA. glycosylase: the rationale for removing undamaged bases from the DNA. Front Genet. 2013;4:18. [PMID: 23450852]
49. Sjolund AB, Senejani AG, Sweasy JB. MBD4 and TDG: multifaceted DNA glycosylases with ever expanding biological roles. Mutat Res. 2013;743-744:12–25. [PMID: 23195996]
50. Mandal SM, Hegde ML, Chatterjee A, Hegde PM, Szczesny B, Banerjee D, Boldogh I, Gao R, Falkenberg M, Gustafsson CM, et al. Role of human DNA glycosylase Nei-like 2 (NEIL2) and single strand break repair protein polynucleotide kinase 3′-phosphatase in maintenance of mitochondrial genome. J Biol Chem. 2012;287:2819–29. [PMID: 22130663]
51. Saad-Hussein A, Noshy M, Taha M, El-Shorbagy H, Shahy E, Abdel-Shafy EA. GSTP1 and XRCC1 polymorphisms and DNA damage in agricultural workers exposed to pesticides. Mutat Res. 2017;819:20–5. [PMID: 28622826]
52. Bhakat KK, Mantha AK, Mitra S. Transcriptional regulatory functions of mammalian AP-endonuclease (APE1/Ref-1), an essential multifunctional protein. Antioxid Redox Signal. 2009;11:621–38. [PMID: 18715144]
53. Li M, Wilson DM 3rd. Human apurinic/apyrimidinic endonuclease 1. Antioxid Redox Signal. 2014;20:678–707. [PMID: 23834463]
54. SL Y, Lee SK, Johnson RE, Prakash L, Prakash S. The stalling of transcription at abasic sites is highly mutagenic. Mol Cell Biol. 2003;23:382–8. [PMID: 12482989]
55. Roychoudhury S, Nath S, Song H, Hegde ML, Bellot LJ, Mantha AK, Sengupta S, Ray S, Natarajan A, Bhakat KK. Human apurinic/apyrimidinic endonuclease (APE1) is acetylated at DNA damage sites in chromatin, and acetylation modulates its DNA repair activity. Mol Cell Biol. 2017;37:e00401-16. [PMID: 27994014]
56. Copeland WC, Kasiviswanathan R, Longley MJ. Analysis of Translesion DNA Synthesis by the Mitochondrial DNA Polymerase gamma. Methods Mol Biol. 2016;1351:19–26. [PMID: 26530671]
57. Sykora P, Kanno S, Akbari M, Kulikowicz T, Baptiste BA, Leandro GS, Lu H, Tian J, May A, Becker KA, et al. DNA polymerase beta participates in mitochondrial DNA repair. Mol Cell Biol. 2017;37:e00237-17. [PMID: 28559431]
58. Simsek D, Furda A, Gao Y, Artus J, Brunet E, Hadjantonakis AK, Van Houten B, Shuman S, McKinnon PJ, Jasin M. Crucial role for DNA ligase III in mitochondria but not in Xrcc1-dependent repair. Nature. 2011;471:245–8. [PMID: 21390132]
59. Opperman CM, Sishi BJ. Tumor necrosis factor alpha stimulates p62 accumulation and enhances proteasome activity independently of ROS. Cell Biol Toxicol. 2015;31:83–94. [PMID: 25761618]
60. Hoch NC, Hanzlikova H, Rulten SL, Tetreault M, Komulainen E, Ju L, Hornyak P, Zeng Z, Gittens W, Rey SA, et al. XRCC1 mutation is associated with PARP1 hyperactivation and cerebellar ataxia. Nature. 2017;541:87–91. [PMID: 28002403]
61. Caldecott KW. DNA single-strand break repair. Exp Cell Res. 2014;329:2–8. [PMID: 25176342]
62. London RE. The structural basis of XRCC1-mediated DNA repair. DNA Repair (Amst). 2015;30:90–103. [PMID: 25795425]

63. Abdou I, Poirier GG, Hendzel MJ, Weinfeld M. DNA Ligase III acts as a DNA strand break sensor in the cellular orchestration of DNA strand break repair. Nucleic Acids Res. 2015;43:875–92. [PMID: 25539916]
64. Lakshmipathy U, Campbell C. Mitochondrial DNA ligase III function is independent of Xrcc1. Nucleic Acids Res. 2000;28:3880–6. [PMID: 11024166]
65. Gredilla R, Garm C, Holm R, Bohr VA, Stevnsner T. Differential age-related changes in mitochondrial DNA repair activities in mouse brain regions. Neurobiol Aging. 2010;31:993–1002. [PMID: 18701195]
66. Fang Q, Inanc B, Schamus S, Wang XH, Wei L, Brown AR, Svilar D, Sugrue KF, Goellner EM, Zeng X, et al. HSP90 regulates DNA repair via the interaction between XRCC1 and DNA polymerase beta. Nat Commun. 2014;5:5513. [PMID: 25423885]
67. Anderson CT, Friedberg EC. The presence of nuclear and mitochondrial uracil-DNA glycosylase in extracts of human KB cells. Nucleic Acids Res. 1980;8:875–88. [PMID: 6253928]
68. Chiang SC, Meagher M, Kassouf N, Hafezparast M, McKinnon PJ, Haywood R, El-Khamisy SF. Mitochondrial protein-linked DNA breaks perturb mitochondrial gene transcription and trigger free radical-induced DNA damage. Sci Adv. 2017;3:e1602506. [PMID: 28508041]
69. Das BB, Dexheimer TS, Maddali K, Pommier Y. Role of tyrosyl-DNA phosphodiesterase (TDP1) in mitochondria. Proc Natl Acad Sci. 2010;107:19790–5. [PMID: 21041670]
70. Kazak L, Reyes A, Holt IJ. Minimizing the damage: repair pathways keep mitochondrial DNA intact. Nat Rev Mol Cell Biol. 2012;13:659–71. 22992591
71. de Souza-Pinto NC, Eide L, Hogue BA, Thybo T, Stevnsner T, Seeberg E, Klungland A, Bohr VA. Repair of 8-oxodeoxyguanosine lesions in mitochondrial dna depends on the oxoguanine DNA glycosylase (OGG1) gene and 8-oxoguanine accumulates in the mitochondrial DNA of OGG1-defective mice. Cancer Res. 2001;61:5378–81. [PMID: 11454679]
72. Karahalil B, de Souza-Pinto NC, Parsons JL, Elder RH, Bohr VA. Compromised incision of oxidized pyrimidines in liver mitochondria of mice deficient in NTH1 and OGG1 glycosylases. J Biol Chem. 2003;278:33701–7. [PMID: 12819227]
73. Hu J, de Souza-Pinto NC, Haraguchi K, Hogue BA, Jaruga P, Greenberg MM, Dizdaroglu M, Bohr VA. Repair of formamidopyrimidines in DNA involves different glycosylases: role of the OGG1, NTH1, and NEIL1 enzymes. J Biol Chem. 2005;280:40544–51. [PMID: 16221681]
74. Dodson ML, Lloyd RS. Mechanistic comparisons among base excision repair glycosylases. Free Radic Biol Med. 2002;32:678–82. [PMID: 11937293]
75. Svilar D, Goellner EM, Almeida KH, Sobol RW. Base excision repair and lesion-dependent subpathways for repair of oxidative DNA damage. Antioxid Redox Signal. 2011;14:2491–507. [PMID: 20649466]
76. Liu P, Qian L, Sung JS, de Souza-Pinto NC, Zheng L, Bogenhagen DF, Bohr VA, Wilson DM 3rd, Shen B, Demple B. Removal of oxidative DNA damage via FEN1-dependent long-patch base excision repair in human cell mitochondria. Mol Cell Biol. 2008;28:4975–87. [PMID: 18541666]
77. Kalifa L, Beutner G, Phadnis N, Sheu SS, Sia EA. Evidence for a role of FEN1 in maintaining mitochondrial DNA integrity. DNA Repair. 2009;8:1242–9. [PMID: 19699691]
78. Zheng L, Zhou M, Guo Z, Lu H, Qian L, Dai H, Qiu J, Yakubovskaya E, Bogenhagen DF, Demple B, et al. Human DNA2 is a mitochondrial nuclease/helicase for efficient processing of DNA replication and repair intermediates. Mol Cell. 2008;325-36:32. [PMID: 18995831]
79. Tann AW, Boldogh I, Meiss G, Qian W, Van Houten B, Mitra S, Szczesny B. Apoptosis induced by persistent single-strand breaks in mitochondrial genome: critical role of EXOG (5′-EXO/endonuclease) in their repair. J Biol Chem. 2011;286:31975–83. [PMID: 21768646]
80. Singh B, Li X, Owens KM, Vanniarajan A, Liang P, Singh KK. Human REV3 DNA Polymerase Zeta Localizes to Mitochondria and Protects the Mitochondrial Genome. PLoS One. 2015;10:e0140409. [PMID: 26462070]
81. Yi C, He C. DNA repair by reversal of DNA damage. Cold Spring Harb Perspect Biol. 2013;5:a012575. [PMID: 23284047]

82. Takahashi M, Teranishi M, Ishida H, Kawasaki J, Takeuchi A, Yamaya T, Watanabe M, Makino A, Hidema J. Cyclobutane pyrimidine dimer (CPD) photolyase repairs ultraviolet-B-induced CPDs in rice chloroplast and mitochondrial DNA. Plant J. 2011;66:433–42. [PMID: 21251107]
83. Ringvoll J, Moen MN, Nordstrand LM, Meira LB, Pang B, Bekkelund A, Dedon PC, Bjelland S, Samson LD, Falnes PO, et al. AlkB homologue 2-mediated repair of ethenoadenine lesions in mammalian DNA. Cancer Res. 2008;68:4142–9. [PMID: 18519673]
84. Bohr VA, Stevnsner T, de Souza-Pinto NC. Mitochondrial DNA repair of oxidative damage in mammalian cells. Gene. 2002;286:127–34. [PMID: 11943468]
85. Pohjoismaki JL, Boettger T, Liu Z, Goffart S, Szibor M, Braun T. Oxidative stress during mitochondrial biogenesis compromises mtDNA integrity in growing hearts and induces a global DNA repair response. Nucleic Acids Res. 2012;40:6595–607. [PMID: 22508755]
86. Martin SA, McCabe N, Mullarkey M, Cummins R, Burgess DJ, Nakabeppu Y, Oka S, Kay E, Lord CJ, Ashworth A. DNA polymerases as potential therapeutic targets for cancers deficient in the DNA mismatch repair proteins MSH2 or MLH1. Cancer Cell. 2010;17:235–48. [PMID: 20227038]
87. Kadyrov FA, Genschel J, Fang Y, Penland E, Edelmann W, Modrich P. A possible mechanism for exonuclease 1-independent eukaryotic mismatch repair. Proc Natl Acad Sci U S A. 2009;106:8495–500. [PMID: 19420220]
88. Wang W, Wang X. Single-cell CRISPR screening in drug resistance. Cell Biol Toxicol. 2017;33:207–10. [PMID: 28474250]
89. Hegde ML, Izumi T, Mitra S. Oxidized base damage and single-strand break repair in mammalian genomes: role of disordered regions and posttranslational modifications in early enzymes. Prog Mol Biol Transl Sci. 2012;110:123–53. [PMID: 22749145]
90. Ahel I, Rass U, El-Khamisy SF, Katyal S, Clements PM, McKinnon PJ, Caldecott KW, West SC. The neurodegenerative disease protein aprataxin resolves abortive DNA ligation intermediates. Nature. 2006;443:713–6. [PMID: 16964241]
91. Stracker TH, Petrini JH. The MRE11 complex: starting from the ends. Mol Cell Biol. 2011;12:90–103. [PMID: 21252998]
92. Tadi SK, Sebastian R, Dahal S, Babu RK, Choudhary B, Raghavan SC. Microhomology-mediated end joining is the principal mediator of double-strand break repair during mitochondrial DNA lesions. Mol Biol Cell. 2016;27:223–35. [PMID: 26609070]
93. Mileshina D, Koulintchenko M, Konstantinov Y, Dietrich A. Transfection of plant mitochondria and in organello gene integration. Nucleic Acids Res. 2011;39:e115. [PMID: 21715377]
94. Davila JI, Arrieta-Montiel MP, Wamboldt Y, Cao J, Hagmann J, Shedge V, YZ X, Weigel D, Mackenzie SA. Double-strand break repair processes drive evolution of the mitochondrial genome in Arabidopsis. BMC Biol. 2011;9:64. [PMID: 21951689]
95. Pohjoismaki JL, Goffart S, Tyynismaa H, Willcox S, Ide T, Kang D, Suomalainen A, Karhunen PJ, Griffith JD, Holt IJ, et al. Human heart mitochondrial DNA is organized in complex catenated networks containing abundant four-way junctions and replication forks. J Biol Chem. 2009;284:21446–57. [PMID: 19525233]
96. Chapman JR, Taylor MR, Boulton SJ. Playing the end game: DNA double-strand break repair pathway choice. Mol Cell. 2012;47:497–510. [PMID: 22920291]
97. Coffey G, Campbell C. An alternate form of Ku80 is required for DNA end-binding activity in mammalian mitochondria. Nucleic Acids Res. 2000;28:3793–800. [PMID: 11000272]
98. Dmitrieva NI, Malide D, Burg MB. Mre11 is expressed in mammalian mitochondria where it binds to mitochondrial DNA. Am J Physiol Regul Integr Comp Physiol. 2011;301:R632–40. [PMID: 21677273]
99. Mita S, Rizzuto R, Moraes CT, Shanske S, Arnaudo E, Fabrizi GM, Koga Y, DiMauro S, Schon EA. Recombination via flanking direct repeats is a major cause of large-scale deletions of human mitochondrial DNA. Nucleic Acids Res. 1990;18:561–7. [PMID: 2308845]

100. Yuzefovych LV, Kahn AG, Schuler MA, Eide L, Arora R, Wilson GL, Tan M, Rachek LI. Mitochondrial DNA Repair through OGG1 Activity Attenuates Breast Cancer Progression and Metastasis. Cancer Res. 2016;76:30–4. [PMID: 26586787]
101. Akbari M, Skjelbred C, Folling I, Sagen J, Krokan HE. A gel electrophoresis method for detection of mitochondrial DNA mutation (3243 tRNA(Leu (UUR))) applied to a Norwegian family with diabetes mellitus and hearing loss. Scand J Clin Lab Invest. 2004;64:86–92. [PMID: 15115244]
102. Fang EF, Scheibye-Knudsen M, Chua KF, Mattson MP, Croteau DL, Bohr VA. Nuclear DNA damage signalling to mitochondria in ageing. Nat Rev Mol Cell Biol. 2016;17:308–21. [PMID:26956196]
103. Saki M, Prakash A. DNA damage related crosstalk between the nucleus and mitochondria. Free Radic Biol Med. 2017;107:216–27. [PMID: 27915046]

Linlin Zhang, MD, is a graduate student co-educated by Wenzhou Medical University and Fudan University. Her main research is focused on lung cancer, mitochondrial diseases and metabolism.

Dr. Aurelio Reyes, PhD, is a Senior Scientist in the MRC Mitochondrial Biology Unit (formerly Dunn Human Nutrition Unit) of Cambridge University and his research has been focused on discovering new genes involved in mitochondrial diseases and elucidating their mechanism of action. Dr. Reyes has published more than 70 scientific papers in well-recognized international journals and 11 book chapters. He is also member of the Editorial board of five international journals.

Xiangdong Wang, MD, PhD, is a Distinguished Professor of Medicine, Director of Shanghai Institute of Clinical Bioinformatics, Executive Director of Clinical Science Institute of Fudan University Zhongshan Hospital, Director of Fudan University Center of Clinical Bioinformatics, Deputy Director of Shanghai Respiratory Research Institute, and visiting professor of King's College of London. His main research is focused on clinical bioinformatics, disease-specific biomarkers, lung chronic diseases, cancer immunology, and molecular & cellular therapies. He is the author of more than 200 scientific publications with the impact factor about 900, citation number about 6918, h-index 46, i10-index 181, and cited journal impact factor about 7000.

Chapter 8
Is Mitochondrial Cell Fragility a Cell Weakness?

William Wang, Jiayuan Hou, Zhenghua Zhu, and Hao Fang

Abstract Mitochondrial dysfunction has historically been linked to the cessation of cell function and ageing. Downstream effects such as reduced calcium buffering capacity, elevated levels of reactive oxygen species, and alterations in adenosine-5′-triphosphate are linked to a wide variety of pathological diseases. The importance of the mitochondria has increasingly been highlighted due to its potential as a therapeutic target for drug intervention and cell elimination in cancer. In addition, due to its origin, drugs targeting bacteria are required to be thoroughly tested prior to administration to prevent toxicity for the mitochondria. In this chapter, we will discuss a variety of factors that could influence mitochondrial dysfunction and highlight potential solutions to these. A comprehensive understanding regarding the mechanisms underlying mitochondrial dysfunction could aid in developing future therapeutic targets in multiple pathologies such as cancer and liver diseases.

Keywords Mitochondria • Dysfunction • Therapy • Fragility • Cancer

1 Introduction

Mitochondria exist within cells as a subcellular organelle involved in the generation of energy via different complex metabolic pathways such as the Krebs cycle and β-oxidation chain. Like its bacterial ancestor, a eukaryotic progenitor, which has engulfed an alpha-proteobacterium, the mitochondrium is divided into an inner membrane (IM) and outer membrane (OM), both with distinctive functions and features [1]. Both an intermembrane space (IMS) and multiple matrix compartments

W. Wang (✉) • H. Fang (✉)
Department of Anaesthesiology, Jinshan Hospital, Fudan University, Shanghai, China

Department of Anaesthesiology, Zhongshan Hospital, Fudan University, Shanghai, China
e-mail: William_95a@hotmail.com; drfanghao@163.com

J. Hou • Z. Zhu
Zhongshan Hospital Institute of Clinical Science, Fudan University, Shanghai, China

H. Sun, X. Wang (eds.), *Mitochondrial DNA and Diseases*, Advances in Experimental Medicine and Biology 1038,
https://doi.org/10.1007/978-981-10-6674-0_8

are encapsulated within IM and OM. Throughout evolution the circular mitochondrial DNA (mtDNA) has progressively been condensed through transfers between the mtDNA and nucleus. Its bacterial ancestry is responsible for mtDNA susceptibility to antimicrobial drugs such as tetracyclines and amino-glycosides. In addition, antiviral nucleoside analogues are likely to affect the maintenance mechanisms of the mtDNA, although the exact mechanisms by which these drugs are challenging to pinpoint remain unclear. Reactive oxygen species (ROS) generation is associated with mutations in NAD/NADPH, AMP/ATP, etc. [2]. Another major contributor for ROS is electron leakage from the mitochondrial electron transport chain. These factors will impact both the regulation and homeostasis of stem cells, but it will also result in mitochondrial damage resulting in dysfunctions. Thus, mitochondrial damage can be directly linked to age-related pathologies and syndromes [3]. An understanding of the mechanisms involved in mitochondrial activities during phases as well as its signalling pathways is crucial to discover treatments for a wide variety of pathologies. In this chapter, we will highlight potential causes to mitochondrial damage as well as what therapeutic treatments are suggested.

2 Mitochondrial Structure and Autophagy

Histone deacetylase inhibitors (HDACi) such as quisinostat can function to reverse non-genotoxic phenotypical alterations. Quisinostat inhibits A549 cell proliferation which prevents migration of cells by inhibiting the epithelial-mesenchymal transition (EMT) process; additionally it induces up-acetylation in a variety of both histone and nonhistone proteins. More significantly, it causes the destruction of mitochondrial membrane potential ($\Delta\Psi m$) and increases reactive oxygen species (ROS) production resulting in extrinsic and intrinsic apoptosis of mitochondrial cell, of which the regulatory details of the apoptosis are explained in Fig. 8.1. p. 53 acetylation was also increased in the presence of quisinostat resulting in the arrest of the G1 phase [4]. A sesquiterpene lactone, gaillardin, has similar properties as quisinostat, resulting in increased ROS production and $\Delta\Psi m$ degradation, and induces apoptotic mechanism activation in mitochondria. However, in contrast to quisinostats, gaillardin affects caspase-3, caspase-6, and caspase-9 activations. Upregulation of pro-apoptotic proteins such as Bax and p53 and downregulation of anti-apoptotic proteins like Bcl-2 occurred when gaillardin was introduced to cancer cell lines [5].

Autophagy occurs via protein complex recruitment at the phagophore assembly site (PAS) where autophagosomes are degraded via endolysosomal interactions. The process of autophagy is furthermore categorised into three subcategories, including macroautophagy, microautophagy, and chaperone-mediated autophagy (Fig. 8.2). Due to its role as a stress responder against infections or environmental fluctuations, it is an important participant in cytoprotective functions. Thus, deficiencies and impairment are frequently indicators for human diseases [6]. Additionally, quinocetone involved in autophagy can induce stress to the endoplasmic reticulum and trigger cell trafficking related to MRLC-mediated mAtg9, known for being a key component in the formation of autophagosomes through death-

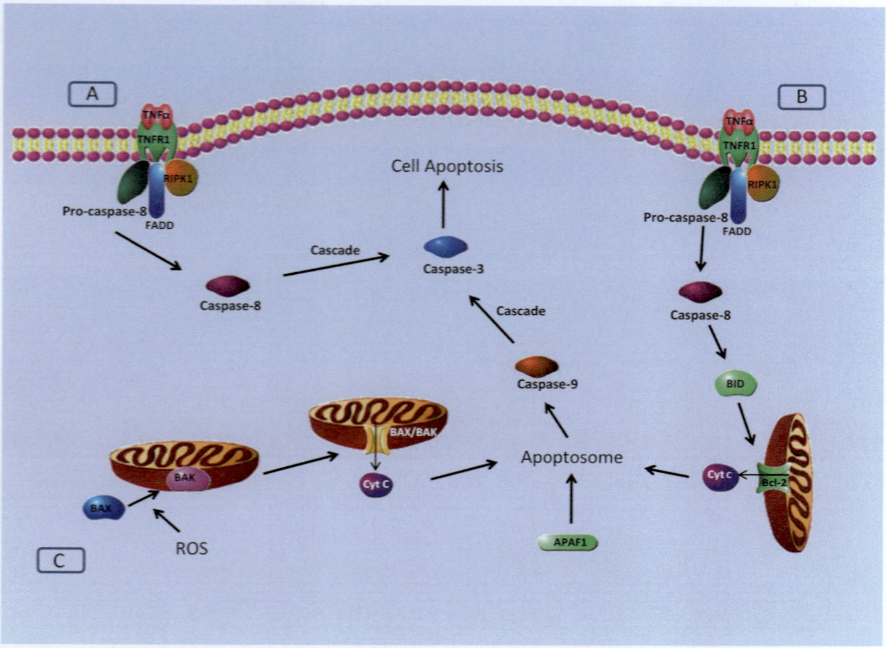

Fig. 8.1 Regulatory mechanism of extrinsic and intrinsic apoptosis. (**a**) Extrinsic apoptosis can be ignited by ligand-death receptor superfamily, such as tumour necrosis factor-α (TNF-α) and tumour necrosis factor receptor 1 (TNFR1). TNFR1 drives the assembly of a complex consisting of Fas-associated via death domain (FADD), receptor-interacting serine/threonine kinase 1 (RIPK1), and pro-caspase-8, which activates the caspase-8 → caspase-3 cascade. The pro-apoptotic cascade eventually leads to cell apoptosis in type I cells, such as lymphocyte, without the contribution of mitochondria. (**b**) Active caspase-8 cleaves the apoptosis regulator Bcl-2 family member BID in hepatocyte, a type II cell, to form truncated BID, leading to mitochondrial outer membrane permeabilisation. (**c**) Intrinsic apoptosis can be ignited by multiple intracellular stress conditions, including oxidative stress and DNA damage. The effector proteins, BAX and BAK, belong to the Bcl-2 family members and form pores in the outer mitochondrial membrane. In both BID and BAX/BAK cases, toxic proteins, such as cytochrome c (Cyt c), are released from permeabilised mitochondria to cytoplasm. The apoptosome can be assembled by Cyt c, apoptotic peptidase activating factor 1 (APAF-1), and dATP, which activates the caspase-9 → caspase-3 cascade, resulting in cell apoptosis

associated protein kinases (DAPK1) overexpression traggered by ATF6. Other effects include ATF6 cleavage and MRLC phosphorylation. Cell lines which were exposed to stable ATF6 and DAPK1 also experience a reduced level of LC3 conversions when introduced to quinocetone [7] (Fig. 8.3).

Ubiquitin-proteasome system (UPS) inhibitors such as MG132, epoxomicin, and lactacystin could be important to consider as potential therapeutic agents in treating amelanotic melanoma (Ab cells). Epoxomicin, the most cytotoxic of the three, induces apoptosis via the mitochondrial pathway by activating caspase-9, cytochrome c, and AIF. In addition, Noxa and Mcl-1, part of the Bcl-2 family are targets

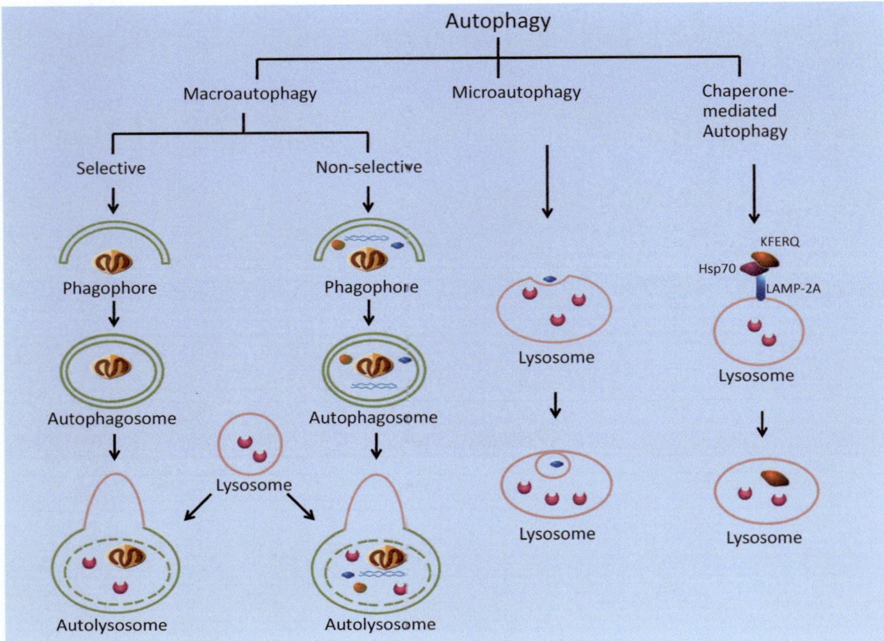

Fig. 8.2 Forming processes from autophagy into three subcategories, including macroautophagy, microautophagy, and chaperone-mediated autophagy. Macroautophagy includes selective autophagy and non-selective autophagy. Selective autophagy can clear damaged or unneeded organelles, such as mitochondria, in a selective manner. Non-selective macroautophagy occurs and mediates the generation of essential nutrients to needed cells. In the process of macroautophagy, a phagophore surrounds a damaged or unneeded organelle or some cytoplasm to form an autophagosome harbouring double membranes. The fusion of autophagosome and lysosome makes the formation of autophago-lysosome, which mediates the selective destruction of the autophagosome contents. The microautophagy is the process from the lysosomal membrane to engulf some cytoplasmic components. The chaperone-mediated autophagy is the process of degradation of KFERQ-amino-acid motif-containing target proteins. The substrate proteins are recognised by the chaperone heat shock protein 70 (Hsp70) complex, bound to lysosomal-associated membrane protein 2 (LAMP-2A) lysosomal protein, and subsequently targeted for lysosomal membrane

for the drug bortezomib. Pan-caspase inhibitor BAF was not only discovered to inhibit the cell cycle post-inactivation of caspases, but it also proved that caspase inhibition is only partially responsible for cell apoptosis [8].

Indole-based chalcones can simulate methuosis to trigger cell death within cancer cell lines and glioblastomas. These molecules (which contain 3-(5-methoxy-2-methyl-1H-indol-3-yl)-1-(4-pyridinyl)-2-propen-1-one (MOMIPP)) can cause expansion and vacuolisation of macropinosome compartments in the endocyte and are also able to destabilise and prevent the trafficking of vesicles in lysosomal nexuses through the EGF and LDL receptor degradation, mutating procathepsin processing, and excessive accumulation of autophagosomes. MOMIPP analogues exert a similar effect with vacuolisation and the disruption of cathepsin processing,

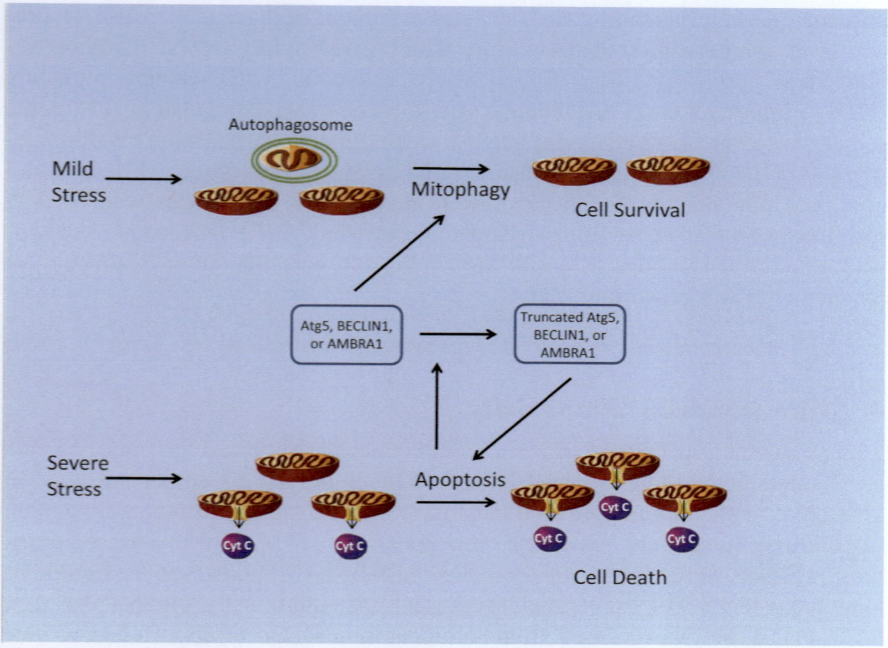

Fig. 8.3 The balance between apoptosis and mitophagy. The mitochondria impaired by mild stress are cleared by mitophagy, leading to cell survival. The mitochondria compromised by severe stress may cause the occurrence of cell apoptosis. Some autophagy-related proteins, such as autophagy-related 5 (Atg5), beclin-1, and autophagy and beclin 1 regulator 1 (AMBRA1), will be cleaved during apoptosis. The truncated Atg5, beclin-1, and AMBRA1 translocate to mitochondria, leading to enhanced apoptosis and reduced mitophagy

while these analogues do not impact cell viability to a large extent, and they do not cause significant receptor degradations either, suggesting that although indole-based chalcones have cytotoxic properties, they vary in response to how it interacts with the pathways for endolysosomal trafficking [9].

3 Mitochondrial DNA

Mitochondrial DNA (mtDNA) content is frequently measured as the ratio between the mitochondrial and nuclear genome using quantitative PCR in real time. With defensive antioxidant properties, free curcumin is a promising ionising radiation (IR) alternative, whereas dendrosomal nanoformulation of curcumin (DNC) expresses THP-1 apoptosis induction by caspase-3 activation and caspase-independent apoptosis using mitochondrial cytochrome c activation when used in tandem with IR [10]. The introduction of BAPTA-AM, a calcium chelator, and the SERCA inhibitor, thapsigargin, could potentially limit cancer progression by influencing Akt levels through cell survival reduction. Both resulted in HIF-1α protein level upregulation; however, while BAPTA-AM mildly inhibited Akt activation, thapsigargin acted as a trigger for Akt activation. Thus, calcium signalling could be important in understanding cancer metastasis and deregulation [11]. In the presence of mono(2-ethylhexyl) phthalate (MEHP) and rosiglitazone, PPARγ pathways involving energy metabolism and signalling were upregulated, whereas adipokine and inflammation signal genes were deregulated. Adipocytes with MEHP/rosiglitazone had elevated lipolysis and glycolysis and MEHP induced nuclear activation and translocations in the PPARγ indicating gene alteration treatments for bioenergetics as a possibility [12].

4 Mitochondrial Cytotoxicity

Extracellular adenosine-5′-triphosphate (ATP) could reduce ethanol toxicity in cells. P2Y1 receptors in epithelial cells have an immunological function and elevated intracellular cyclic adenosine monophosphate (cAMP) levels intracellularly. Similar effects were noted using forskolin and 8-Br-cAMP; however these could be inhibited using an H-89 protein kinase A inhibitor. Additionally, intracellular toxic mechanisms that were caused from significantly elevated levels of ethanol were found to be reversed with micromolar ranges of ATP present, and an increase in structural integrity within the monolayers of cells was observed preventing leakage. These observations highlight the potential use of purinergic receptors in treating patients with ethanol toxicity [13]. Methanol has similar properties to ethanol and contains formic acid and formaldehyde, causing neurotoxicity when ingested. In particular, formaldehyde has multiple toxic effects and in contrast to formic acid is able to exert cytotoxic effects in a large amount of cell types. In neuronal SK-N-SH cells, formaldehyde was noted to induce oxidative stress, significantly reduce the availability of ATP intracellularly, reduce the mitochondrial membrane potential (MMP), and cause a dose-dependent inhibition of NADH dehydrogenase (complex I), aconitases sensitive to oxidative fluctuations, and other mitochondrial respiratory enzymes [14]. Similarly, to formaldehyde, haloalkenes such as trichloroethylene and tetrachloroethylene are known to cause nephrotoxicity via the glutathione conjugation pathway. Cysteine conjugate β-lyase activation resulted from the presence of haloalkenes has varying degrees of toxicity such as *S*-(1,2,2-trichlorovinyl)-L-cysteine being more toxic in contrast to *S*-(1,2,3,4,4-pentachloro-1:3-butadienyl)-L-cysteine with *N*-acetyl-L-cysteine conjugates also possessing toxic characteristics.

These cysteine-c conjugates are also involved in oxidative pathways forming sulfoxides which in vitro and in vivo were more toxic compared to cysteines through mediators such as cytochrome P-450 3A and flavin-containing monooxygenase 3 which are also toxic [15]. Non-nucleoside reverse transcriptase inhibitors (NNRTI) are often considered as therapeutic targets for antiretroviral treatments. However, the mechanisms are poorly understood and frequently have significant negative effects in tandem with treatment. An example is efavirenz (EFV) which is suggested to cause dysfunction and has significant toxic effects towards the mitochondria. The toxic effects are stemmed from depolarisation in $\Delta\psi m$ due to OM alterations in the mitochondrium. Important events in the cascade are the secretion of cytochrome c, morphological alterations, and activation of apoptosis via the mitochondrial pathway, which ultimately results in the reduction of mitochondrial DNA and RNA levels and apoptosis [16].

5 Mitochondrial Stem Cells

Mitochondrial dysfunction is increasingly present in aged cells suggesting a direct relationship; however pinpointing the mechanisms for cell ageing is challenging. The free radical theory of ageing is suggested to be one major factor, as elevated ROS levels result in mitochondrial DNA (mtDNA) mutations causing respiratory chain dysfunction, and mitochondrial DNA polymerase γ which mediates errors in replication inducing mtDNA mutations is suggested as well [17]. Cellular and metabolic alterations are often considered to be participants, such as the homeostasis of energy and the detection and sensitivity to nutrients [18]. Studies suggest that mitochondrial decay is linked with NAD^+ reduction which disrupts the process of oxidative phosphorylation. In contrast, enhanced function of mitochondrium is associated with increased stem cell functions through inducing oxidative metabolism and increasing content within the mitochondria.

An emerging way to investigate how drugs interact with cells and potentially could result in toxicology is via stem cells. Human-induced pluripotent stem cells (hiPSC) or embryonic stem cells (ESC) are differentiated into target cells, and drugs such as acetaminophen (AAP) and aflatoxin B1 (AFB1) are administered to test for toxic effects to hepatocytes. Functional markers such as secretion of albumin or urea, calcium influx, and oxidative stress were reduced in the presence of AAP and AFB1 [19].

6 Mitochondrial Immunity

Tests using caudatin, derived from *Cynanchum bungei decne* and evaluated using U251 and U87 cells, resulted in inhibition of cell production and maturation. This was caused by the stimulation of cell apoptosis shown from PARP cleavage, activation of caspase-3, caspase-7, and caspase-9, and the maximal increase of sub-G1 peak. Additionally, Bcl-2 members were un-stabilised resulting in dysfunction in mitochondrial functions. The apoptosis was linked to the induction of ROS production through redox homeostasis disturbance. Cell angiogenesis was noted to be inhibited in xenografted in vivo tumour U251 cells, and reduced mitochondrial functionality was present [20].

Cancer stem cells (CSC) possess strong resistance mechanisms to drug therapy complicating therapeutic drug regiments. One factor affecting drug-induced stress is the endoplasmic reticulum (ER) and its structural integrity. Clusterin (CLU), a component of regulating the integrity, is a potential target to inhibit and reduce the efficacy of drug resistances in CSC and preventing recurrences of metastatic tumours. Camptothecin (CPT)-treated cells had reduced EC_{50} values in both CSC and MCF-7 cells with CLU suppression increasing its anticancer effects through elevating induction signals for necrosis and apoptosis. Increased values for mitochondrial permeability transition pores (MPTP) and Ca^{2+} cytosolic were also observed highlighting CLU suppression methods to be a beneficial tool for anticancer agents [21].

Elevated ROS levels trigger the activation of the autophagosomal pathway involved in degradation of lysosomes intracellularly. Hepatitis C virus (HCV) inhibits enzymes involved in detoxifying ROS that are Nrf2/ARE-dependent with a poorly understood mechanism. Theories propose that endosomal pathways and their derivatives, exosomes, are involved when released virally. Due to the connection between the endocytic pathway and autophagy, ROS elevated levels caused by HCV could stimulate viral particle releases via autophagy activation [22].

TNF-α, responsible for mediating immunological responses via membrane receptors, is frequently elevated in a variety of pathologies. The severity of TNF-α in pathology such as congestive heart failure (CHF) is often attributed to how the endothelium and the heart itself respond to the downstream effects. Protein breakdown in tissues is elevated with an increased TNF-α presence due to its effect on cultured myotubes where it directly stimulates catabolism. Increased ubiquitination of protein is also present due to poor pathway clearance and an induction of autophagy. Together with an increase in E3 ubiquitin ligase expression and a reduction in proteasome levels, it is suggested that TNF-α results in an increased myofibrillar protein degradation rate as well as a disturbance in autophagy resulting in sarcomeric dysfunction [23]. Additionally, inflammatory mediators that impact glucocorticoids (GCs) have been shown to be influenced by mucin-1 (MUC1) through testing done with tumour necrosis factor-alpha (TNF-α) or dexamethasone (DEX), indicating the use of MUC1 as a target for adenocarcinoma (ADC) treatment [24].

7 Conclusion

Mitochondria participate in multiple key roles in a wide variety of metabolic processes within the cell and are integral for the structural and functional features of the cell. However, due to this mitochondrial dysfunction is also a major factor for cell failure, ageing, and pathologies. Thus, it is vital that the complex interactions and signalling networks in the mitochondria act as a focal point for drug research. Currently, a multitude of studies regarding the functional characteristics of mitochondria have been researched, and multiple potential targets have been identified, yet both the knowledge regarding the interactions within the mitochondria itself and how drug interacts with this organelle remains superficial. This is partially due to the barriers that need to be overcome to succeed in selectively target sites but also due to a lack of optimal methods. Although advancements have been made, it is crucial to continue researching how to optimise therapeutic outcomes for different pathogens, increase drug efficiency, and discover applicable analytics regarding the accumulation of drug concentration and what the detailed physiological effects these drugs have on the mitochondria and on the human body.

Acknowledgements The work was supported by Zhongshan Distinguished Professor Grant (XDW), National Nature Science Foundation of China (91230204, 81270099, 81320108001, 81270131, 81300010), Shanghai Committee of Science and Technology (12JC1402200, 12431900207, 11410708600, 14431905100), Operation funding of Shanghai Institute of Clinical Bioinformatics, Ministry of Education for Academic Special Science and Research Foundation for PhD Education (20130071110043), and National Key Research and Development Program (2016YFC0902400, 2017YFSF090207).

References

1. Lane N, Martin W. The energetics of genome complexity. Nature. 2010;467(7318):929–34.
2. Shyh-Chang N, Daley GQ, Cantley LC. Stem cell metabolism in tissue development and aging. Development. 2013;140(12):2535–47.
3. Sahin E, DePinho RA. Linking functional decline of telomeres, mitochondria and stem cells during ageing. Nature. 2010;464(7288):520–8.
4. Bao L, Diao H, Dong N, Su X, Wang B, Mo Q, Yu H, Wang X, Chen C. Histone deacetylase inhibitor induces cell apoptosis and cycle arrest in lung cancer cells via mitochondrial injury and p53 up-acetylation. Cell Biol Toxicol. 2016;32(6):469–82.
5. Fallahian F, Aghaei M, Abdolmohammadi MH, Hamzeloo-Moghadam M. Molecular mechanism of apoptosis induction by Gaillardin, a sesquiterpene lactone, in breast cancer cell lines. Cell Biol Toxicol. 2015;31(6):295–305.
6. Lippai M, Szatmári Z. Autophagy—from molecular mechanisms to clinical relevance. Cell Biol Toxicol. 2017;33(2):145–68.
7. Zhou Y, Zhang S, Dai C, Tang S, Yang X, Li D, Zhao K, Xiao X. Quinocetone triggered ER stress-induced autophagy via ATF6/DAPK1-modulated mAtg9a trafficking. Cell Biol Toxicol. 2016;32(2):141–52.
8. Sidor-Kaczmarek J, Cichorek M, Spodnik JH, Wójcik S, Moryś J. Proteasome inhibitors against amelanotic melanoma. Cell Biol Toxicol. 2017;33:1–17.
9. Mbah NE, Overmeyer JH, Maltese WA. Disruption of endolysosomal trafficking pathways in glioma cells by methuosis-inducing indole-based chalcones. Cell Biol Toxicol. 2017;33(3):263–82.
10. Soltani B, Ghaemi N, Sadeghizadeh M, Najafi F. Curcumin confers protection to irradiated THP-1 cells while its nanoformulation sensitizes these cells via apoptosis induction. Cell Biol Toxicol. Dec. 2016;32(6):543–61.
11. Divolis G, Mavroeidi P, Mavrofrydi O, Papazafiri P. Differential effects of calcium on PI3K-Akt and HIF-1α survival pathways. Cell Biol Toxicol. 2016;32(5):437–49.
12. Chiang H-C, Wang C-H, Yeh S-C, Lin Y-H, Kuo Y-T, Liao C-W, Tsai F-Y, Lin W-Y, Chuang W-H, Tsou T-C. Comparative microarray analyses of mono(2-ethylhexyl)phthalate impacts on fat cell bioenergetics and adipokine network. Cell Biol Toxicol. 2017:1–16.

13. Seo JB, Jung S-R, Hille B, Koh D-S. Extracellular ATP protects pancreatic duct epithelial cells from alcohol-induced damage through P2Y1 receptor-cAMP signal pathway. Cell Biol Toxicol. Jun. 2016;32(3):229–47.
14. Zerin T, Kim J-S, Gil H-W, Song H-Y, Hong S-Y. Effects of formaldehyde on mitochondrial dysfunction and apoptosis in SK-N-SH neuroblastoma cells. Cell Biol Toxicol. Dec. 2015;31(6):261–72.
15. Cristofori P, Sauer AV, Trevisan A. Three common pathways of nephrotoxicity induced by halogenated alkenes. Cell Biol Toxicol. Feb. 2015;31(1):1–13.
16. Ganta KK, Mandal A, Chaubey B. Depolarization of mitochondrial membrane potential is the initial event in non-nucleoside reverse transcriptase inhibitor efavirenz induced cytotoxicity. Cell Biol Toxicol. Feb. 2017;33(1):69–82.
17. Trifunovic A, Wredenberg A, Falkenberg M, Spelbrink JN, Rovio AT, Bruder CE, Bohlooly-Y M, Gidlöf S, Oldfors A, Wibom R, Törnell J, Jacobs HT, Larsson N-G. Premature ageing in mice expressing defective mitochondrial DNA polymerase. Nature. 2004;429(6990):417–23.
18. Bonawitz ND, Chatenay-Lapointe M, Pan Y, Shadel GS. Reduced TOR signaling extends chronological life span via increased respiration and upregulation of mitochondrial gene expression. Cell Metab. 2007;5(4):265–77.
19. Kang S-J, Lee H-M, Park Y-I, Yi H, Lee H, So B, Song J-Y, Kang H-G. Chemically induced hepatotoxicity in human stem cell-induced hepatocytes compared with primary hepatocytes and HepG2. Cell Biol Toxicol. 2016;32(5):403–17.
20. Zhu L, Hou Y, Zhao M, Yang M, Fu X, Sun J, Fu X, Shao L, Zhang H, Fan C, Gao H, Sun B. Caudatin induces caspase-dependent apoptosis in human glioma cells with involvement of mitochondrial dysfunction and reactive oxygen species generation. Cell Biol Toxicol. 2016;32(4):333–45.
21. Arumugam P, Samson A, Ki J, Song JM. Knockdown of clusterin alters mitochondrial dynamics, facilitates necrosis in camptothecin-induced cancer stem cells. Cell Biol Toxicol. 2017;33(3):307–21.
22. Medvedev R, Hildt E, Ploen D. Look who's talking—the crosstalk between oxidative stress and autophagy supports exosomal-dependent release of HCV particles. Cell Biol Toxicol. 2017;33(3):211–31.
23. Opperman CM, Sishi BJN. Tumor necrosis factor alpha stimulates p62 accumulation and enhances proteasome activity independently of ROS. Cell Biol Toxicol. 2015;31(2):83–94.
24. Xu M, Wang X. Critical roles of mucin-1 in sensitivity of lung cancer cells to tumor necrosis factor-alpha and dexamethasone. Cell Biol Toxicol. 2017;33(4):361–71.

Chapter 9
Mitochondrial DNA Methylation and Related Disease

Danyan Gao, Bijun Zhu, Hongzhi Sun, and Xiangdong Wang

Abstract Most researchers focused on methylation of genomic DNA, while methylation of mitochondrial DNA (mtDNA) is rarely tauched, and there is still controversy about the existence of mtDNA methylation. The study of cytosine methylation in mtDNA is limited. The mtDNA was recently found to exist CpG hypomethylation, and more studies provided evidence that mtDNA methylation plays an important role in mitochondrial gene regulation. In present review, we will overview recent studies of mitochondrial DNA methylation and potential influencing factors in diseases that are involved in mtDNA methylation. Thus, the further studies on mtDNA methylation will provide more evidence to explain the mechanism of mtDNA methylation and an advantageous approach for human clinical diagnosis and prevention.

Keywords Mitochondrial DNA • Methylation

D. Gao • B. Zhu
Zhongshan Hospital Institute of Clinical Science, Fudan University Medical School, Shanghai, China

Shanghai Institute of Clinical Bioinformatics, Shanghai, China

H. Sun (✉)
Liaoning Province Key Laboratory of Cancer Metabolomics, Jinzhou Hospital of Jinzhou Medical University, Jinzhou, Liaoning, China
e-mail: 752816371@qq.com; cmushz@163.com

X. Wang (✉)
Zhongshan Hospital Institute of Clinical Science, Fudan University, Shanghai Medical College, Shanghai, China
e-mail: xiangdong.wang@clintransmed.org

H. Sun, X. Wang (eds.), *Mitochondrial DNA and Diseases*, Advances in Experimental Medicine and Biology 1038,
https://doi.org/10.1007/978-981-10-6674-0_9

1 Introduction

Mitochondria exist in almost all eukaryotic cells and are one of, if not, the major organelles tasked with cellular energy production. The nucleus DNA (nDNA) and the mitochondrial DNA (mtDNA), two genetic systems, control mitochondrial function. MtDNA is found in the matrix of mitochondria or the inner membrane of mitochondria, with the loop structure of heavy chains (H) or light chains (L). More specifically, mtDNA is 16,569 base pairs in length and encodes 37 genes. These genes include 2 rRNA molecules, 22 tRNA molecules, and 13 protein subunits of the electron transport chain/oxidative phosphorylation system (OXPHOS) [1], as explained in Fig. 9.1. The mtDNA has unique genetic characteristics, including

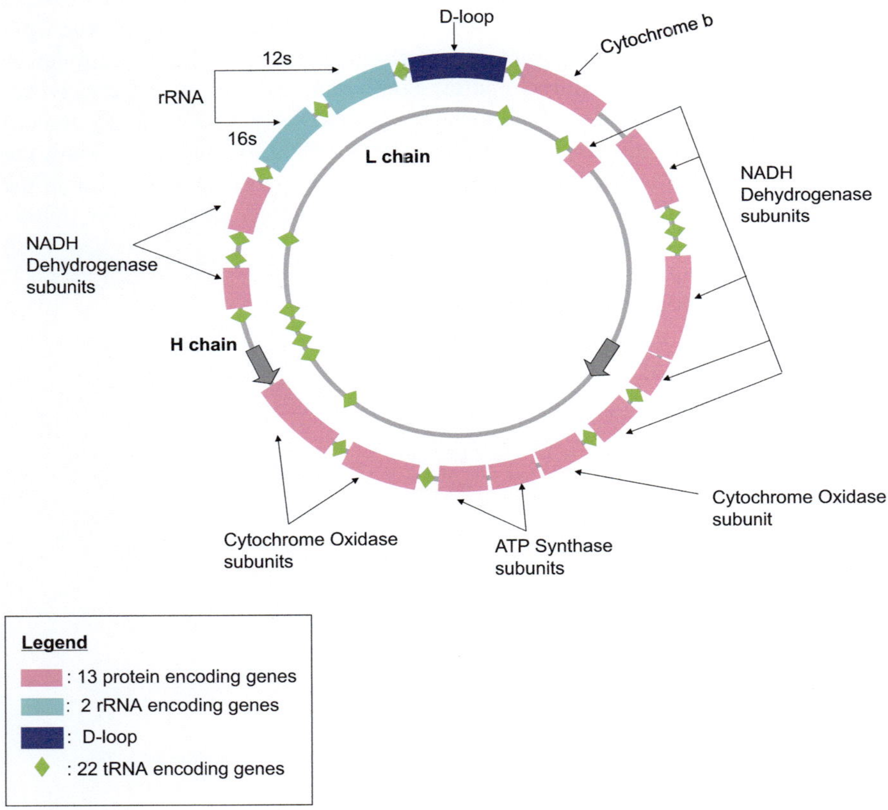

Fig. 9.1 Mitochondrial DNA. Mitochondrial DNA (mtDNA) molecules in mammals generally are 15,000 ~ 18,000 base pairs in size, double stranded, circular, and divided into two chains (a guanine-rich outer heavy chain [H] and a cytosine-rich inner light chain [L]). mtDNA contains a total of 37 genes: two ribosomal RNA-encoding genes, 13 genes encoding polypeptides (including cytochrome b, three subunits of cytochrome oxidase, two subunits of ATP synthase, and seven subunits of NADH dehydrogenase), and 22 genes encoding transfer RNAs. mtDNA also contains a noncoding region known as the D-loop. The D-loop is rich with transcription promoters

heterogeneity, maternal inheritance, high mutation, and threshold effect, especially in stem cells [2, 3]. In response to changes of biological environments and genetic status, mtDNA can be mutated without binding with histone ten times higher than nDNA. The ATP levels, redox state, and membrane potential can be varied in the depletion of mtDNA, although it is not clear whether individual changes in the cell are related with mitochondrial dysfunction or in combination with signal methylation changes. The current article will be specially focused on the biological significance of mtDNA mutation.

2 Biological Significance of mtDNA Mutation

The mutation in mtDNA may be involved in important functional areas of the human genome [2], of which mtDNA methylation can be affected by dietaries, drugs, oxidative stresses, or environmental factors, while others have no methylation. The existence of mtDNA methylation was found in umbilical cord blood which is correlated with the level of free thyroid hormones (FT3 and FT4) in MT-RNR1 (12S ribosomal RNA). In 12S ribosomal RNA and the D-loop control region, using bisulfite pyrosequencing to measure the level of mtDNA methylation, Bram et al. found that the placental mtDNA methylation was negatively correlated with D-loop regions while positively with two types of hormones [4]. Three regions of mtDNA are methylated by sodium bisulfite or high concentrations of metals. 12S ribosomal RNA (MT-RNR1) and the transfer RNA phenylalanine (MT-TF) were significantly associated with airborne particulate matter (PM1) exposure [5]. DNA methyltransferase 1 (DNMT1) – one major enzyme catalyzing DNA methylation – can translocate to the mitochondria, bind to mtDNA, and possibly facilitate mtDNA cytosine methylation, which exists in the mitochondrial matrix and can be influenced by 5-hydroxymethylcytosine [6]. However, the existence of mtDNA methylation was not detected in the mitochondria of frogs and Hela cells 30 years ago [7]. Further evidence for the lack of mtDNA methylation was spurred by an inability to detect histone proteins in the mitochondria. Given that histone proteins are a major component of chromatin and play a large role in epigenetic processes, their absence seemed to further suggest that mtDNA methylation was unlikely [8]. More recent study demonstrates that there are very low levels of DNA methylation in HTC116 cells and normal male blood samples or little methylation [9]. Such disagreement or different findings may be resulted from the variation of applied methods and their sensitivities, e.g., mass spectrometry, restriction enzyme digestion, bisulfite conversion, affinity enrichment, and others. It is also associated with various distributions of enzymes among species and tissues during the mtDNA methylation.

3 Regulations of mtDNA in Biological Processes

DNA methylation is one of the three major epigenetic modifications – histone modifications and noncoding RNAs are the other two – known as mitochondrial epigenetics (mitoepigenetics), representing interactions between the mitochondrial and nuclear genome [10]. The mitoepigenetics take the action to reverse gene function, heritable changes, DNA methylation, or histone modification, when nDNA sequence changes are absent. The mtDNA methylation can affect gene transcription activity without changes in DNA sequence. In mammals, DNA methylation almost exclusively occurs on cytosine residues that are followed by guanine residues (CpG sites). Clusters of CpGs near transcriptional start sites are called CpG islands. The CpG dinucleotide sequence in mitochondria is altered with challenge of chemicals and multifactors and contributes to cell aging [6, 11]. The amount of CpG dinucleotides and 1 typical CpG islands in mtDNA are less than those in noncoding DNA, since the primary targets of internal and external challenges may mediate gene regulation in the noncoding region of DNA.

DNA methylation regulates a number of biological processes, e.g., gene expression, development, genomic imprinting, or inactivation of X chromosome. DNA methylation is catalyzed by enzymes called DNA methyltransferases (DNMTs). Mammals have three major DNMTs: DNMT1, DNMT3A, and DNMT3B. DNMT1 is the most abundant methyltransferase and active throughout adulthood. DNMT1 is active in maintenance methylation which involves maintaining methylation patterns throughout an organism's life. When DNA is replicated and the template strand is methylated, DNMT1 methylates the newly synthesized strand accordingly. In most organisms, S-adenosylmethionine serves as a methyl donor in methylation processes [6]. In addition, DNMT3A and DNMT3B are recently found in mitochondria. S-Adenosylmethionine is found in living cells and organisms, by methionine and ATP synthesis, which plays an important role in the regulation of mitochondrial function for cytotoxicity and autophagy [12]. In addition to being methyl group donor in methylation pathway, S-adenosylmethionine also contributes to the metabolic pathway of sulfur-containing metabolites for synthesis of cysteine and glutathione, as well as the aminopropylation pathway for polyamine synthesis of precursor molecules, in early gene expression, protein metabolism, and disease occurrence [13–15]. Deficiency of S-adenosylmethionine may lead to mitochondrial dysfunction and finally lead to insulin resistance. Of the two forms of mtDNA methylation, CpG methylation occurs in mtDNA without changes of mtDNA gene expression, while mtDNA expression is downregulated during the process of GpC methylation [16].

DNA methylation can change the molecular structure and protein binding with DNA, e.g., transcription factor, topoisomerase, RNA polymerase, or inhibitory protein, leading to the loss and dysfunction of gene expression and transcription. mtDNMT1 plays an important role in the epigenetic mechanism of mtDNA methylation in the mitochondrial matrix and in the maintenance of mtDNA methylation and regulation of mitochondrial gene expression [6]. MtDNMT1 regulates the expression of mitochondrial gene through mtDNMT1 binding to the D-loop control region and carrying the promoters to drive transcription of both strands. The central roles of mtDNMT1 in mtDNA methylation and expression can be influenced by a large number of intra- or extramitochondrial factors. MtDNMT1 binding to the mitochondrial genome is associated with the density of CpG dinucleotides. DNMT3A and DNMT3B are involved in de novo methylation and are most active during embryogenesis and early life [17].

Cytosine methylation is essential and critical in normal development during which the lethality and mitosis of the embryos occur if DNMT1 is deleted. CpG methylation patterns of human mtDNA depend upon the number of the promoter region and first exon of multiple genes. Methylation of CpG islands with enriched regions of CpG will lead to condensed chromatin, delayed replication, and inhibition of transcription initiation by which new drugs can be discovered and developed [18–20]. The genome function can be dependent upon the location of DNA methylation, e.g., methylation occurs in the promoter region to inhibit gene expression. In fact, approximately 80% of CpGs in mammalian genomes are methylated. Many CpG sites exist in clusters near transcriptionally integral regions of the genome like promoters and enhancers, controlling DNA chain synthesis during cell division. Dysfunction of mtDNA could increase genetic heterogeneity, instability, fragile, and strand break. Aberrance of methylation in CpG islands causes the inactivation of gene. The development of human tumor is often related with the decreased level of total genomic DNA methylation or hypermethylation or oncogene hypomethylation of tumor suppressor gene promoter [21].

4 mtDNA Methylation in Late-Onset Alzheimer's Disease

Changes of mtDNA methylation and mtDNMT1 expression are influenced by levels of mtDNA-encoded RNAs. The second heavy-strand promoter of human mtDNA is repressed by transcription factor A by altering bind sites and subsequent transcription reaction as epigenetic markers [22]. Alzheimer's disease (AD) is a chronic neurodegenerative disease that accounts for 60–70% of all dementia cases. AD symptoms can be clustered into three primary groups: The first group includes cognitive dysfunction such as language barriers, memory loss, and executive dysfunction; the second group includes psychiatric symptoms, depression, and behavioral disorders; and the third group includes difficulties implementing activities of daily living. Mitochondrial dysfunction including issues with ion permeability and oxidative stress has been shown to play an important role in neurodegenerative diseases such as AD [23, 24]. A study by Stoccoro and colleagues found significantly lower mtDNA D-loop methylation in the blood of 133 patients with late-onset Alzheimer's disease patients compared to the blood of 130 normal controls (2.3% vs. 3.1%) [20]. D-loop is a noncoding region of mitochondrial DNA with 1.1 Dalton and plays an important role in the replication and transcription of mtDNA. Methylation of D-loop region regulates transcription of mitochondrial genes. The late onset of Alzheimer's disease is a complex and critical phase caused by the interaction of more factors. Of those, DNA methylation and histone modification play an important role in the pathogenesis of late-onset Alzheimer's disease [25]. The methylation level in the mitochondrial D-loop region is considered as a critical measure in the clinical stage of a healthy individual; a greater understanding of how methylation in this region affects mtDNA genes could provide further insight to the pathophysiology of AD [26].

5 mtDNA in Metabolic Diseases

Approximately 422 million people worldwide have diabetes mellitus (DM), and obesity is one of the major risk factors for type 2 diabetes mellitus (T2DM), which is characterized by insulin resistance. Given its involvement in cellular energy dynamics, it has long been appreciated that mitochondrial dysfunction is related to DM, insulin resistance, and obesity. Furthermore, the mitochondrial genome and qualitative changes in mtDNA (including mtDNA methylation) are widely appreciated genetic contributors of DM pathophysiology [22]. The mtDNA methylation is associated with early insulin sensitivity and BMI with a complex molecular mechanisms [23–25]. Insulin signaling regulates the signaling axis of genes on mitochondria as the main metabolic platforms and responses to therapies. mtDNA methylation in D-loop region occurs under with the help from mitochondrial NADH dehydrogenase 6, which is a major contributor to insulin sensitivity [26–28]. It seems clear that insulin resistance is associated with DNA methylation in mitochondrial NADH dehydrogenase 6 and D loop-region. Insulin resistance can affect sirtuin-1 activity and sirtuin-1-mediated deacetylation of DNMT1, leading to alternations in DNMT1 activity and gene expression [29–32]. Patients with type 2 diabetes and insulin resistance have mutation or deletion of mitochondrial dysfunction. Altered S-adenosylmethionine is associated with mitochondrial dysfunction and insulin resistance. This is evidenced by the fact that S-adenosylmethionine is correlated with skeletal muscle mitochondrial density and insulin sensitivity [33]. S-Adenosylmethionine can increase the density of mtDNA in skeletal muscle and furthermore improve the insulin sensitivity. Mitochondrial function may be related with age which easily develops insulin resistance to reduce the effect of insulin on glucose [34]. Decreased number of mtDNA copies is associated with insulin resistance in obese individuals, possibly due to increased methylation in the D-loop region [35]. The mitochondrial D region controls the replication and transcription of mtDNA, revealing an increase in DNA methylation and a decrease in the number of mtDNA replication in obese patients. It is unclear whether and how abnormal methylation in D-loop region can regulate mtDNA replication and transcription, to change cell biological roles, evolution, and pharmacologic manipulation [36–38].

6 mtDNA and Neuro-Diseases

Amyotrophic lateral sclerosis (ALS) is an adult-onset neurodegenerative disease that results in the death of motor neurons (brain, brainstem, spinal cord) ultimately manifesting in paralysis and death. To date there is no cure for ALS, and most individuals die 3–5 years following symptom onset. The expression levels of DNMT1 and DNMT3A in neuronal mitochondria increased in amyotrophic lateral sclerosis [39]. Levels of DNMT1 protein increases in nuclear fraction, activity and amount of transferase and 5-methylcytosine and DNMT expression are upregulated in the motor cortex of sporadic patients with amyotrophic lateral sclerosis. Increased activity and amount of 5-methylcytosine in DNA is an important feature of pathology in amyotrophic lateral sclerosis and can be a therapeutic target for drug development. The upregulation of DNMTs and 5mC in human ALS motor neuron mitochondria suggests that aberrant regulation of DNA methylation is an important feature of ALS pathology and may serve as a target for drug development.

Trisomy 21 or Down syndrome (DS) is closely associated with Alzheimer's disease and has common genetic risk factors. Down syndrome is the most common trisomy syndrome caused by all parts or critical parts of trisomy 21 of human chromosomes. Mitochondrial dysfunction has been widely described in DS, and mtDNA defects have also been reported in DS. Infantino and colleagues investigated the role of methylation of mitochondrial DNA in Down syndrome patients' cells. They found that the level of S-adenosylmethionine was decreased in DS subjects when compared to the normal control group [40], leading to the imbalance in mitochondrial methylation. The abnormalities in the mtDNA methylation donors and mtDNA methylation status may impact the mitochondrial dysfunction that is widely documented in Down syndrome.

7 Influencing Factors of mtDNA Methylation

mtDNA has a molecular weight of 16.5 kb, without the protection of histone and loss of relative high-efficiency repair system [41]. It is easy for mtDNA to be attacked by a large number of influencing factors from intra- and extra-cell, organ, and body. For example, reactive oxygen species can play an important role in alterations of mtDNA methylation, the air pollutants [42] can increase mtDNA copy number, and chemicals can damage mtDNA and change mitochondrial gene expression (Fig. 9.2) [43]. Long-term exposure of alcohol could change the expression of more than 700 genes in the promoter region of the h3k4me3 [44]. From one early in vitro study of human HTC 116 cells, alcohol exposure was found to induce oxidative stress and increase levels of mtDNAMT1 [45]. From Bram G. Janssen's study, we got the idea that fetal thyroid hormones play an important role in mitochondrial methylation and mitochondrial DNA (mtDNA) content.

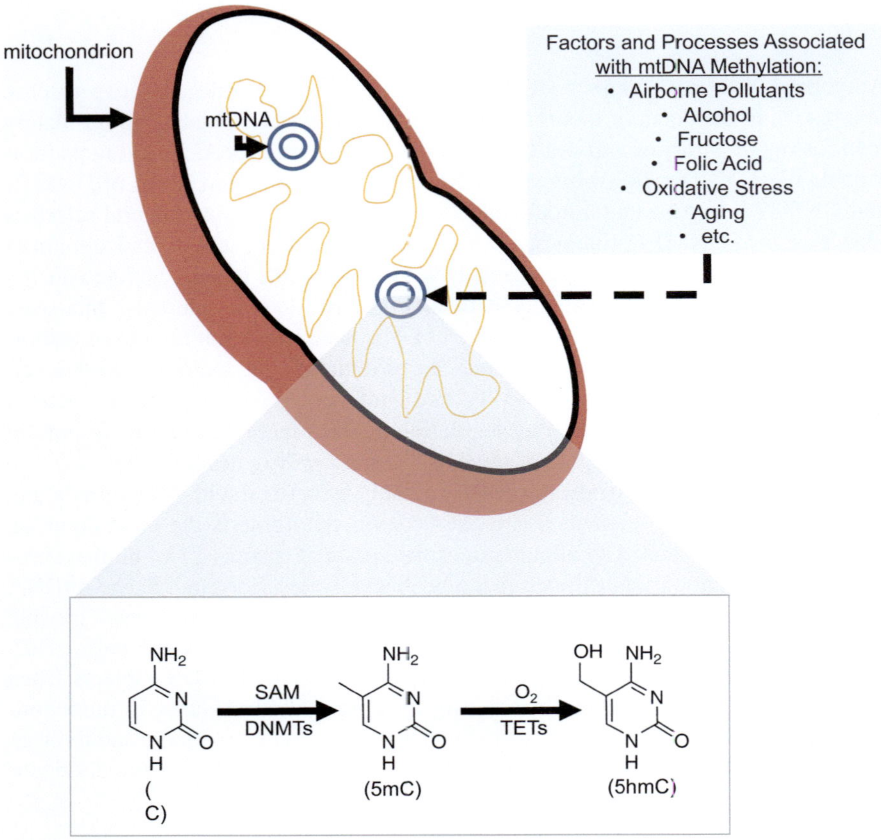

Fig. 9.2 Mitochondrial DNA methylation. This figure depicts the enzymes involved in mitochondrial DNA methylation. Cytosine residues are converted to 5-methylcytosine (5mC) and then 5 hydroxymethylcytosine (5hmC). Existing research demonstrates that the process of mtDNA methylation may be influenced by a number of factors including air pollution, diet, aging, and oxidative stress

Although the mechanism by which the fructose induces metabolic disorders remains unclear, the role of fructose in the apparent modification of mtDNA in cell biology and toxicology should be paid attentions to [46, 47]. One existing study of fructose-fed rats reported that fructose increased mtDNA content and mtDNA transcription in the rat liver cells [48]. It is possible that metabolic syndrome caused by fructose is dependent upon the status of DNA methylation. Folic acid is one of the necessary substances for methylation [49]. The deficiency of folic acid can damage mtDNA and induce deletions of mtDNA. Recent study demonstrates that mtDNA methylation is a new form of epigenetic mechanism, with a special sensitivity to environmental exposures or signals from nuclear genome. Changes of mtDNA methylation lead to dysregulation of gene expression in the development of nervous system and brain activity. A number of toxicants can perform toxic effects by influencing mitochondrial function, including membrane, mtDNA, mtRNA, and metabolisms [50, 51]. The oxidative stress as one of the most important factors can regulate mtDNMT1 and increase mtDNA susceptibility to reactive oxygen species much more than nDNA [52]. The aging also plays an important role in the maintenance of mtDNA methylation [53]. For example, 12SrRNA methylation was found to be associated with aging, by analyzing the methylation of cytosine residues of two mitochondrial genes, 12S and 16S rRNA [54].

8 mtRNA and nDNA

mtDNA and nuclear genes have the close communication and interaction. The nuclear genome-dominated epigenetic alterations can affect the mitochondrial function by regulating the expression of mitochondrial genes. mtDNA changes influence patterns of nuclear gene expression and levels of nDNA methylation. The specific mtDNA copy number and activity in the cell can determine the methylation pattern of nuclear genes. Like noncoding DNA, mtDNA is also subject to epigenetic modifications, particularly in the 5-methylcytosine and 5-hydroxymethylcytosine marks. The relationship between the mitochondrial and nuclear genomes is a bidirectional effect in the regulation of metabolism and apoptosis pathway [55]. Besides the 37 genes which are encoded by mtDNA, the remaining genes that are involved in mitochondrial structure and function are encoded by the nDNA, synthesized in cytoplasm. nDNA-directed genes then enter into mitochondria to play a role in mitochondria, nDNA-encoding polymerase, and ribosomal protein and protein structure. Mitochondrial transcription factor A (TFAM) is a key molecule encoded by nDNA to regulate mtDNA replication. Moreover, the TFAM gene contains many potentially methylated CpG islands in its promoter region. Under the control of TSH, the thyroid gland produces FT3 and FT4, and there are two pathways T3 plays in the methylation of mitochondrial DNA: (1) FT3 binds to the thyroid receptors (TRs) to promote the expression of peroxisome proliferator-activated receptor-γ coactivator 1-α (PPARGC1A), transcription factors encoded by nuclear genes, such as TFAM, regulating mitochondrial gene; (2) in the mitochondrial matrix, it binds to the specific TRs directly to promote the mitochondrial genome transcription (Fig. 9.3) [56]. TFAM promoter is linked with luciferase reporter gene to perform the methylation under methyltransferase, respectively, in SssI (CG), HpaII (CCGG), and HhaI (GCGC) and be transiently transfected into cells for the detection of fluorescein activity [57]. Binding sites of SssI and HpaII are not involved in the regulation of TFAM promoter activity. Nuclear respiratory factor −1 exists in the promoter of TFAM, a DNA binding domain which contains two HhaI binding sites rather than the HpaII site. mtDNA methylation is related with respiratory factor −1 methylation of TFAM promoter. The methylation of gene silencing can regulate such communication and mitochondrial biogenesis and respiratory function, contributing to the pathogeneses of human diseases.

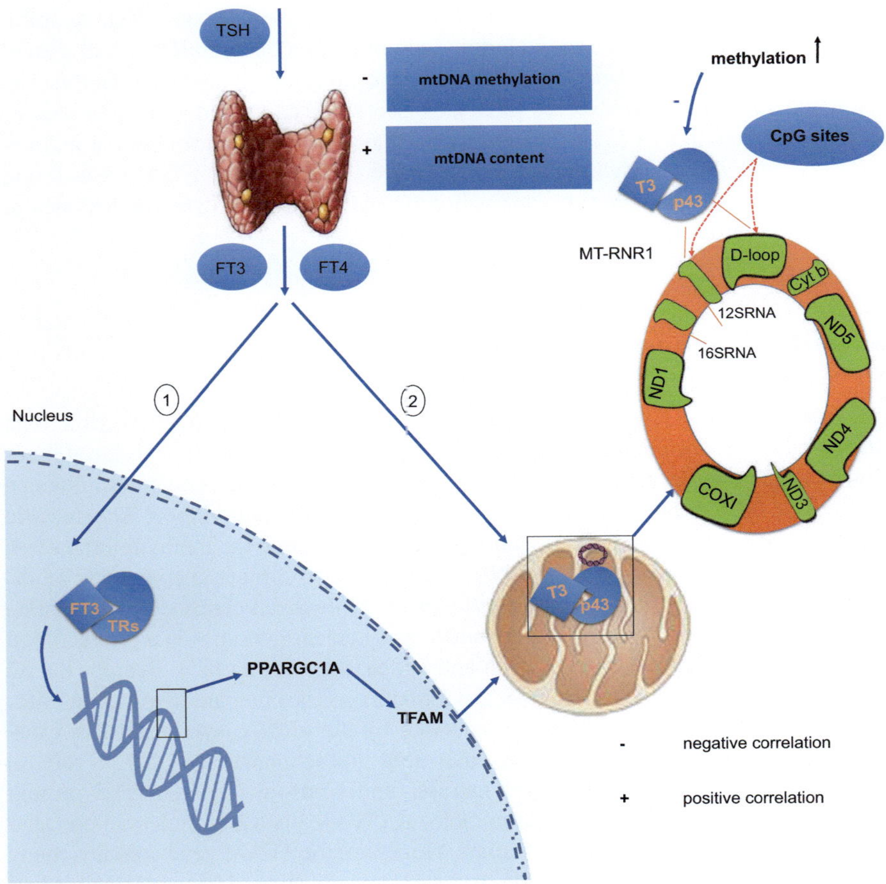

Fig. 9.3 FT3 and mitochondrial DNA methylation. Seventy-seven percent methylation in 12S ribosomal RNA (MT-RNR1) in mtDNA and 47% methylation in D-loop region. FT3 and FT4 were negatively correlated with mtDNA methylation. FT3 controls the gene PPARGC1A and the gene regulating mitochondrial gene expression, mitochondrial transcription factor A (TFAM) which is a nuclear-encoded transcription factors that is responsible for mitochondrial biogenesis. FT3 has positive association between FT3 and placental mtDNA content. While it is negatively correlated with mtDNA methylation, T3 binds to the specific receptors, p43, in the mitochondrial matrix. In addition, 16S rRNA, COXI, ND3, ND4, ND1, and ND5 also exist in the mtDNA. High DNA methylation levels in the mitochondrial genome make the mtDNA less accessible to proteins and p43 [4, 67]

On the other hand, mtDNA damage also affects the methylation status of nDNA. In the condition of anoxia and ischemic, the content of 8-OHdG as a marker of oxidative damage in mtDNA increased, while DNA methylation decreased [58]. 8-OHdG could inhibit the methylation of cytosine and interfere with the function of methyltransferase, to change the DNA methylation and gene expression. Changes in mtDNA copy number could also affect the methylation level of nDNA [59]. Using genomic restriction scanning, 64 low methylation regions and 50 hypermethylation regions were found in the nDNA of tumor cells, and 22 regions were altered after removal of mtDNA, of which 17 or 5 were hypomethylation or hypermethylation, respectively. Those changes can be reversed after they were reintroduced to mtDNA, indicating that mtDNA reduction or deletion can cause epigenetic changes. The copy number of mtDNA can regulate the methylation level of nDNA to control gene expression, differentiation, and development [60]. The depletion of mtDNA leads to genetic changes in the nuclear genome of the muscle and spleen. mtDNA is considered as a genetic material outside the nucleus, controls the basic function of mitochondria, and is regulated by nDNA. The mechanism of mitochondrial transcription is different from that of nuclear transcription [61]. The effect of mtDNA methylation may be hardly associated with nDNA methylation. Mitochondria contain a noncoding region called the D-loop.

9 Conclusions and Perspectives

mtDNA has the genetic function outside of the nucleus, of which the proportion may be limited, while the role of mtDNA in the body cannot be ignored. Mitochondrial dysfunction is a common cause of diseases. The existence of mtDNA methylation can be affected by a large number of intra- or extra-cellular factors. mtDNA methylation plays a critical role in the development of diseases, although the understanding of mechanisms and diseases impacted by mitochondrial DNA methylation is at present relatively rudimentary. As its role in the body is better understood, mtDNA methylation may potentially be a useful biomarker of disease detection and diagnosis. We have to realize that the studies on mtDNA methylation are just excellent starts. At the same time, we also should develop more accurate methodologies to detect mtDNA methylation-oriented signal pathways, interaction, and communications among regions of mtDNA sequences and between mtDNA and nDNA. The ability to understand and predict mitochondrial disease progression is important for clinicians, who can provide guidance on the patient's development of disease symptoms and provide optimal care for the patient [62, 63]. Exploring more about mtDNA in single-cell and mtDNA heterogeneity by measuring single-cell DNA sequencing [64–66] is what we're going to do. On the other hand, comprehensive characterizations of mtDNA and bidirectional effects between mtDNA and 3D genome, instability, and gene editing should also be investigated. In all, more and more attention should be paid to the roles of mtDNA in disease occurrence, duration, severity, response to therapy, as well as prognosis.

Acknowledgments The work was supported by the Zhongshan Distinguished Professor Grant (XDW), the National Nature Science Foundation of China (91230204, 81270099, 81320108001, 81270131, 81300010), the Shanghai Committee of Science and Technology (12JC1402200, 12431900207, 11410708600, 14431905100), the operation funding of Shanghai Institute of Clinical Bioinformatics, the Ministry of Education for Academic Special Science and Research Foundation for PhD Education (20130071110043), and the National Key Research and Development Program (2016YFC0902400, 2017YFSF090207).

References

1. Brandon MC, Lott MT, Nguyen KC, Spolim S, Navathe SB, Baldi P, et al. MITOMAP: a human mitochondrial genome database – 2004 update. Nucleic Acids Res. 2005;33:D611–3. [PMID: 15608272]
2. Wallace DC. A mitochondrial paradigm of metabolic and degenerative diseases, aging, and cancer: a Dawn for evolutionary medicine. Annu Rev Genet. 2005;39:359–407. [PMID: 16285865]
3. Paes B, Moco PD, Pereira CG, Porto GS, de Sousa Russo EM, Reis LCJ, et al. Ten years of iPSC: clinical potential and advances in vitro hematopoietic differentiation. Cell Biol Toxicol. 2017;33:233–50. [PMID: 28039590]
4. Janssen BG, Byun HM, Roels HA, Gyselaers W, Penders J, Baccarelli AA, et al. Regulating role of fetal thyroid hormones on placental mitochondrial DNA methylation: epidemiological evidence from the ENVIRONAGE birth cohort study. Clin Epigenetics. 2017;9:66. [PMID: 28649287]
5. Byun HM, Panni T, Motta V, Hou L, Nordio F, Apostoli P, et al. Effects of airborne pollutants on mitochondrial DNA methylation. Part Fibre Toxicol. 2013;10:18. [PMID: 23656717]
6. Shock LS, Thakkar PV, Peterson EJ, Moran RG, Taylor SMDNA. Methyltransferase 1, cytosine methylation, and cytosine hydroxymethylation in mammalian mitochondria. Proc Natl Acad Sci U S A. 2011;108:3630–5. [PMID: 21321201]
7. Dawid IB. 5-methylcytidylic acid: absence from mitochondrial DNA of frogs and HeLa cells. Science. 1974;184:80–1. [PMID: 4815237]
8. Choi YS, Hoon Jeong J, Min HK, Jung HJ, Hwang D, Lee SW, et al. Shot-gun proteomic analysis of mitochondrial D-loop DNA binding proteins: identification of mitochondrial histones. Mol BioSyst. 2011;7:1523–36. [PMID: 21359316]
9. Hong EE, Okitsu CY, Smith AD, Hsieh CL. Regionally specific and genome-wide analyses conclusively demonstrate the absence of CpG methylation in human mitochondrial DNA. Mol Cell Biol. 2013;33:2683–90. [PMID: 23671186]
10. Manev H, Dzitoyeva S. Progress in mitochondrial epigenetics. Biomol Concepts. 2013;4:381–9. [PMID: 25436587]
11. Kaarniranta K, Tokarz P, Koskela A, Paterno J, Blasiak J. Autophagy regulates death of retinal pigment epithelium cells in age-related macular degeneration. Cell Biol Toxicol. 2017;33:113–28. [PMID: 27900566]
12. Giromini C, Rebucci R, Fusi E, Rossi L, Saccone F, Baldi A. Cytotoxicity, apoptosis, DNA damage and methylation in mammary and kidney epithelial cell lines exposed to ochratoxin A. Cell Biol Toxicol. 2016;32:249–58. [PMID: 27154019]
13. Gomez-Sagasti MT, Becerril JM, Epelde L, Alkorta I, Garbisu C. Early gene expression in Pseudomonas fluorescens exposed to a polymetallic solution. Cell Biol Toxicol. 2015;31:39–81. [PMID: 25754557]
14. Santamaria E, Avila MA, Latasa MU, Rubio A, Martin-Duce A, SC L, et al. Functional proteomics of nonalcoholic steatohepatitis: mitochondrial proteins as targets of S-adenosylmethionine. Proc Natl Acad Sci U S A. 2003;100:3065–70. [PMID: 12631701]
15. Fierbinteanu-Braticevici C, Sinescu C, Moldoveanu A, Petrisor A, Diaconu S, Cretoiu D, et al. Nonalcoholic fatty liver disease: one entity, multiple impacts on liver health. Cell Biol Toxicol. 2017;33:5–14. [PMID: 27680752]
16. van der Wijst MG, van Tilburg AY, Ruiters MH, Rots MG. Experimental mitochondria-targeted DNA methylation identifies GpC methylation, not CpG methylation, as potential regulator of mitochondrial gene expression. Sci Rep. 2017;7:177. [PMID: 28282966]
17. Rinaldi L, Avgustinova A, Martin M, Datta D, Solanas G, Prats N, et al. Loss of Dnmt3a and Dnmt3b does not affect epidermal homeostasis but promotes squamous transformation through PPAR-gamma. elife. 2017;6:e21697. [PMID: 28425913]
18. Guney Eskiler G, Cecener G, Tunca B, Egeli U. An in vitro model for the development of acquired tamoxifen resistance. Cell Biol Toxicol. 2016;32:563–81. [PMID: 27585693]

19. Chen T, Hevi S, Gay F, Tsujimoto N, He T, Zhang B, et al. Complete inactivation of DNMT1 leads to mitotic catastrophe in human cancer cells. Nat Genet. 2007;39:391–6. [PMID: 17322882]
20. Wang XCBT. Profiles of cabozantinib approved for advanced renal cell carcinomas. Cell Biol Toxicol. 2016;32:259–61. [PMID: 27383755]
21. Mohsenzadeh M, Sadeghi RN, Vahedi M, Kamani F, Hashemi M, Asadzadeh H, et al. Promoter hypermethylation of RAR-beta tumor suppressor gene in gastric carcinoma: association with histological type and clinical outcomes. Cancer Biomark. 2017;20:7–15. [PMID: 28759951]
22. Lodeiro MF, Uchida A, Bestwick M, Moustafa IM, Arnold JJ, Shadel GS, et al. Transcription from the second heavy-strand promoter of human mtDNA is repressed by transcription factor A in vitro. Proc Natl Acad Sci U S A. 2012;109:6513–8. [PMID: 22493245]
23. Zerin T, Kim JS, Gil HW, Song HY, Hong SY. Effects of formaldehyde on mitochondrial dysfunction and apoptosis in SK-N-SH neuroblastoma cells. Cell Biol Toxicol. 2015;31:261–72. [PMID: 26728267]
24. Lippai M, Szatmari Z. Autophagy-from molecular mechanisms to clinical relevance. Cell Biol Toxicol. 2017;33:145–68. [PMID: 27957648]
25. Stoccoro A, Siciliano G, Migliore L, Coppede F. Decreased methylation of the mitochondrial D-loop region in late-onset Alzheimer's disease. J Alzheimers Dis. 2017;59:559–64. [PMID: 28655136]
26. Cacabelos R, Torrellas C. Epigenetics of aging and Alzheimer's disease: implications for pharmacogenomics and drug response. Int J Mol Sci. 2015;16:30483–543. [PMID: 26703582]
27. Zhou Y, Ning Z, Lee Y, Hambly BD, McLachlan CS. Shortened leukocyte telomere length in type 2 diabetes mellitus: genetic polymorphisms in mitochondrial uncoupling proteins and telomeric pathways. Clin Transl Med. 2016;5:8. [PMID: 26951191]
28. Zheng LD, Linarelli LE, Brooke J, Smith C, Wall SS, Greenawald MH, et al. Mitochondrial epigenetic changes link to increased diabetes risk and early-stage prediabetes indicator. Oxidative Med Cell Longev. 2016;2016:5290638. [PMID: 27298712]
29. Miko E, Vida A, Bai P. Translational aspects of the microbiome-to be exploited. Cell Biol Toxicol. 2016;32:153–6. [PMID: 27098154]
30. Cao T, Yang D, Zhang X, Wang Y, Qiao Z, Gao L, et al. FAM3D inhibits glucagon secretion via MKP1-dependent suppression of ERK1/2 signaling. Cell Biol Toxicol. 2017;33:457–66. [PMID: 28247283]
31. Kuznetsova T, Knez J. Peripheral blood mitochondrial DNA and myocardial function. Adv Exp Med Biol. 2017;982:347–58. [PMID: 28551797]
32. Hulman A, Simmons RK, Brunner EJ, Witte DR, Faerch K, Vistisen D, et al. Trajectories of glycaemia, insulin sensitivity and insulin secretion in South Asian and white individuals before diagnosis of type 2 diabetes: a longitudinal analysis from the Whitehall II cohort study. Diabetologia. 2017;60:1252–60. [PMID: 28409212]
33. Tabak AG, Herder C, Rathmann W, Brunner EJ, Kivimaki M. Prediabetes: a high-risk state for diabetes development. Lancet. 2012;379:2279–90. [PMID: 22683128]
34. Vedi M, Sabina EP. Assessment of hepatoprotective and nephroprotective potential of withaferin A on bromobenzene-induced injury in Swiss albino mice: possible involvement of mitochondrial dysfunction and inflammation. Cell Biol Toxicol. 2016;32:373–90. [PMID: 27250656]
35. O'Hagan HM, Wang W, Sen S, Destefano Shields C, Lee SS, Zhang YW, et al. Oxidative damage targets complexes containing DNA methyltransferases, SIRT1, and polycomb members to promoter CpG Islands. Cancer Cell. 2011;20:606–19. [PMID: 22094255]
36. Seo JB, Jung SR, Hille B, Koh DS, Extracellular ATP. Protects pancreatic duct epithelial cells from alcohol-induced damage through P2Y1 receptor-cAMP signal pathway. Cell Biol Toxicol. 2016;32:229–47. [PMID: 27197531]
37. Amodio G, Sasso E, D'Ambrosio C, Scaloni A, Moltedo O, Franceschelli S, et al. Identification of a microRNA (miR-663a) induced by ER stress and its target gene PLOD3 by a com-

bined microRNome and proteome approach. Cell Biol Toxicol. 2016;32:285–303. [PMID: 27233793]
38. Peng L, Yuan Z, Ling H, Fukasawa K, Robertson K, Olashaw N, et al. SIRT1 deacetylates the DNA methyltransferase 1 (DNMT1) protein and alters its activities. Mol Cell Biol. 2011;31:4720–34. [PMID: 21947282]
39. Engin AB, Engin A. The interactions between Kynurenine, folate, methionine and Pteridine pathways in obesity. Adv Exp Med Biol. 2017;960:511–27. [PMID: 28585214]
40. Petersen KF, Befroy D, Dufour S, Dziura J, Ariyan C, Rothman DL, et al. Mitochondrial dysfunction in the elderly: possible role in insulin resistance. Science. 2003;300:1140–2. [PMID: 12750520]
41. Infantino V, Castegna A, Iacobazzi F, Spera I, Scala I, Andria G, et al. Impairment of methyl cycle affects mitochondrial methyl availability and glutathione level in Down's syndrome. Mol Genet Metab. 2011;102:378–82. [PMID: 21195648]
42. Peng C, Bind MC, Colicino E, Kloog I, Byun HM, Cantone L, et al. Particulate air pollution and fasting blood glucose in nondiabetic individuals: associations and epigenetic mediation in the normative aging study, 2000–2011. Environ Health Perspect. 2016;124:1715–21. [PMID: 27219535]
43. Zheng LD, Linarelli LE, Liu L, Wall SS, Greenawald MH, Seidel RW, et al. Insulin resistance is associated with epigenetic and genetic regulation of mitochondrial DNA in obese humans. Clin Epigenetics. 2015;7:60. [PMID: 26110043]
44. Thomas PD, Kahn M. Kat3 coactivators in somatic stem cells and cancer stem cells: biological roles, evolution, and pharmacologic manipulation. Cell Biol Toxicol. 2016;32:61–81. [PMID: 27008332]
45. Bellizzi D, D'Aquila P, Scafone T, Giordano M, Riso V, Riccio A, et al. The control region of mitochondrial DNA shows an unusual CpG and non-CpG methylation pattern. DNA Res. 2013;20:537–47. [PMID: 23804556]
46. He J, Mao CC, Reyes A, Sembongi H, Di Re M, Granycome C, et al. The AAA+ protein ATAD3 has displacement loop binding properties and is involved in mitochondrial nucleoid organization. J Cell Biol. 2007;176:141–6. [PMID: 17210950]
47. Sidor-Kaczmarek J, Cichorek M, Spodnik JH, Wojcik S, Morys J. Proteasome inhibitors against amelanotic melanoma. Cell Biol Toxicol. 2017, March 9. doi:https://doi.org/10.1007/s10565-017-9390-0. [PMID: 28281027]
48. Ganta KK, Mandal A, Chaubey B. Depolarization of mitochondrial membrane potential is the initial event in non-nucleoside reverse transcriptase inhibitor efavirenz induced cytotoxicity. Cell Biol Toxicol. 2017;33:69–82. [PMID: 27639578]
49. Bao L, Zhang Y, Wang J, Wang H, Dong N, Su X, et al. Variations of chromosome 2 gene expressions among patients with lung cancer or non-cancer. Cell Biol Toxicol. 2016;32:419–35. [PMID: 27301951]
50. Ohashi K, Munetsuna E, Yamada H, Ando Y, Yamazaki M, Taromaru N, et al. High fructose consumption induces DNA methylation at PPARalpha and CPT1A promoter regions in the rat liver. Biochem Biophys Res Commun. 2015;468:185–9. [PMID: 26519879]
51. Gao X, Liu YA. Transcriptomic study suggesting human iPSC-derived hepatocytes potentially offer a better in vitro model of hepatotoxicity than most hepatoma cell lines. Cell Biol Toxicol. 2017;33:407–21. [PMID: 28144825]
52. Yamazaki M, Munetsuna E, Yamada H, Ando Y, Mizuno G, Murase Y, et al. Fructose consumption induces hypomethylation of hepatic mitochondrial DNA in rats. Life Sci. 2016;149:146–52. [PMID: 26869391]
53. Crott JW, Choi SW, Branda RF, Mason JB. Accumulation of mitochondrial DNA deletions is age, tissue and folate-dependent in rats. Mutat Res. 2005;570:63–70. [PMID: 15680403]
54. Bao L, Diao H, Dong N, Su X, Wang B, Mo Q, et al. Histone deacetylase inhibitor induces cell apoptosis and cycle arrest in lung cancer cells via mitochondrial injury and p53 up-acetylation. Cell Biol Toxicol. 2016;32:469–82. [PMID: 27423454]

55. Kang SJ, Lee HM, Park YI, Yi H, Lee H. So B, et al. chemically induced hepatotoxicity in human stem cell-induced hepatocytes compared with primary hepatocytes and HepG2. Cell Biol Toxicol. 2016;32:403–17. [PMID: 27287938]
56. Blanch M, Mosquera JL, Ansoleaga B, Ferrer I, Barrachina M, Altered Mitochondrial DNA. Methylation pattern in Alzheimer disease-related pathology and in Parkinson disease. Am J Pathol. 2016;186:385–97. [PMID: 26776077]
57. Domcke S, Bardet AF, Adrian Ginno P, Hartl D, Burger L, Schubeler D. Competition between DNA methylation and transcription factors determines binding of NRF1. Nature. 2015;528:575–9. [PMID: 26675734]
58. Kornicka K, Marycz K, Maredziak M, Tomaszewski KA, Nicpon J. The effects of the DNA methyltranfserases inhibitor 5-Azacitidine on ageing, oxidative stress and DNA methylation of adipose derived stem cells. J Cell Mol Med. 2017;21:387–401. [PMID: 27998022]
59. Gu J, Wang X. New future of cell biology and toxicology: thinking deeper. Cell Biol Toxicol. 2016;32:1–3. [PMID: 26874518]
60. Kelly RD, Mahmud A, McKenzie M, Trounce IA, St John JC. Mitochondrial DNA copy number is regulated in a tissue specific manner by DNA methylation of the nuclear-encoded DNA polymerase gamma A. Nucleic Acids Res. 2012;40:10124–38. [PMID: 22941637]
61. Bestwick ML, Shadel GS. Accessorizing the human mitochondrial transcription machinery. Trends Biochem Sci. 2013;38:283–91. [PMID: 23632312]
62. Liu T, Liu WH, Zhao JS, Meng FZ, Wang H. Lutein protects against beta-amyloid peptide-induced oxidative stress in cerebrovascular endothelial cells through modulation of Nrf-2 and NF-kappa b. Cell Biol Toxicol. 2017;33:57–67. [PMID: 27878403]
63. Zhu LZ, Hou YJ, Zhao M, Yang MF, XT F, Sun JY, et al. Caudatin induces caspase-dependent apoptosis in human glioma cells with involvement of mitochondrial dysfunction and reactive oxygen species generation. Cell Biol Toxicol. 2016;32:333–45. [PMID: 27184666]
64. Wang W, Wang X, Single-cell CRISPR. Screening in drug resistance. Cell Biol Toxicol. 2017;33:207–10. [PMID: 28474250]
65. Wang W, Gao D, Wang X. Can single-cell RNA sequencing crack the mystery of cells? Cell Biol Toxicol. 2017, July 21. doi:https://doi.org/10.1007/s10565-017-9404-y. [PMID: 28733864]
66. Wang W, Zhu B, Wang X. Dynamic phenotypes: illustrating a single-cell odyssey. Cell Biol Toxicol. 2017;33:423–7. [PMID: 28638956]
67. Liu B, Du Q, Chen L, Fu G, Li S, Fu L, et al. CpG methylation patterns of human mitochondrial DNA. Sci Rep. 2016;6:23421. [PMID: 26996456]

Danyan Gao, is a student in graduate school in Wenzhou Medical University and now is doing experiment in Clinical Science Institute of Fudan University Zhongshan Hospital, Shanghai, China. She participates in publishing two articles: “Can single-cell RNA sequencing crack the mystery of cells?” and “Genomic mechanisms of transformation from chronic obstructive pulmonary disease to lung cancer.” Her main research direction is lung cancer.

Bijun Zhu, BS, is a project leader of electron microscope technique in Clinical Science Institute of Fudan University Zhongshan Hospital, Shanghai, China. She has strong background of biomedical techniques and biotechnology. She has published ten articles in the past 5 years and participated in three funded projects, including National Natural Science Fund. Now, she is focusing on the field of oncology and iconography.

Hongzhi Sun, MD, PhD, is a famous surgical specialist in China and the director of the First Affiliated Hospital of Liaoning Medical University. He is the chairman of Surgical Biomaterials Specialized Committee, the vice chairman of Tumor Cell Biology Specialized Committee and Surgical Biological Cell Technology Specialized Committee in Liaoning Cell Biology Society, a member of the Breast Sciences Section of Liaoning Medical Association, the executive director of Liaoning Anticancer Association, the associate editor of the international journal *Molecular and Cellular Therapies*, and the editorial board member of *Chinese Journal of Tissue Engineering Research* and *Clinical Rehabilitation* and Liaoning Medical University newspaper. He hosted seven fund projects and published more than ten scientific papers with the impact factor over 20. His major research is focused on the mechanism of tumorigenesis and progression, invasion and metastasis, and the basic and clinical research of biotherapy of tumor.

Xiangdong Wang, MD, PhD, is a distinguished professor of Medicine, director of Shanghai Institute of Clinical Bioinformatics, executive director of Clinical Science Institute of Fudan University Zhongshan Hospital, director of Fudan University Center of Clinical Bioinformatics, deputy director of Shanghai Respiratory Research Institute, and visiting professor of King's College London. His main research is focused on clinical bioinformatics, disease-specific biomarkers, lung chronic diseases, cancer immunology, and molecular and cellular therapies. He is the author of more than 200 scientific publications with the impact factor about 900, citation number about 6918, h-index 46, i10-index 181, and cited journal impact factor about 7000.

Chapter 10
Beyond Deubiquitylation: USP30-Mediated Regulation of Mitochondrial Homeostasis

Jiayun Hou, Mohmmad Eldeeb, and Xiangdong Wang

Abstract Mutations or sequence aberrations in the Parkin gene are among the most common causes of autosomal recessive Parkinson's disorder (PD). Parkin, a cytoplasmic E3 ubiquitin ligase, is involved in mitochondrial quality control pathways, including mitochondrial fission and mitophagy by autophagy-related genes. Parkin mediates the covalent addition of ubiquitin (Ub) chains to Lys 6, Lys 11, and Lys 63 on diverse mitochondrial-related target proteins. USP30, a mitochondrial deubiquitinase, promotes mitochondrial fusion by mediating the deubiquitination of ubiquitylated forms of mitofusins, such as Mfn1 and Mfn2. USP30 preferentially mediates the removal of Ub chains from Lys 6 and Lys 11 on mitochondria-derived proteins. USP30 mediates the removal of the ubiquitin chains added by Parkin. It was demonstrated that overexpression of USP30 triggers the mitochondrial dynamic signaling toward elevated fusion and reduced fission and halts mitochondrial clearance via mitophagy. Although mounting lines of evidences reveal the pivotal role of Parkin in mitochondrial quality control pathways, the crucial role of deubiquitinases including the USP30 deubiquitinase is emerging. Herein, we review briefly the role of USP30 in the dynamic networks of mitochondrial quality control and its physiological implications.

Keywords Mitochondrial fusion • Mitochondrial fission • Mitophagy • Deubiquitylation • USP30

J. Hou (✉)
Zhongshan Hospital Institute of Clinical Science, Fudan University, Shanghai, China

Shanghai Institute of Clinical Bioinformatics, Biomedical Research Center, Shanghai, China
e-mail: hjyfs@hotmail.com

M. Eldeeb
Department of Biochemistry, University of Alberta, Edmonton, AB, Canada
e-mail: Eldeeb@ualberta.ca

X. Wang (✉)
Zhongshan Hospital Institute of Clinical Science, Fudan University, Shanghai Medical College, Shanghai, China
e-mail: xiangdong.wang@clintransmed.org

H. Sun, X. Wang (eds.), *Mitochondrial DNA and Diseases*, Advances in Experimental Medicine and Biology 1038,
https://doi.org/10.1007/978-981-10-6674-0_10

1 Introduction

Mitochondria as dynamic organelles are constantly undergoing fusion and fission, culminating in mitochondrial elongation or fragmentation. Among the crucial roles that mitochondrial fusion serves are mixing of mitochondrial components and safeguarding the mitochondrial DNA integrity, generating more ATP production. Mitochondrial fission contributes to the proper distribution of mitochondria to meet local ATP demands and selective clearance of any otherwise abnormal or damaged mitochondria [1]. The fission of mitochondria is essential to progression of mitochondrial autophagy, i.e., mitophagy [2]. Two major intracellular protein degradation pathways exist. Autophagy-lysosomal degradation pathway provides a way to clean organelles and supramolecular components in the cytoplasm [3]. Ubiquitylation-mediated (also known as ubiquitination) proteasomal degradation involves the degradation of most of the intracellular proteins [4].

2 USP30 Deubiquitylation Counteracts Parkin-Mediated Ubiquitylation

Ubiquitin-specific peptidase 30 (USP30), a member of the ubiquitin-specific protease family, is a mitochondrial deubiquitinase [5]. UPS-mediated protein degradation relies on the first recognition of the substrate and covalent conjugation to Ub by specific Ub E3 ligases. After the initial ubiquitination, further Ub molecules are conjugated in series at lysine 48 of the Ub to form a polyubiquitin structure, which is then recruited to the 26S proteasome for degradation [6]. The canonical ubiquitination of target substrates of the UPS system proceeds via a three-enzyme cascade: E1 (ubiquitin-activating enzyme, using ATP), E2 (ubiquitin-conjugating enzyme or carrier enzyme), and E3 (ubiquitin-protein ligase). The ubiquitin-conjugated substrate is degraded into short peptides via the 26S proteasome. Most Ub molecules of the Ub chain are depolymerized by mitochondrial deubiquitinases, and the rest are often degraded by the 26S proteasome [7]. Deubiquitylation provides the counteracting process of ubiquitylation, which requires deubiquitylating enzymes, leading to the reversible regulation [8].

USP30 can mediate the removal of the Ub chains added by Parkin E3 Ub ligase [9] (Fig. 10.1). Parkin is a member of RING-between-RING family. Similar to HECT domain E3 Ub protein ligase, RING-between-RING E3 Ub ligases have a thiol ester intermediate with ubiquitin, differentiating them from RING finger E3 Ub-protein ligase without intermediate with ubiquitin. The Ser/Thr kinase PINK1 triggers Parkin self-association and subsequently triggers its ubiquitination activity to target protein substrates [10]. USP30 plays the crucial role in mitochondrial quality control and in counteracting the ubiquitylation mediated by Parkin in response to carbonyl cyanide-3-chlorophenylhydrazone (CCCP) treatment in HEK293 cells [9]. CCCP-mediated mitochondrial damage stimulates Parkin to add Ub chains to

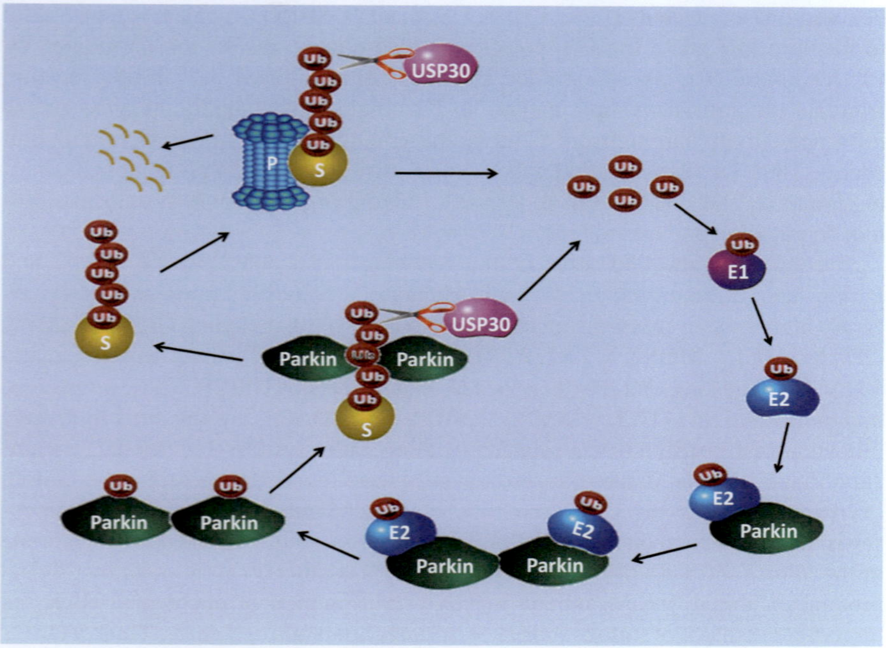

Fig. 10.1 Model of USP30 deubiquitylating Parkin substrates.
Ubiquitin (Ub) is activated by E1 ubiquitin-activating enzyme (E1), which forms an E1 ubiquitin-activating enzyme-thiol ester intermediate via ATP. The activated Ub by E1 transferred to an E2 ubiquitin-conjugating enzyme (E2) to generate a second thiol ester intermediate. And the activated Ub conjugated with E2 can be transferred to Parkin E3 ubiquitin ligase to generate a third labile thiol ester intermediate. PINK1 induces Parkin self-association. Then, the activated Ub is transferred to the substrate (S), where forms a s isopeptide bond. Additional ubiquitin is then added in the same way to generate a polyubiquitin chain. The ubiquitin-conjugated substrate is degraded to short peptides through the 26S proteasome (P). Most Ub molecules of the Ub chain are depolymerized by USP30 enzyme, and the rest are often degraded by the 26S proteasome

Lys 6, Lys 11, and Lys 63 on mitochondria-derived proteins. USP30 mediates the removal of both noncanonical (Lys 6 and Lys 11))- linked and canonical (Lys 48 and Lys 63)-linked Ub chains [11]. The ablation of USP30 induces the specific accumulation of Lys 6-linked Ub chains [9], strongly suggesting that Lys 6-linked Ub chains is the primary target of USP30.

Although phosphorylation impacts ubiquitylation and subsequent degradation of diverse signaling proteins [12, 13], phosphorylation of ubiquitin chains may affect deubiquitylation by mitochondrial deubiquitinases. For instance, phosphorylation of Ub (pUb) at Ser65 impacts Ub polymerization, culminating in mitigation of ubiquitylated protein levels [14], while the pUb chains don't impact the expression of USP30 in cells [11]. Recombinant USP30 appears to cleave non-phosphorylated Ub chains in ubiquitin-conjugated substrates and Lys 6-linked diubiquitin [14]. USP30 reduced activity toward pUb chains, which is common among other mitochondrial deubiquitinases, such as USP2, USP5, USP8, and USP15 [14, 15]. It is noteworthy to mention that all of the Ub molecules in Ub chains can be phosphorylated by PINK1 kinase. The phosphorylated Ub seems to not impact the Ub translocation from E1 ubiquitin-activating enzyme to E2 ubiquitin-conjugating enzyme and to influence Ub discharging from E2 ubiquitin-conjugating enzyme to form ubiquitin chains. Thus, Ub phosphorylation has a regulatory effect on ubiquitination and deubiquitination and downstream processes of ubiquitin-proteasome system like protein degradation and others [15].

Recent work unveiled some of the common target substrates of USP30 and Parkin. The deubiquitylation of 12 mitochondrially associated proteins was significantly high upon overexpression of recombinant USP30, including CISD1, FKBP8, GDAP1, MUL1, MOS2, MTX1, PTRH2, TOM20, TDRKH, VDAC1, VDAC2, and VDAC3, suggesting that they may be target substrates of USP30. These substrates, including MUL1, MTX1, VDAC1, VDAC2, and VDAC3, are involved in diverse mitochondrial quality control-related signaling pathways [16–18]. MUL1, a mitochondrial E3 ligase, is the most prominently deubiquitylated protein upon USP30 overexpression [16, 19]. USP30 contributes to the balance of mitochondrial homeostasis. MTX1, a component of a preprotein, imports complex in the outer membrane of the mitochondrion [18]. Cells lacking VDAC2, a mitochondrial outer membrane protein, exhibited the monomeric inactive conformation of augmented BAK, an important mediator of mitochondrial or intrinsic apoptotic cell death. Thus, VDAC2 enhances the activity of BAK and plays a crucial role in the regulation of intrinsic apoptotic pathway [17].

3 USP30 Promotes Mitochondrial Fusion by Deubiquitylating Mitofusins

Mitochondria can be converted from state of fragmentation into a hyper-fused, giant network known as elongation, upon G1-S transition in cell cycle. The mitochondrial fusion has greater ATP outputs than other cell cycle stages, where depolarizing mitochondria at the early stage of G1 phase halts the blocking of G1 phase progression into S phase. Prolonged mitochondria by chronic inhibition of dynamin-related protein (DRP)-1, culminating in cyclin E overexpression, can entry into S phase and mitotic chromosome alignment [20]. Mitochondrial fusion, like endoplasmic reticulum fusion, relies on DRPs [21, 22]. DRPs belong to a family of GTPases, which provide the structural and mechanical support for membrane remodeling [23, 24]. DRPs play an important role in membrane scission, particularly in the process of endocytosis and some membrane budding, and participate in fission of mitochondria and peroxisome biogenesis [25, 26]. Mitochondrial fusion occurs upon binding of the two tips [27].

DRPs, known as mitofusins, are anchored to the mitochondrial outer membranes (OM) by two transmembrane domains, and the N-terminal end and C-terminal end are all exposed to the cytoplasm [28]. DRPs, including Fzo1 in budding yeast, Fzo and Marf/Dmfn in flies, Mfn1 and Mfn2 in mammals, and FZO-1 in the worm, mediate fusion of mitochondrial OM [29]. FZO, shortly expressed in the development of fly, is an early indicator for mitochondrial fusion [29, 30]. All the other DRPs, excluding FZO, are constitutively expressed in mitochondria [31, 32]. In flies, the loss of Fzo leads to male sterility [33]. Fzo1 mutation causes mitochondrial DNA loss and subsequent mitochondrial translation loss, culminating in respiratory incompetence [34]. In mice, knockout of Mfn1 or Mfn2 is embryonic lethal because of a placental defect features [35]. In addition, Mfn2 loss occurs after placental formation, but not Mfn1 loss, resulting in impaired cerebellar development and lethality soon after birth [36]. In vitro, partial ablations of both Mfn1 and Mfn2, neither Mfn1 nor Mfn2 alone, culminate in many severe cellular defects, such as poor cell growth, decreased cellular respiration, and high heterogeneity of mitochondrial membrane potential [31]. The overexpression of Mfn2 in cells devoid of Mfn1 and overexpression of Mfn1 in Mfn2-deficient cells can both rescue mitochondrial morphology, due to the abundant homotypic and heterotypic complexes assembled by Mfn1 and Mfn2 [1, 35]. The Mfn1 has different roles in OM fusion with respect to Mfn2, because the absence of Mfn1 or Mfn2 makes cells harbor different mitochondrial morphologies [37]. In mammals, knockout of Mfn1 results in numerous scattering small mitochondria throughout the cytoplasm, and loss of Mfn2 culminates in mitochondrial fragmentation whose fragments are larger than Mfn1 [35, 38]. The BCL-2-like protein CED-9 of *C. elegans* can promote mitochondrial fusion of both the outer and inner mitochondrial membranes via FZO-1 and EAT-3 [39]. 15-Oxospiramilactone, a diterpenoid derivative identified from plants, can trigger the mitochondrial fusion via mitofusins (Mfns) or OPA1 on both the outer and inner mitochondrial membranes. 15-Oxospiramilactone directly interacts with the catalytic domain of USP30, resulting in USP30 inactivation. USP30 acts on ubiquitylated forms of Mfn1 and Mfn2, facilitating mitochondrial fusion [40].

Fzo1 and Mfn1/Mfn2 are involved in mitochondrial docking, mitofusins anchor to the OM by two transmembrane domains, and the N- and C-terminal endings are exposed to the cytosol. Mitochondrial fusion interacts between the tips of two mitochondria. Monomeric form of mitofusins assembles a dimer in the same mitochondria, i.e., in cis, when the binding of GTP under the help of Ugo1. Interactions between two mitochondria occur via the higher oligomerization state of mitofusins [41, 42]. Subsequently, GTP hydrolysis mediates mitofusin ubiquitylation and ultimately degradation, to facilitate membrane approximation. Ugo1 and Mito-PLD in mammals mediate the merging of the associated OM membrane between two mitochondria [43].

Mitochondrial motility is related to mitochondrial fusion, which means that the stationary mitochondria have less probability to connect and fuse with another mitochondrion [44]. Mitochondrial axonal transport is mediated by motor/adaptor complex, which anchors in the OM of mitochondria and encompasses Miro, Milton, and kinesin-1 heavy chain. PINK1 senses mitochondrial depolarization upon CCCP treatment, mediates Parkin binding and the subsequent degradation of Miro/Milton, and culminates in motility arrest of mitochondria [5].

4 USP30 Dampens Mitochondrial Fission by Deubiquitylating Mitofusins

Ubiquitylation impacts the degradation of target protein substrates and regulates the dynamics of target protein functions. This is exemplified by the observation that ubiquitylation of mitofusins can culminate in two opposite outcomes, increased and decreased OM fusions, which are profusion and antifusion. In the profusion aspect, the mitofusins, Fzo in flies or Mfns in mammals, are assembled, ubiquitylated, and degraded to facilitate the OM fusion, when the OM fusion occurs. In the antifusion aspect, nonfunctional mitochondria lose membrane potential, leading to fusion incompetency and fragmentation of mitochondrial tubules by ongoing fission [45]. These damaged mitochondria, which cannot mix together or protect mitochondrial DNA integrity, become target autophagy substrates [46]. Crucial mitofusins of mitochondrial fusion are inactivated, resulting in mitochondrial fragmentation. The Ub E3 ligase APC/C ubiquitylates Mfn1 and OPA1, leading to the proteasomal-dependent degradation [47]. Fzo1 degradation takes place in the context of the mitochondrial fission promoted by cellular growth arrest [48]. Huwe1, an E3 Ub protein ligase, is recruited to the OM for apoptosis via diverse apoptotic stimuli or mitochondrial depolarization, ubiquitylates, and subsequently degrades Mfn1 and Mfn2, leading to mitochondrial fission. Parkin is also involved in the mitochondrial fission, similar in flies and mammalian cellular models [49, 50].

The mitochondrial dynamic network encompasses fusion and fission related to mitophagy and is pivotal to distribute mtDNA and proteins [51]. Fission often occurs after fusion, leading to produce mitochondrial fragmentation with uneven mitochondrial constituents. The mitochondrial fragmentations participate in fusion and fission dynamic network, whereas the depolarized mitochondrial fragmentation is degraded by mitophagy [22]. Depolarized mitochondria undergo fission, when Mfn1 and Mfn2 lost in outer membrane of mitochondria and Opa1 lost in inner and outer membrane of mitochondria [1]. Mitochondrial fission is essential to progression of mitophagy [2]. It was revealed that Drp1 is involved in the induction of mitophagy [52] and inhibits mitophagy by downregulating overexpression of Drp1 or Opa1 [53].

5 USP30 Halts Mitochondrial Mitophagy

Autophagy can be categorized by macroautophagy, microautophagy, and chaperone-mediated autophagy [3]. Macroautophagy includes selective autophagy and nonselective autophagy. Mitophagy is one of the selective macroautophagy pathways to contribute to clearing damaged or unneeded organelles in a selective manner. Nonselective macroautophagy occurs and mediates the generation of essential nutrients to needed cells. In macroautophagy, a phagophore surrounds a damaged or unneeded organelle or cytoplasm to form an autophagosome harboring double

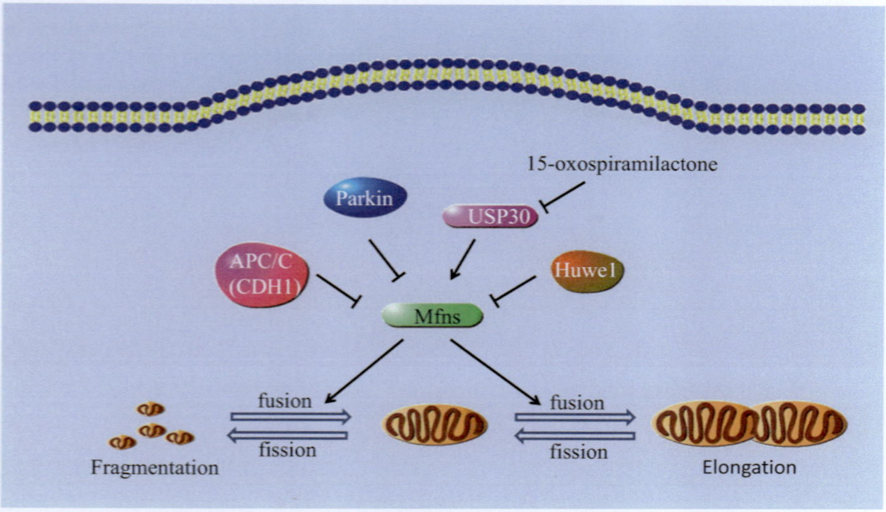

Fig. 10.2 Ubiquitylation and deubiquitylation of mitofusins in regulating dynamic mitochondria of fusion and fission. Mitochondria constantly occur in fusion and fission, leading to mitochondrial elongation or fragmentation. E3 ubiquitin ligases, including Parkin, APC/C, or Huwe1, inhibit mitochondrial fusion to ubiquitylate Mfn1 and Mfn2 (Mfns). Overexpression of deubiquitylase USP30 makes the mitochondrial dynamic network toward increased fusion and decreased fission by removing ubiquitin from ubiquitylated Mfns. 15-Oxospiramilactone, a small molecule inhibitor of USP30, inhibits mitochondrial fusion

membranes. The fusion of autophagosome and lysosome forms autophago-lysosome and mediates the selective destruction of the autophagosome contents [54, 55]. The microautophagy dictates that lysosomal membrane invaginates to engulf cytoplasmic components. The chaperone-mediated autophagy degrades KFERQ amino acid motif-containing target proteins. The substrate proteins are recognized by the chaperone Hsp70 complex, bound to LAMP-2A lysosomal protein, and subsequently targeted for lysosomal membrane [55] (Fig. 10.2).

Mitophagy in mammalian cellular system is regulated, at least in part, by PINK1- and Parkin-mediated pathway (Fig. 10.3). PINK1 is transferred through the outer membrane of mitochondria by TOM; located in the inner membrane of mitochondria by TIM; cleaved by proteases, such as MPP and PARL proteases; and proteolytically degraded through the N-end rule degradation machinery [1, 50]. In damaged mammalian mitochondria, PINK1 is hardly delivered to the inner membrane of mitochondria probably via the abnormal mitochondrial membrane potential and thus just localized to the outer membrane of mitochondria. PINK1 recruits the Parkin and initiates mitophagy to selectively target the tagged mitochondria for destruction [56, 57]. However, in flies, Parkin overexpression can rescue mitochondrial defects without the need of PINK1 kinase, indicating another initiator or mediator molecules in addition to PINK1 [58, 59].

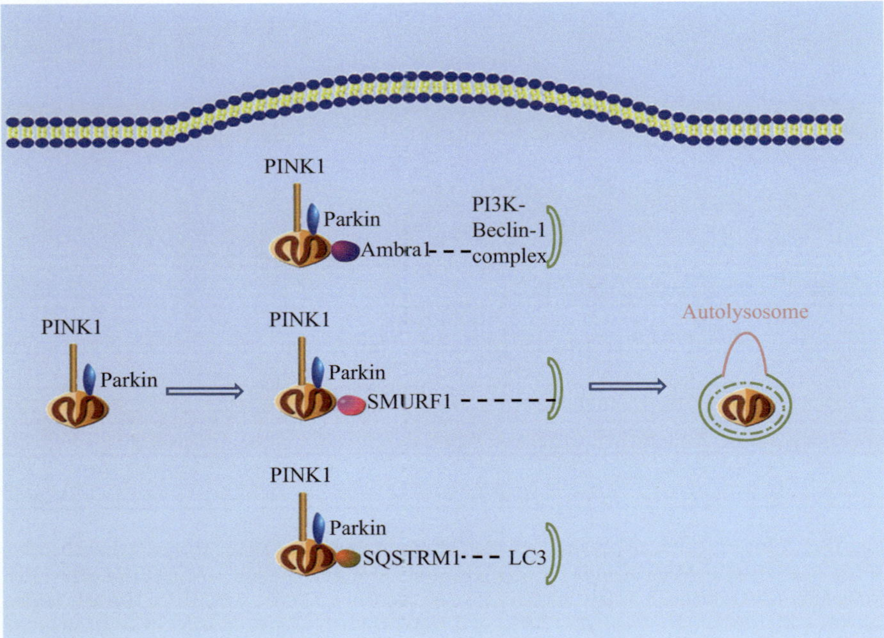

Fig. 10.3 Mitophagy is relative to PINK1- and Parkin-mediated pathway. In depolarized mitochondria, ubiquitin-binding adaptor SQSTRM1 localizes in mitochondria and binds the mitochondria to autophagosomes by LC3. Parkin recruits Ambra1 to damaged mitochondria, which activates the PI3K-Beclin-1 complex and recruits pre-autophagosomal membranes, leading to mitophagy. The Parkin-dependent mitophagy also needs the E3 ubiquitin ligase SMURF1, which might deliver mitochondria to the autophagosome, because of its membrane-targeting domain rather than its ubiquitin ligase activity

In neuronal-derived cellular models, knockdown of PINK1 stimulates mitophagy [60]. Parkin overexpression promotes CCCP-induced mitophagy [61]. In unwanted or damaged mitochondria, degradation of mitofusin dampens mitochondrial fusion and augments mitochondrial fission and subsequent mitophagy. The loss of mitofusin fails to induce mitophagy *per se*, while PINK1 and Parkin can induce mitophagy without the need for mitofusin [5, 62]. Parkin-mediated mitophagy involves the autophagy-related genes (ATGs), including ATG1–10, ATG12–16, and ATG18, to form the core machinery of autophagy. In macroautophagy, ATGs initiate the selective removal of organelles or abnormal cytoplasmic components. About 30 ATGs were identified in *Saccharomyces cerevisiae* by genetic screens [63]. The CCCP-treated mitochondria are rarely degraded via shutting function of class III phosphatidylinositol-3-kinase (PI3K) Vps34, ATG3, ATG5, and ATG5 [1]. In depolarized mitochondria, ubiquitin-binding adaptor p62/SQSTRM1 is localized in mitochondria and binds the mitochondria to autophagosomes by LC3 [64]. The Parkin-mediated mitophagy still takes place without the p62-mediated clustering [65, 66]. Recent work has revealed the direct interaction between PINK1/Parkin pathway and Beclin-1 PI3K complex [67, 68]. Parkin recruits Ambra1 to damaged mitochondria and activates the Beclin-1 complex, recruiting pre-autophagosomal membranes and culminating in mitophagy. The Parkin-dependent mitophagy also needs the E3 Ub ligase SMURF1 in the cytoplasm [69]. It was postulated that SMURF1 might deliver mitochondria to the autophagosome, because of its membrane-targeting domain rather than its Ub ligase activity.

The starvation-induced macroautophagy is initiated by a complex ATG1 and ATG13 [70]. In unstressed condition, ATG1 and ATG13 are separated by rapamycin complex 1. Upon starvation, ATG1 and ATG13 are gathered to induce macroautophagy. Rapamycin complex 1 also regulates the ortholog of ATG1 and ATG13, the ULK1 complex. The Hsp90-Cdc37 complex induces mitophagy via the ULK1-ATG13 pathway in response to stress [70]. The ATG1/ULK1 complex, rather than ATG13, is involved in other pathways in the autophagosome assembly by ATG14, Beclin1, and Vps34 [71]. Vps34 mediates the induction of the nucleation of the isolation membrane by PIP3-rich membrane domains [54].

USP30 regulates Parkin-mediated mitophagy. Lys 6 and Lys 11 Ub chains are involved in mitophagic signaling pathway. Mitochondrial damage by CCCP stimulates Parkin to covalently add Ub chains to Lys 6, Lys 11, and Lys 63 on mitochondria-derived proteins, and USP30 selectively removes Ub chains from Lys 6- and Lys 11-containing protein substrates [9]. USP30 overexpression culminates in deubiquitinating target substrates, including TOM20, TOM70, and VDACs, and thus can rescue the Parkin-mediated mitophagy [72]. The downregulation of USP30 rescues abrogated mitophagy via the overexpression of autophagy receptor SQSTM1 [72]. The overexpression of USP30 shifts the mitochondrial dynamic network toward augmented fusion and reduced fission and curbs mitochondrial clearance via mitophagy.

6 The Relative Diseases

The abrogation in mitochondrial quality control pathways contributes to neurodegenerative disorder. The deregulation of mitochondrial fusion contributes to Charcot-Marie-Tooth neuropathy type 2A, autosomal dominant optic atrophy, cardiopathies, and neurodegenerative disorders. E3 Ub ligase Parkin ubiquitylates mitofusins, leading to damaged mitochondria to degrade. Parkin mutation is often found in Parkinson's disease patients [50], suggesting that the invalid Parkin might not effectively ubiquitylate mitofusins, leading to noneffective clearance of damaged mitochondria. Because neuronal function needs high energy, fusion of mitochondria takes important part in neurons [73]. MUL1 is the most prominently deubiquitylated protein by USP30, among the substrates identified in the shared substrates of USP30 and Parkin. MUL1 is relative to some mitochondrial diseases, including Parkinson's disease and muscle wasting [19].

Parkin contributes to a large number of mitochondrial and cellular dysfunctions. Parkin is involved in the development of neurotoxicity and lipotoxicity through the autophagy-dependent pathway [74–76]. USP30 may stimulate the overproduction of inflammatory mediators and proteasome activity dependently or independently of reactive oxygen species or of endolysosomal trafficking pathways [77–79]. It is also questioned whether USP30 and Parkin are involved in mitochondria-driven metabolism and bioenergy generation [80–82], change cellular sensitivity and response to drug [83–86], and promote the development of drug resistance [87–89]. With the development of technology, it is also interesting to explore biological functions of Parkin and USP30 in single-cell biology by measuring single-cell RNA sequencing and using CRISPR [90–93].

7 Conclusions and Perspectives

Parkin adds Ub chains to Lys 6, Lys 11, and Lys 63 on mitochondria,while USP30 preferentially removes Ub chains from Lys 6 and Lys 11 on mitochondria. This suggests some other deubiquitinating enzymes corresponding to Parkin ubiquitination. There are about 100 deubiquitinating enzymes in humans [94], but only USP30, USP15, and USP35 deubiquitinate substrates of Parkin. Otherwise, USP8 directly targets Parkin itself. USP15 opposes Parkin-mediated mitophagy on damaged mitochondria. Moreover, USP15 knockdown rescues mitophagy in PD patient fibroblasts with Parkin defect [95]. Similar to USP30 and USP15, USP35 overexpression impairs mitophagy, and knockdown enhances mitophagy [96]. But there is no data showing which Ub chain can be removed by USP15 or USP35. Moreover, only several deubiquitinating enzymes have been investigated for small inhibitor molecules, such as 15-oxospiramilactone and VLX1570 [94].

Acknowledgments The work was supported by Zhongshan Distinguished Professor Grant (XDW), the National Nature Science Foundation of China (91230204, 81270099, 81320108001, 81270131, 81300010), the Shanghai Committee of Science and Technology (12JC1402200, 12431900207, 11410708600, 14431905100), the operation funding of Shanghai Institute of Clinical Bioinformatics, the Ministry of Education for Academic Special Science and Research Foundation for PhD Education (20130071110043), and the National Key Research and Development Program (2016YFC0902400, 2017YFSF090207).

References

1. Bordi M, Nazio F, Campello S. The close interconnection between mitochondrial dynamics and mitophagy in cancer. Front Oncol. 2017;7:81. [PMID:28512624]
2. Gugnoni M, Sancisi V, Manzotti G, Gandolfi G, Ciarrocchi A. Autophagy and epithelial-mesenchymal transition: an intricate interplay in cancer. Cell Death Dis. 2016;7:e2520. [PMID:27929542]
3. Mizushima N, Komatsu M. Autophagy: renovation of cells and tissues. Cell. 2011;147:728–41. [PMID:22078875]

4. Takayama K, Matsuura A, Itakura E. Dissection of ubiquitinated protein degradation by basal autophagy. FEBS Lett. 2017;591:1199–211. [PMID:28369861]
5. Bragoszewski P, Turek M, Chacinska A. Control of mitochondrial biogenesis and function by the ubiquitin-proteasome system. Open Biol. 2017;7:170007. [PMID:28446709]
6. Eldeeb M, Fahlman R. The-N-end rule: the beginning determines the end. Protein Pept Lett. 2016;23:343–8. [PMID:26743630]
7. Weissman AM, Shabek N, Ciechanover A. The predator becomes the prey: regulating the ubiquitin system by ubiquitylation and degradation. Nat Rev Mol Cell Biol. 2011;12:605–20. [PMID:21860393]
8. Escobar-Henriques M. Mitofusins: ubiquitylation promotes fusion. Cell Res. 2014;24:387–8. [PMID:24556809]
9. Cunningham CN, Baughman JM, Phu L, Tea JS, Yu C, Coons M, et al. USP30 and parkin homeostatically regulate atypical ubiquitin chains on mitochondria. Nat Cell Biol. 2015;17:160–9. [PMID:25621951]
10. Lazarou M, Narendra DP, Jin SM, Tekle E, Banerjee S, Youle RJ. PINK1 drives Parkin self-association and HECT-like E3 activity upstream of mitochondrial binding. J Cell Biol. 2013;200:163–72. [PMID:23319602]
11. Ordureau A, Sarraf SA, Duda DM, Heo JM, Jedrychowski MP, Sviderskiy VO, et al. Quantitative proteomics reveal a feedforward mechanism for mitochondrial PARKIN translocation and ubiquitin chain synthesis. Mol Cell. 2014;56:360–75. [PMID:25284222]
12. Hunter T. The age of crosstalk: phosphorylation, ubiquitination, and beyond. Mol Cell. 2007;28:730–8. [PMID:18082598]
13. Eldeeb MA, Fahlman RP. Phosphorylation impacts N-end rule degradation of the proteolytically activated form of BMX kinase. J Biol Chem. 2016;291:22757–68. [PMID:27601470]
14. Swaney DL, Rodriguez-Mias RA, Villen J. Phosphorylation of ubiquitin at Ser65 affects its polymerization, targets, and proteome-wide turnover. EMBO Rep. 2015;16:1131–44. [PMID:26142280]
15. Wauer T, Swatek KN, Wagstaff JL, Gladkova C, Pruneda JN, Michel MA, et al. Ubiquitin Ser65 phosphorylation affects ubiquitin structure, chain assembly and hydrolysis. EMBO J. 2015;34:307–25. [PMID:25527291]
16. Lokireddy S, Wijesoma IW, Teng S, Bonala S, Gluckman PD, McFarlane C, et al. The ubiquitin ligase Mul1 induces mitophagy in skeletal muscle in response to muscle-wasting stimuli. Cell Metab. 2012;16:613–24. [PMID:23140641]
17. Hosoi KI, Miyata N, Mukai S, Furuki S, Okumoto K, Cheng EH, et al. The VDAC2-BAK axis regulates peroxisomal membrane permeability. J Cell Biol. 2017;216:709–22. [PMID:28174205]
18. Ott C, Dorsch E, Fraunholz M, Straub S, Kozjak-Pavlovic V. Detailed analysis of the human mitochondrial contact site complex indicate a hierarchy of subunits. PLoS One. 2015;10:e0120213. [PMID:25781180]
19. Yun J, Puri R, Yang H, Lizzio MA, Wu C, Sheng ZH, et al. MUL1 acts in parallel to the PINK1/parkin pathway in regulating mitofusin and compensates for loss of PINK1/parkin. elife. 2014;3:e01958. [PMID:24898855]
20. Syu JS, Baba T, Huang JY, Ogawa H, Hsieh CH, JX H, et al. Lysosomal activity maintains glycolysis and cyclin E1 expression by mediating Ad4BP/SF-1 stability for proper steroidogenic cell growth. Sci Rep. 2017;7:240. [PMID:28325912]
21. Riba A, Deres L, Eros K, Szabo A, Magyar K, Sumegi B, et al. Doxycycline protects against ROS-induced mitochondrial fragmentation and ISO-induced heart failure. PLoS One. 2017;12:e0175195. [PMID:28384228]
22. Prieto J, Torres J. Mitochondrial dynamics: in cell reprogramming as it is in cancer. Stem Cells Int. 2017;2017:8073721. [PMID:28484497]
23. Li Z, Ding B, Zhou X, Wang GL. The Rice Dynamin-related protein OsDRP1E negatively regulates programmed cell death by controlling the release of cytochrome c from mitochondria. PLoS Pathog. 2017;13:e1006157. [PMID:28081268]

24. Ulengin I, Park JJ, Lee TH. ER network formation and membrane fusion by atlastin1/SPG3A disease variants. Mol Biol Cell. 2015;26:1616–28. [PMID:25761634]
25. Santoro A, Campolo M, Liu C, Sesaki H, Meli R, Liu ZW, et al. DRP1 suppresses leptin and glucose sensing of POMC neurons. Cell Metab. 2017;25:647–60. [PMID:28190775]
26. Moustaq L, Smaczynska-de R II, Palmer SE, Marklew CJ, Ayscough KR. Insights into dynamin-associated disorders through analysis of equivalent mutations in the yeast dynamin Vps1. Microb Cell. 2016;3:147–58. [PMID:28357347]
27. Dalmasso G, Marin Zapata PA, Brady NR, Hamacher-Brady A. Agent-based modeling of mitochondria links sub-cellular dynamics to cellular homeostasis and heterogeneity. PLoS One. 2017;12:e0168198. [PMID:28060865
28. Cao YL, Meng S, Chen Y, Feng JX, DD G, Yu B, et al. MFN1 structures reveal nucleotide-triggered dimerization critical for mitochondrial fusion. Nature. 2017;542:372–6. [PMID:28114303]
29. Sawyer EM, Brunner EC, Hwang Y, Ivey LE, Brown O, Bannon M, et al. Testis-specific ATP synthase peripheral stalk subunits required for tissue-specific mitochondrial morphogenesis in Drosophila. BMC Cell Biol. 2017;18:16 [PMID:28335714]
30. Shirozu R, Yashiroda H, Murata S. Proteasome impairment induces recovery of mitochondrial membrane potential and an alternative pathway of mitochondrial fusion. Mol Cell Biol. 2015;36:347–62. [PMID:26552703]
31. Pendin D, Filadi R, Pizzo P. The concerted action of mitochondrial dynamics and positioning: new characters in cancer onset and progression. Front Oncol. 2017;102:7. [PMID:28589083]
32. Qi Y, Yan L, Yu C, Guo X, Zhou X, Hu X, et al. Structures of human mitofusin 1 provide insight into mitochondrial tethering. J Cell Biol. 2016;215:621–9. [PMID:27920125]
33. Del Campo A, Jaimovich E, Tevy MF. Mitochondria in the aging muscles of flies and mice: new perspectives for old characters. Oxidative Med Cell Longev. 2016;2016:9057593. [PMID:27630760]
34. Norris KL, Hao R, Chen LF, Lai CH, Kapur M, Shaughnessy PJ, et al. Convergence of Parkin, PINK1, and alpha-Synuclein on stress-induced mitochondrial morphological remodeling. J Biol Chem. 2015;290:13862–74. [PMID:25861987]
35. Fang CL, Sun DP, Chen HK, Lin CC, Hung ST, Uen YH, et al. Overexpression of mitochondrial GTPase MFN2 represents a negative prognostic marker in human gastric cancer and its inhibition exerts anti-cancer effects. J Cancer. 2017;8:1153–61. [PMID:28607589]
36. Beresewicz M, Boratynska-Jasinska A, Charzewski L, Kawalec M, Kabzinska D, Kochanski A, et al. The effect of a novel c.820C>T (Arg274Trp) mutation in the Mitofusin 2 gene on fibroblast metabolism and clinical manifestation in a patient. PLoS One. 2017;12:e0169999. [PMID:28076385]
37. Trotta AP, Chipuk JE. Mitochondrial dynamics as regulators of cancer biology. Cell Mol Life Sci. 2017;74:1999–2017. [PMID:28083595]
38. Bannerman P, Burns T, Xu J, Miers L, Pleasure D. Mice hemizygous for a pathogenic Mitofusin-2 allele exhibit hind limb/foot gait deficits and phenotypic perturbations in nerve and muscle. PLoS One. 2016;11:e0167573. [PMID:27907123]
39. Hardwick JM, Soane L. Multiple functions of BCL-2 family proteins. Cold Spring Harb Perspect Biol. 2013;5(2):pii: a008722. https://doi.org/10.1101/cshperspect.a008722. [PMID:23378584]
40. Yue W, Chen Z, Liu H, Yan C, Chen M, Feng D, et al. A small natural molecule promotes mitochondrial fusion through inhibition of the deubiquitinase USP30. Cell Res. 2014;24:482–96. [PMID:24513856]
41. Hu C, Huang Y, Li L. Drp1-dependent mitochondrial fission plays critical roles in physiological and pathological progresses in mammals. Int J Mol Sci. 2017;18:144. [PMID:28098754]
42. Romanello V, Sandri M. Mitochondrial quality control and muscle mass maintenance. Front Physiol. 2015;6:422. [PMID:26793123]
43. Cohen MM, Amiott EA, Day AR, Leboucher GP, Pryce EN, Glickman MH, et al. Sequential requirements for the GTPase domain of the mitofusin Fzo1 and the ubiquitin

ligase SCFMdm30 in mitochondrial outer membrane fusion. J Cell Sci. 2011;124:1403–10. [PMID:21502136]
44. Novak EA, Mollen KP. Mitochondrial dysfunction in inflammatory bowel disease. Front Cell Dev Biol. 2015;3:62. [PMID:26484345]
45. Van Laar VS, Berman SB. The interplay of neuronal mitochondrial dynamics and bioenergetics: implications for Parkinson's disease. Neurobiol Dis. 2013;51:43–55. [PMID:22668779]
46. Tang YC, Tian HX, Yi T, Chen HB. The critical roles of mitophagy in cerebral ischemia. Protein Cell. 2016;7:699–713. [PMID:27554669]
47. Garedew A, Andreassi C, Moncada S. Mitochondrial dynamics, biogenesis, and function are coordinated with the cell cycle by APC/C CDH1. Cell Metab. 2012;15:466–79. [PMID:22482729]
48. Braganza A, Li J, Zeng X, Yates NA, Dey NB, Andrews J, et al. UBE3B is a Calmodulin-regulated, mitochondrion-associated E3 ubiquitin ligase. J Biol Chem. 2017;292:2470–84. [PMID:28003368]
49. Leboucher GP, Tsai YC, Yang M, Shaw KC, Zhou M, Veenstra TD, et al. Stress-induced phosphorylation and proteasomal degradation of mitofusin 2 facilitates mitochondrial fragmentation and apoptosis. Mol Cell. 2012;47:547–57. [PMID:22748923]
50. Mouton-Liger F, Jacoupy M, Corvol JC, Corti O. PINK1/Parkin-dependent mitochondrial surveillance: from pleiotropy to Parkinson's disease. Front Mol Neurosci. 2017;10:120. [PMID:28507507]
51. Kuhlbrandt W. Structure and function of mitochondrial membrane protein complexes. BMC Biol. 2015;13:89. [PMID:26515107]
52. Bin-Umer MA, McLaughlin JE, Butterly MS, McCormick S, Tumer NE. Elimination of damaged mitochondria through mitophagy reduces mitochondrial oxidative stress and increases tolerance to trichothecenes. Proc Natl Acad Sci U S A. 2014;111:11798–803. [PMID:25071194]
53. Knorre DA, Sokolov SS, Zyrina AN, Severin FF. How do yeast sense mitochondrial dysfunction? Microb Cell. 2016;3:532–9. [PMID:28357322]
54. Tanida I. Autophagy basics. Microbiol Immunol. 2011;55:1–11. [PMID:21175768]
55. Singh V, Finke-Isami J, Hopper-Chidlaw AC, Schwerk P, Thompson A, Tedin K. Salmonella co-opts host cell chaperone-mediated autophagy for intracellular growth. J Biol Chem. 2017;292:1847–64. [PMID:27932462]
56. Budini M, Buratti E, Morselli E, Criollo A. Autophagy and its impact on neurodegenerative diseases: new roles for TDP-43 and C9orf72. Front Mol Neurosci. 2017;10:170. [PMID:28611593]
57. Lazarou M, Jin SM, Kane LA, Youle RJ. Role of PINK1 binding to the TOM complex and alternate intracellular membranes in recruitment and activation of the E3 ligase Parkin. Dev Cell. 2012;22:320–33. [PMID:22280891]
58. Pathak D, Berthet A, Bendor JT, Yu K, Sellnow RC, Orr AL, et al. Loss of alpha-Synuclein does not affect mitochondrial bioenergetics in rodent neurons. eNeuro. 2017;4(2). pii: ENEURO.0216-16. doi:https://doi.org/10.1523/ENEURO.0216-16. [PMID:28462393]
59. Anding AL, Baehrecke EH. Cleaning house: selective autophagy of organelles. Dev Cell. 2017;41:10–22. [PMID:28399394]
60. Packiriswamy N, Coulson KF, Holcombe SJ, Sordillo LM. Oxidative stress-induced mitochondrial dysfunction in a normal colon epithelial cell line. World J Gastroenterol. 2017;23:3427–39. [PMID:28596679]
61. Gong Y, Schumacher SE, WH W, Tang F, Beroukhim R, Chan TA. Pan-cancer analysis links PARK2 to BCL-XL-dependent control of apoptosis. Neoplasia. 2017;19:75–83. [PMID:28038320]
62. Han K, Hassanzadeh S, Singh K, Menazza S, Nguyen TT, Stevens MV, et al. Parkin regulation of CHOP modulates susceptibility to cardiac endoplasmic reticulum stress. Sci Rep. 2017;7:2093. [PMID:28522833]
63. Kankuan W, Wanichanon C, Titone R, Engsusophon A, Sumpownon C, Suphamungmee W, et al. Starvation promotes autophagy-associated maturation of the ovary in the giant freshwater prawn, Macrobrachium rosenbergii. Front Physiol. 2017;300:8. [PMID:28553234]

64. Dagda RK, Rice M. Protocols for assessing mitophagy in neuronal cell lines and primary neurons. NeuroMethods. 2017;123:249–77. [PMID:28603343]
65. Reina S, Guarino F, Magri A, De Pinto V. VDAC3 as a potential marker of mitochondrial status is involved in cancer and pathology. Front Oncol. 2016;264:6. [PMID:28066720]
66. Bartolome F, Esteras N, Martin-Requero A, Boutoleau-Bretonniere C, Vercelletto M, Gabelle A, et al. Pathogenic p62/SQSTM1 mutations impair energy metabolism through limitation of mitochondrial substrates. Sci Rep. 2017;7:1666. [PMID:28490746]
67. Strappazzon F, Cecconi F. AMBRA1-induced mitophagy: a new mechanism to cope with cancer? Mol Cell Oncol. 2015;2:e975647. [PMID:27308437]
68. Gelmetti V, De Rosa P, Torosantucci L, Marini ES, Romagnoli A, Di Rienzo M, et al. PINK1 and BECN1 relocalize at mitochondria-associated membranes during mitophagy and promote ER-mitochondria tethering and autophagosome formation. Autophagy. 2017;13:654–69. [PMID:28368777]
69. Lai SC, Devenish R. Peering into the 'black box' of pathogen recognition by cellular autophagy systems. Microb Cell. 2015;2:322–8 [PMID:28357309]
70. Wesselborg S, Stork B. Autophagy signal transduction by ATG proteins: from hierarchies to networks. Cell Mol Life Sci. 2015;72:4721–57. [PMID:26390974]
71. Itakura E, Kishi-Itakura C, Koyama-Honda I, Mizushima N. Structures containing Atg9A and the ULK1 complex independently target depolarized mitochondria at initial stages of Parkin-mediated mitophagy. J Cell Sci. 2012;125:1488–99. [PMID:22275429]
72. Bingol B, Tea JS, Phu L, Reichelt M, Bakalarski CE, Song Q, et al. The mitochondrial deubiquitinase USP30 opposes parkin-mediated mitophagy. Nature. 2014;510:370–5. [PMID:24896179]
73. Rugarli EI, Langer T. Mitochondrial quality control: a matter of life and death for neurons. EMBO J. 2012;31:1336–49. [PMID:22354038]
74. Lippai M, Szatmari Z. Autophagy-from molecular mechanisms to clinical relevance. Cell Biol Toxicol. 2017;33:145–68. [PMID:27957648]
75. Hsu HC, Li SJ, Chen CY, Chen MF. Eicosapentaenoic acid protects cardiomyoblasts from lipotoxicity in an autophagy-dependent manner. Cell Biol Toxicol. 2017, July 24. doi:https://doi.org/10.1007/s10565-017-9406-9.[PMID:28741157]
76. Poindexter KM, Matthew S, Aronchik I, Firestone GL. Cooperative antiproliferative signaling by aspirin and indole-3-carbinol targets microphthalmia-associated transcription factor gene expression and promoter activity in human melanoma cells. Cell Biol Toxicol. 2016;32:103–19. [PMID:27055402]
77. Opperman CM, Sishi BJ. Tumor necrosis factor alpha stimulates p62 accumulation and enhances proteasome activity independently of ROS. Cell Biol Toxicol. 2015;31:83–94. [PMID:25761618]
78. Mbah NE, Overmeyer JH, Maltese WA. Disruption of endolysosomal trafficking pathways in glioma cells by methuosis-inducing indole-based chalcones. Cell Biol Toxicol. 2017;33:263–82. [PMID:27822587]
79. Venkatesan T, Choi YW, Mun SP, Kim YK. Pinus radiata bark extract induces caspase-independent apoptosis-like cell death in MCF-7 human breast cancer cells. Cell Biol Toxicol. 2016;32:451–64. [PMID:27400986]
80. Zerin T, Kim JS, Gil HW, Song HY, Hong SY. Effects of formaldehyde on mitochondrial dysfunction and apoptosis in SK-N-SH neuroblastoma cells. Cell Biol Toxicol. 2015;31:261–72. [PMID:26728267]
81. Soltani B, Ghaemi N, Sadeghizadeh M, Najafi F. Curcumin confers protection to irradiated THP-1 cells while its nanoformulation sensitizes these cells via apoptosis induction. Cell Biol Toxicol. 2016;32:543–61. [PMID:27473378]
82. Seo JB, Jung SR, Hille B, Koh DS, Extracellular ATP. Protects pancreatic duct epithelial cells from alcohol-induced damage through P2Y1 receptor-cAMP signal pathway. Cell Biol Toxicol. 2016;32:229–47. [PMID:27197531]

83. Ganta KK, Mandal A, Chaubey B. Depolarization of mitochondrial membrane potential is the initial event in non-nucleoside reverse transcriptase inhibitor efavirenz induced cytotoxicity. Cell Biol Toxicol. 2017;33:69–82. [PMID:27639578]
84. Fathi H, Ebrahimzadeh MA, Ziar A, Mohammadi H. Oxidative damage induced by retching; antiemetic and neuroprotective role of Sambucus ebulus L. Cell Biol Toxicol. 2015;31:231–9. [PMID:26493312]
85. Zhu LZ, Hou YJ, Zhao M, Yang MF, XT F, Sun JY, et al. Caudatin induces caspase-dependent apoptosis in human glioma cells with involvement of mitochondrial dysfunction and reactive oxygen species generation. Cell Biol Toxicol. 2016;32:333–45. [PMID:27184666]
86. Bao L, Diao H, Dong N, Su X, Wang B, Mo Q, et al. Histone deacetylase inhibitor induces cell apoptosis and cycle arrest in lung cancer cells via mitochondrial injury and p53 up-acetylation. Cell Biol Toxicol. 2016;32:469–82. [PMID:27423454]
87. Gu J, Wang X. New future of cell biology and toxicology: thinking deeper. Cell Biol Toxicol. 2016;32:1–3. [PMID:26874518]
88. Divolis G, Mavroeidi P, Mavrofrydi O, Papazafiri P. Differential effects of calcium on PI3K-Akt and HIF-1alpha survival pathways. Cell Biol Toxicol. 2016;32:437–49. [PMID:27344565]
89. Guney Eskiler G, Cecener G, Tunca B, Egeli U. An in vitro model for the development of acquired tamoxifen resistance. Cell Biol Toxicol. 2016;32:563–81. [PMID:27585693]
90. Chu MP, Kriangkum J, Venner CP, Sandhu I, Hewitt J, Belch AR, et al. Addressing heterogeneity of individual blood cancers: the need for single cell analysis. Cell Biol Toxicol. 2017;33:83–97. [PMID:27761761]
91. Wang W, Gao D, Wang X. Can single-cell RNA sequencing crack the mystery of cells? Cell Biol Toxicol. 2017, July 21. doi:https://doi.org/10.1007/s10565-017-9404-y. [PMID:28733864]
92. Wang W, Zhu B, Wang X. Dynamic phenotypes: illustrating a single-cell odyssey. Cell Biol Toxicol. 2017;33:423–7. [PMID:28638956]
93. Wang W, Wang X, Single-cell CRISPR. Screening in drug resistance. Cell Biol Toxicol. 2017;33:207–10. [PMID:28474250]
94. He M, Zhou Z, Shah AA, Zou H, Tao J, Chen Q, et al. The emerging role of deubiquitinating enzymes in genomic integrity, diseases, and therapeutics. Cell Biosci. 2016;6:62. [PMID:28031783]
95. Cornelissen T, Haddad D, Wauters F, Van Humbeeck C, Mandemakers W, Koentjoro B, et al. The deubiquitinase USP15 antagonizes Parkin-mediated mitochondrial ubiquitination and mitophagy. Hum Mol Genet. 2014;23:5227–42. [PMID:24852371]
96. Wang Y, Serricchio M, Jauregui M, Shanbhag R, Stoltz T, Di Paolo CT, et al. Deubiquitinating enzymes regulate PARK2-mediated mitophagy. Autophagy. 2015;11:595–606. [PMID:25915564]

Jiayun Hou, a doctoral candidate, is a research assistant of Clinical Science Institute of Fudan University Zhongshan Hospital. Her main research is focused on immunoelectron microscopy technique, disease-specific biomarkers, and lung chronic diseases. She is the author of 13 scientific publications.

Mohamed Eldeeb, just finished his PhD studies in Biochemistry (2017) from the University of Alberta, Canada. His main research is related to the regulation of N-end rule protein degradation, ubiquitin-dependent signaling, apoptotic cell death, proteolysis, and molecular and cellular biological studies. He is the author of more than six scientific publications.

Xiangdong Wang, MD, PhD, is a distinguished professor of Medicine, director of Shanghai Institute of Clinical Bioinformatics, executive director of Clinical Science Institute of Fudan University Zhongshan Hospital, director of Fudan University Center of Clinical Bioinformatics, deputy director of Shanghai Respiratory Research Institute, and visiting professor of King's College of London. His main research is focused on clinical bioinformatics, disease-specific biomarkers, lung chronic diseases, cancer immunology, and molecular and cellular therapies. He is the author of more than 200 scientific publications with the impact factor about 900, citation number about 6918, h-index 46, i10-index 181, and cited journal impact factor about 7000.

Chapter 11
Metabolic Regulation in Mitochondria and Drug Resistance

Yue Pan, Min Cao, Jianzhou Liu, Qing Yang, Xiaoyu Miao, Vay Liang W. Go, Paul W.N. Lee, and Gary Guishan Xiao

Abstract Mitochondria are generally considered as a powerhouse in a cell where the majority of the cellular ATP and metabolite productions occur. Metabolic rewiring and reprogramming may be initiated and regulated by mitochondrial enzymes. The hypothesis that cellular metabolic rewiring and reprogramming processes may occur as cellular microenvironment is disturbed, resulting in alteration of cell phenotype, such as cancer cells resistant to therapeutics seems to be now acceptable. Cancer metabolic reprogramming regulated by mitochondrial enzymes is now one of the hallmarks of cancer. This chapter provides an overview of cancer metabolism and summarizes progress made in mitochondria-mediated metabolic regulation in cancer drug resistance.

Keywords Mitochondrial • Metabolism • Drug resistance • Signaling pathway

---

Y. Pan • M. Cao • J. Liu • Q. Yang • X. Miao
School of Pharmaceutical Science and Technology, Dalian University of Technology, Dalian, China

V.L.W. Go
Agi Hirshberg UCLA Center for Pancreas Diseases, University of California, Los Angeles, CA, USA

P.W.N. Lee
Harbor University of California Los Angeles Medical Center, Torrance, CA, USA

G.G. Xiao (✉)
Agi Hirshberg UCLA Center for Pancreas Diseases, University of California, Los Angeles, CA, USA

Functional Genomics and Proteomics Laboratories, Creighton University Medical Center, Omaha, NE, USA

School of Pharmaceutical Science and Technology, Dalian University of Technology, Dalian, China

Harbor University of California Los Angeles Medical Center, Torrance, CA, USA
e-mail: gxiao@dlut.edu.cn

H. Sun, X. Wang (eds.), *Mitochondrial DNA and Diseases*, Advances in Experimental Medicine and Biology 1038,
https://doi.org/10.1007/978-981-10-6674-0_11

1 Introduction

Mitochondria are generally considered as the powerhouse in a cell where the majority of the cellular ATP and metabolite productions occur. Mitochondrial metabolism is directly or indirectly connected to many cellular metabolic pathways through common substrates and enzymes serving numerous cellular functions (Figs. 11.1 and 11.2). Energy production in a cell via oxidative phosphorylation is an essential function of mitochondria, which is also involved in Ca^{2+} homoeostasis and cellular apoptosis. Studies showed that mitochondrial dysfunction resulted from mutations of nuclear DNA or mitochondrial DNA (mtDNA) occurred in diseases [1, 2]. Efficient energy production by the mitochondria is required for rapid proliferation in cancer [3–5]. Mitochondria dysfunction may be a common marker for tumorigenesis. In this chapter, altered regulations of metabolic functions and specific pathways by mitochondrial metabolism will be reviewed and their relevance to the development of cancer chemoresistance discussed.

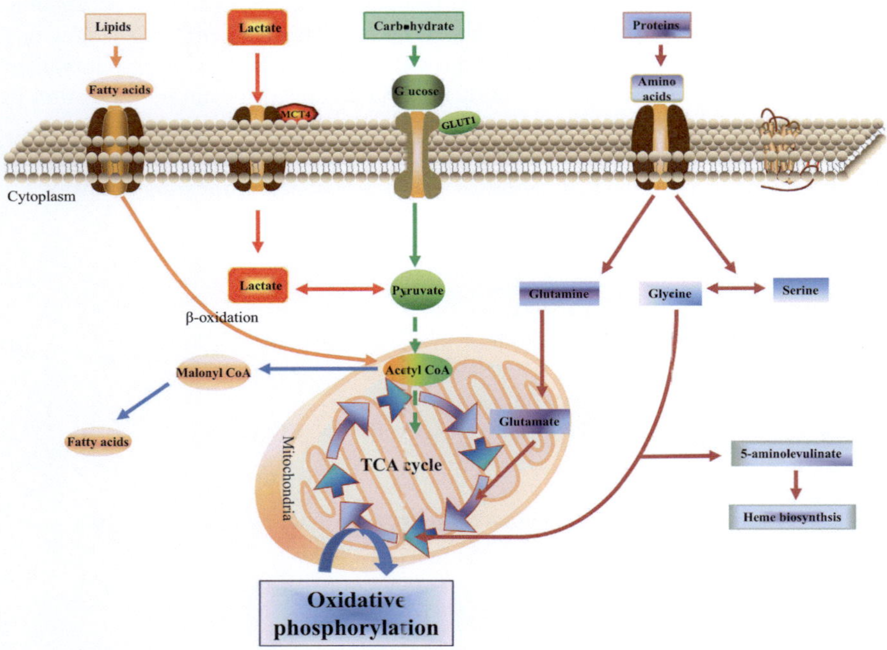

Fig. 11.1 Schematic representation of metabolic pathways involving mitochondrial metabolites. An overview of the main catabolic and anabolic metabolic pathways supporting tumor cell growth and survival.
Abbreviations: GLUT, glucose transporter; MCT, monocarboxylate transporter; TCA, tricarboxylic acid cycle

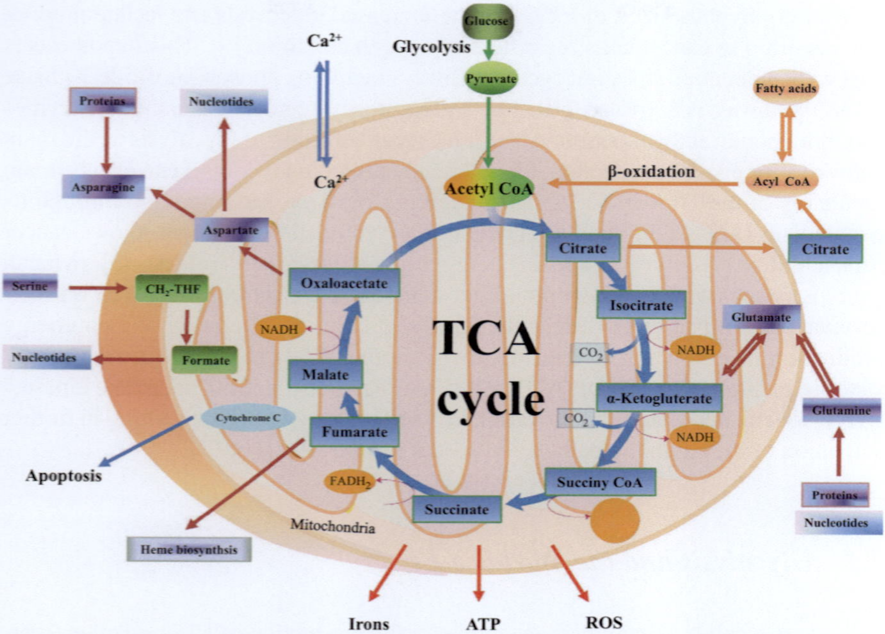

Fig. 11.2 Overview of mitochondrial function and catabolic and anabolic metabolic pathways. The TCA cycle uses substrates from glycolysis, fatty acid oxidation, and amino acid catabolism to generate building blocks and high-energy electrons (NADH and FADH2) to power the ETC. Major biosynthetic products produced by mitochondria and signaling pathways regulated by mitochondrial function are indicated
Abbreviations: NADH, nicotinamide adenine dinucleotide reduced form; ROS, reactive oxygen species; $FADH_2$, flavin adenine dinucleotide; ATP, adenosine triphosphate; 5,10-CH_2-THF, 5,10 methylenetetrahydrofolate

2 Mitochondrial Metabolism in Tumor Growth

Cellular bioenergetics and biosynthesis are maintained by mitochondria-coordinated anabolic and catabolic reactions for energy metabolism. Acetyl-CoA, produced by metabolism of glucose, fatty acids, and amino acids, is an important component in energy production and mitochondrial metabolism. Although mitochondrial metabolism is necessary for cancer cell proliferation, the exact mechanism underlying mitochondria-mediated regulation in cancer metabolism is still unknown. A number of studies showed that well-functioned mitochondria are required for rapid growth of tumor cells as they are not only an energy-producing factory but also a supplier of metabolic components. Active metabolic intermediates resulting from a myriad of glucose metabolic networks maintain high proliferation rate in a cancer cell. Intermediates from tricarboxylic acid (TCA) cycle can be utilized as carbon sources, such as α-ketoglutarate (α-KG) from glutamine metabolism [6–8].

Warburg hypothesized in 1956 that the increased glycolysis and lactate production occurred in cancer cells regardless of oxygen condition [9]. This hypothesis is now well understood since oncogenic pathways including phosphoinositide 3-kinase (PI3K) pathway is activated [10, 11], leading to dysregulated metabolic enzymes and mitochondrial dysfunction. Although cancer cells utilize glycolysis as the main pathways for energy production, the intermediates from glycolysis can also be main regulators of the TCA cycle for cancer cell proliferation and migration through its metabolism [12]. Additionally, mitochondrial dysfunction in cancer allows cancer cells to utilize glycolysis as a main energy resource for its survival [10, 13]. In harsh conditions such as hypoxia and nutrient deficiency, cancer cells can turn on their survival mechanism through mitochondrial regulation by alteration of their energy production in adaptation to their microenvironment [14, 15]. In these low-energy states (e.g., high ratio of ADP/ATP or AMP/ATP), mitochondrial adenylate kinase 4 (AK4) is activated leading to activation of AMP kinase (AMPK), resulting in further activation of the catabolic pathway to produce more ATP [16, 17].

2.1 Glycolysis and the Warburg Effect

Glucose in a cell is a major carbon source for the production of ATP and other intermediates by oxygen-independent glycolysis [18]. The intermediate pyruvate from glycolysis can be used by mitochondria in the cell to produce energy through oxidative phosphorylation (OXPHOS), as well as potential production of reactive oxygen species (ROS) [19]. However, in a hypoxic environment, cancer cells can use glycolysis instead of oxidative phosphorylation to generate energy regardless of oxygen condition [20, 21]. Therefore, the majority of glucose in cancer cells is used for production of lactate through non-OXPHOS pathway, which is independent on mitochondria [22]. Presumably, cancer cells utilize glucose through glycolysis for energy production; thus it keeps less ROS production and incremental production of metabolic intermediates [23, 24]. Because less ATP is produced through glycolysis than from OXPHOS in a cell [25], cancer cells maintain their rapid growth and redox homeostasis by increasing glucose uptake and glycolysis [26]. Conveniently, detection of glycolytic rates in cancer cells by ^{18}F-deoxyglucose positron emission tomography (FDG-PET) is used for detection, diagnosis, and therapy of cancer [27].

Over the years, the Warburg effect, aerobic glycolysis, is believed to result from mutations in a variety of signaling pathways [24, 28–32]. Mitochondrial dysfunction and activation of crucial enzymes in glycolysis such as hexokinase 2 and M2 pyruvate kinase alter cancer cell metabolic phenotype [32]. Studies showed that the increased glucose uptake induced by glucose transporter overexpression [33] and the crucial glycolytic enzymes [34] regulated by transcription factors including hypoxia-inducible factor 1 (HIF-1), Myc, and PI3K may contribute to the Warburg effect and maintain cancer cell proliferation [24, 28, 30]. On the other hand, tumor suppressor gene p53 represses the transcription of glucose transporters GLUT1 and GLUT4, activates the cytochrome c oxidase assembly protein (SCO2) to promote the OXPHOS, and upregulates glycolysis and apoptosis regulator (TIGAR), which can reduce the glycolytic activator fructose-2,6-bisphosphate in cancer cells. These studies suggest that the tumor suppressor gene p53 may play a crucial role in modifying Warburg effect [19, 35].

Many metabolic enzymes are key regulators in cancer metabolic phenotypes that regulate cell apoptosis and glucose homeostasis [36, 37]. For instance, the rate-limiting enzyme of glycolysis hexokinase 2 (HK2) isoform plays a crucial role in the Warburg effect. HK2 binds competitively with the voltage-dependent anion channel (VDAC) in mitochondria, which prevent VDAC from binding with pro-apoptotic factor Bax, resulting in inhibition of proliferation, prevention of mitochondrial apoptosis, and enhancement of autophagy [38]. Thus, multifunctional metabolic enzymes can be exploited as potential targets for cancer therapy.

2.2 Pentose Phosphate Pathway

Pentose phosphate pathway (PPP) is another key energy-producing pathway in metabolism, which supports DNA and RNA synthesis, to maintain cancer cell growth by regulating the lipogenesis and pentose synthesis. The pentose phosphate and nicotinamide adenine dinucleotide phosphate (NADPH) are mainly generated by PPP. NADPH is a necessary cofactor for lipid and amino acid biosynthesis, and it can modulate the oxidative stress by maintaining reduction of glutathione (GSH) [39]. Studies showed close association of PPP with cancer cell proliferation, indicating that PPP can counteract oxidative stress in maintaining cellular redox homeostasis in cancer cells.

PPP oxidative branch pathway is mainly regulated by rate-limiting enzymes of glucose oxidation-producing NADPH and ribose-5-phosphate (R5P). PPP non-oxidative branch pathway is driven by mass action through a system of transaldolase and transketolase. Glucose-6-phosphate dehydrogenase (G6PD) is the first enzyme and main regulator in the oxidative branch, and the key enzyme in the non-oxidative branch is transketolase (TKT) [40]. Several oncogenic signaling pathways may activate G6PD, such as tumor suppressor p53 which can directly activate G6PD by causing K403 acetylation. Activation of G6PD affects late G1 and S phases of cell cycle. Moreover, G6PD interacted with the protein-b-transduction repeat-containing protein (b-TrCP) to activate SCF ubiquitin ligase, which degrades PFKFB3 during S phase [41]. Degradation of PFKFB3 increases the generation of NADPH and R5P leading to the shuttling of glycolytic substrates through the PPP.

2.3 Lipid Metabolism

Fatty acids are the principal component of phosphoglycerides, triacylglycerides, sterol esters, and sphingolipids. Plasma membranes are made up of phosphoglycerides and sterols. Triacylglycerides are mainly used for energy storage. Lipids are involved in signal transduction, membrane homeostasis, proliferation, apoptosis, differentiation, and drug resistance of cancer cells [42]. Lipid biosynthesis in cancer cells is increased for cell membrane synthesis to maintain rapid growth of cancer cell. Especially, normal cells rely on extracellular lipids, while proliferation of cancer cells relies on de novo lipid biosynthesis.

In cancer cells, oncogenic signaling pathway enhances fatty acid synthase to accelerate the lipogenesis by activating many enzymes related to fatty acid synthase [43, 44], thus increasing glutamine metabolism and glucose transport and promoting glycolysis and PPP activity. Mitochondrial citrate is the source of acetyl groups used for biosynthesis of fatty acids. The citrate in the mitochondria is first exported to the cytosol, and ATP citrate lyase (ACLY) can then produce oxaloacetate and acetyl-CoA. Malic enzyme (ME) and malate dehydrogenase (MDH) catalyze oxaloacetate into pyruvate and generate NADPH used for biosynthesis of fatty acid. Additionally, acetyl-CoA carboxylase (ACC) catalyzes acetyl-CoA into malonyl-CoA, and both groups of acetyl and malonyl are condensed through cyclical series of reactions by fatty acid synthase (FASN), producing long-chain fatty acids. Saturated fatty acids are synthesized by elongases and desaturases. On the other hand, fatty acids can be degraded by β-oxidation in mitochondria. During this process, large amounts of ROS and ATP from OXPHOS and TCA are generated [42].

Fatty acid and sterol synthesis is also regulated at the transcription level. Sterol regulatory element-binding proteins (SREBPs) are transcriptional factors that regulate enzymes related to cholesterol and fatty acid synthesis. In turn, tumor suppressors pRB, p53, and AMPK can inhibit SREBP expression, and oncogenes PI3K and Akt can activate it [43]. Since enzymes related to lipid biosynthesis are activated in cancer cells and are associated to cell growth, lipogenic pathways are potential targets for cancer therapy.

2.4 Amino Acid Metabolism

Amino acids contain a specific side chain and both amino ($-NH_2$) and carboxyl ($-COOH$) groups, which are basic building blocks for protein synthesis. Catabolism of amino acids provides carbon and nitrogen for biosynthesis of other important metabolites. In fact, amino acids contribute to survival and regeneration of cells. Progressive deficiency of specific amino acids such as leucine, glutamine, and arginine can activate the mTOR pathway, and the levels of amino acids in a cell are regulated by rapamycin (mTOR) signaling [44]. In response to amino acid deficiency, activation of mTOR leads to autophagy and reduced protein synthesis. Nonessential amino acids can be produced from glycolytic and TCA cycle intermediates. For example, 3-phosphoglycerate and pyruvate can be converted to alanine. Oxaloacetate and α-ketoglutarate generated by TCA cycle can be converted to aspartate (asparagine) and glutamate (glutamine). Furthermore, glutamate can be transformed to 1-pyrroline-5-carboxylate (P5C) and L-glutamate-5-semialdehyde (GSA), which could be further converted to proline and ornithine, which produces arginine at urea cycle. Other nonessential amino acids of importance are serine and glycine which are precursors to the synthesis of cysteine and nucleobases [45].

Endogenous amino acid synthesis usually is insufficient to support protein synthesis in cell growth. Exogenous supply of amino acids is required, which are transported across cell membrane by selective transporters on the cell membrane. Accordingly, SLC1A5, SLC7A5, SLC7A11, and SLC6A14 are amino acid transporters and overexpressed in cancer cells regulated by MYC and miR-23a. Overexpression of amino acid transporters facilitates amino acid utilization and cell proliferation [46–48]. Interestingly, the glutamine transporters SLC1A5 and SLC7A5 coupling can increase the use of glutamine and help increase tumor growth regulated by mTOR pathway in cancer cells [49].

In proliferating cells, more amino acids are used than is required for protein biosynthesis. They are the sources of one-carbon units, acetyl units in fatty acids, TCA cycle intermediates, and nucleotides in intermediary metabolism [45]. For instance, glutamine, aspartate, and glycine are necessary for the synthesis of nucleotide, while glycine and serine are crucial in the metabolism of one-carbon unit; it can provide the precursors for lipid, nucleotide, and protein synthesis, taking part in nucleic acid and protein methylation and regulating the redox state [50]. The process of converting serine to glycine can be catalyzed by the cytosolic or mitochondrial serine hydroxymethyltransferase (SHMT1 and SHMT2, respectively). It has been shown that the metabolic activity of SHMT2 is closely related to the proliferation rate in cancer cell. In fact, SHMT2 is thought to be the basis for the maintenance of cancer metabolism by promoting hemoglobin biosynthesis and oxidative phosphorylation [51].

The degradation of proline is closely associated with production of ROS and regulation of redox potential in cancer metabolism. The first step in the degradation of proline is catalyzed by a tumor suppressor, which is the mitochondrial proline dehydrogenase (PRODH) [52]. This mitochondrial enzyme is linked to the electron transport chain through complex III, and it could be as a source of ROS production. Additionally, P5C and proline can be used as a redox couple, through mitochondrial PRODH and cytosolic P5C reductase (PYCR) binding activity to balance mitochondrial reduction potential and oxidative potential, during which process NADPH is more preferred [53]. Since proline is a product of glutamine, many studies have shown that glutamine plays a prominent role in tumor metabolism. Thus, it has been reported that some tumor cells exhibit a survival dependent on glutamine.

2.5 *Glutamine Metabolism*

Glutamine as one of the nonessential amino acids is a highly abundant amino acid compared to others in plasma and intracellular pools, and cancer cells utilize it actively for energy metabolism. Glutamine is a versatile substrate in cancer cells and plays a key role in tumor development [54, 55]. It can be used for synthesizing protein, nucleotide, and lipid, generating energy, and providing the precursor for glutathione and other nonessential amino acids. Glutamine metabolism is regulated by MYC, TP53, KRAS, SIRT4, and mTOR, that is to say, tumor genetics can affect cancer proliferation by regulating glutamine [56]. Additionally, many cancer cells can use glutamine to meet their metabolic needs for proliferation. Glutamine catabolism can generate the precursors for de novo lipid biosynthesis and TCA cycle. The consumption of citrate from the TCA cycle creates a need for anaplerotic replenishment of the cycle, which can be provided through oxidative metabolism of glutamine [57]. Glutaminase (GLS) catalyzing glutamate is the first step of mitochondrial oxidation of glutamine. The crucial enzyme in glutamine metabolism may be the target of cancer therapy. There are three mammalian glutaminase isoforms: kidney (K-type) glutaminase (KGA), liver (L-type) glutaminase (LGA), and glutaminase C (GAC). KGA and GAC are encoded by GLS, while LGA is encoded by GLS2. Transaminases and glutamate dehydrogenase (GDH) can convert glutamate into α-ketoglutarate and used in TCA cycle. Furthermore, glutamate can act as a precursor of GSH and nonessential amino acids. Interestingly, levels of α-ketoglutarate play a key role in the regulation of HIF-1α degradation through prolyl hydroxylase (PHD) sensing pathway [58]. During the activation of mTOR signaling, glutaminolysis and α-ketoglutarate are also involved in. In the form of malic acid, glutamine carbons can leave the TCA cycle, which can be converted to pyruvate by malic enzyme (ME) and product NADPH [59]. Both glutamine-derived NADPH and GSH production can reduce oxidative stress associated with mitochondrial respiration and rapid cell proliferation of tumor cells.

It is worth noting that glutamine is used in TCA cycle as a respiratory substrate that produces NADH and FADH2, which can generate ATP through providing electrons for the mitochondrial electron transport chain. Remarkably, the contribution of glycolytic to total ATP synthesis differs widely depending on cell type in tumor cells, ranging from over 60% to less than 1%, and the average contribution in the tested cell lines is 17 ± 18% [60]. Analysis of a flux balance across the NCI-60 cell lines has confirmed the results, suggesting that OXPHOS contributes to 70–84% of the total cellular ATP production [ref]. Thus, in many cancer cell lines, oxidative metabolism of glutamine can be a major energetic source. Together, glucose and glutamine supply both ATP and key precursors for protein, lipid, and nucleic acid biosynthesis to coordinately fuel the proliferation of tumor cells. In fact, depending on nutrient availability, some cancer cells can switch their carbon source. For example, in MYC-transformed glioblastoma cells, glucose withdrawal increases GDH activity [61] and inhibits the oxidative metabolism of glutamine by impairing glutamine. The new upregulated pathways provide metabolic flexibility and compensatory abilities in some tumor cells. Therefore, these pathways should be carefully considered in cancer combined therapies.

2.6 Glutamine Reductive Carboxylation

Acetyl-CoA for fatty acid biosynthesis can be derived from glutamine via two different pathways. The deamination of glutamate produces α-ketoglutarate which can either be converted to citrate by reductive carboxylation or to fumarate and malate.

Citrate thus generated is then transported to the cytoplasm and dissociated into oxaloacetate and acetyl-CoA for fat synthesis. On the other hand, malate produced in the TCA cycle can generate pyruvate by the role of malic enzyme, which in turn be further converted into acetyl-CoA for fatty acid metabolism. The use of aconitine and reversible transformation of catalytic isocitrate dehydrogenase. The cytosolic NADP+/NADPH-dependent isocitrate dehydrogenase 1 (IDH1) is the main enzyme catalyzing the reversible reductive carboxylation of α-ketoglutarate to isocitrate and NADP+. The reductive carboxylation of α-ketoglutarate is an important source for acetyl-CoA in some cancer cell especially under hypoxia [62]. In fact, the reduction of glutamine carboxylation for biosynthesis of acetyl-CoA is independent of glucose availability, allowing maximum use of glucose for biosynthesis of other metabolic intermediates [62].

2.7 *Mitochondrial Metabolism*

The function of mitochondria is essential to the development of tumor, like ATP production, apoptosis, ROS-mediated signal pathway change, Ca^{2+} channels, and the transport of small molecule metabolites. Indeed, mitochondrial function and mitochondrial biogenesis are increasingly recognized to play an important role in cancer formation and proliferation. Mitochondrial dysfunction is also involved in muscular degeneration and cardiovascular dysfunctions besides cancer [63]. In normal cells with sufficient oxygen, OXPHOS is the main way for energy generation. Mitochondrial respiratory chain includes four complexes (I to IV) and is located in the mitochondrial membrane. Its function is to transfer electrons and H^+ and generate ATP, CO_2, and water [64].

In all metabolic pathways that occurred in the mitochondria, TCA cycle is particularly important, especially for OXPHOS. The main function of TCA cycle is generation of the precursors for lipid, nucleic acid, and protein synthesis. The key enzyme mutations (such as succinate dehydrogenase, isocitrate dehydrogenase, and fumarate hydratase) in the TCA cycle are closely related to the tumor development, the mitochondrial dysfunction, and the accumulation of metabolites. The intermediates succinate and fumarate generated by TCA cycle can affect DNA demethylases and α-ketoglutarate-dependent histone and the stabilization of HIF. Isocitrate dehydrogenases (IDHs) catalyze isocitrate to alpha-ketoglutaric acid by oxidative decarboxylation. There are three kinds of IDH enzyme; IDH1 is found in the cytosol and peroxidase, and IDH2 and IDH3 are located in the mitochondria. Targeted IDH1 and IDH2 mutations can lead to the loss of natural enzyme activity and the production of 2-hydroxyglutaric acid, which can be competitive inhibition of α-ketoglutarate-dependent dioxygenases and be associated with tumorigenesis [65–67].

3 Role of Metabolism Alteration and Anticancer Drug Resistance

Metabolic phenotypes are affected by genetic and epigenetic factors and have been correlated with drug activity and clinical outcome, supporting the hypothesis that chemical resistance results from altered cancer metabolism. Metabolic data from in vitro studies, proteomics platforms, and ^{13}C metabolic flux analysis (MFA) have provided a better understanding of the complex metabolic mechanisms of cancer which could guide selection of therapeutic molecular targets [68]. These issues are discussed further in the following paragraphs from the perspective of PDAC.

3.1 Metabolism-Mediated Modulation of Survival and/or Apoptosis Pathways

Metabolic reprogramming of tumor metabolism can affect cancer cell differentiation, proliferation, and/or apoptosis, as well as its therapeutic responses. In particular, several regulatory enzymes in glycolysis could induce the occurrence of drug resistance in cancer. In the first step of the glycolysis pathway, the enzyme hexokinase (HK) plays a vital catalytic role. In two HK isoforms known, cytoplasmic HK1 and mitochondrial HK2, the latter is found to be upregulated in many cancers and can directly insert in the mitochondrial outer membrane to suppress mitochondrial apoptosis. HK2 is most highly expressed in PDAC metastases, suggesting a link between HK2 and pancreatic aggressive tumor biology [69]. Moreover, in the PI3K/Akt/mTOR pathway and other survival pathways, HK2 not only activates HK in cancer cells but induces drug resistance as well. Because HK2 can regulate cellular apoptosis and bioenergetics, it is considered to be an important anticancer drug target. 3-Bromopyrrole is a HK2 inhibitor, which can reduce ATP reserves and thereby reverse chemoresistance [70]. In contrast, increased levels of ATP due to accelerated glycolysis activate HIF-1 and confer drug resistance. Fructose-bisphosphate aldolase (FBA) catalyzes fructose 1,6-bisphosphate to glyceraldehyde-3-phosphate (G3P), and dihydroxyacetone phosphate is another metabolic enzyme overexpressed in PDAC. Overexpression of FBA delays induction of apoptosis, as does G3P, by suppressing caspase-3 activity. Additionally, glyceraldehyde-3-phosphate dehydrogenase (GAPDH) overexpression may stimulate glycolysis, increase cellular ATP levels, and promote autophagy, thereby preventing caspase-independent cell death. In the glycolytic pathway, PKM2 is one of the rate-limiting enzymes, which converts phosphoenolpyruvate (PEP) and ADP into pyruvate and ATP. PK is a tetrameric enzyme composed of homotetramers or hybrid forms of four isotypes (M1, M2, L, and R) being differentially expressed in different cell types. PKM2 is highly expressed in cancers, conceivably to drive higher glycolytic fluxes because of its high substrate affinity for PEP, and maintains high lactate levels, which may induce carcinogenesis. A recent study has shown that lactic acid in hypoxic conditions can lead to activation of survival pathways. In addition, lactate-dependent stabilization was shown that the combined protein NDRG3 and c-Raf can promote neovascularization and survival [71]. In addition to promoting cell survival, lactate can relieve immune signaling, and in particular, tumor-derived lactate can prevent the response of human T cell, which dominates the immune infiltration in human PDAC. Of further notice, inhibiting the monocarboxylic acid family of transporters including lactate transporters has been considered to be a new cancer treatment method, including PDAC (Fig. 11.3).

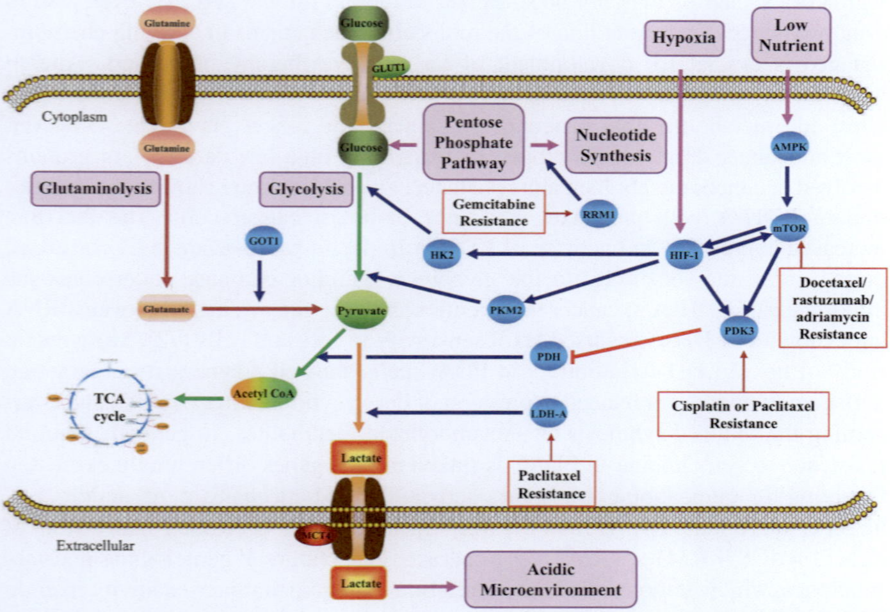

Fig. 11.3 Summary of metabolic reprogramming and drug resistance. Metabolic reprogramming and crucial signaling pathway are characterized by enhanced glycolysis, PPP, and glutaminolysis, among others. These suggested cancer cells with not only essential energy but also important precursors to supply large-scale biosynthesis, rapid proliferation, growth, invasion, metastasis, and resistance to anticancer therapies

Abbreviations: GOT1, cytosolic aspartate transaminase; AMPK, AMP-activated protein kinase; KRAS, V-Ki-ras2 Kirsten rat sarcoma viral oncogene homolog; HK2, hexokinase 2; PPP, pentose phosphate pathway; TCA, tricarboxylic acid cycle; GLS1, glutaminase 1; GLUD, glutamate dehydrogenase; HIF-1a, hypoxia-inducible factor 1-alpha; MCT, monocarboxylic acid transporter; LDH-A, lactate dehydrogenase A; mTOR, mammalian target of rapamycin; SLC1A5, neutral amino acid transporter B; PK, pyruvate kinase; PDH, pyruvate dehydrogenase; RRM1, ribonucleoside-diphosphate reductase large subunit; PDK, pyruvate dehydrogenase kinase

3.2 Metabolism-Mediated Regulation of Drug Targets, Transport, and Catabolism

At present, metabolic adaptation is a distinctive feature of cancer cells, and accumulating studies support that metabolic alterations are linked to cancer drug resistance. Targeting such adaptations has suggested new anticancer strategies to overcome chemoresistance. However, the Warburg effect involves complex control of the expression of multiple genes and pathways. Targeting one target or segment of the regulatory scenarios may not be sufficient to repress tumors and may even lead to drug resistance. Understanding of the molecular mechanisms underlying chemoresistance is crucial for development of an effective therapy of cancers. Altered metabolism comprises a multifactorial process of concerted action of genes, proteins, and metabolites that generates a characteristic cancer phenotype. However, most anticancer drugs are developed focusing only on a few pathways or proteins involved in cancer metabolism and resistance. One recent study showed the association of LDH-A with paclitaxel resistance in breast cancer cells. The fact that increased expression and activity of LDH-A in paclitaxel-resistant cells correlated directly with the sensitivity to the glycolysis inhibitor oxamate underscores the importance of LDH-A in cancer therapeutics and drug sensitivity. Moreover, siRNA knockdown of LDH-A reversed taxol sensitivity in resistant cells [72]. More recent study of novel LDH-A inhibitors in PDAC cells showed a synergistic interaction with gemcitabine by enhanced expression of deoxycytidine kinase (dCK) and overcoming the reduced synthesis of phosphorylated metabolites. In general, acquired resistance to gemcitabine in PDAC is linked to four genes differentially expressed involving in gemcitabine transport, activation, and mechanism of action, i.e., hENT1, dCK, and RRM1 and RRM2. Specifically, a decreased expression of hENT1 × dCK/RRM1 × RRM2 was a characteristic feature of gemcitabine-resistant subclones, which demonstrated the correlation with gemcitabine sensitivity in eight PDAC cell lines, whereas no single gene expression level correlated with enhanced sensitivity [73]. Another enzyme that has a critical role in metabolism-mediated resistance is pyruvate dehydrogenase kinase 3 (PDK3), which catalyzes the first step of OXPHOS. PDK3 upregulation may contribute to hypoxia-induced drug resistance in cervical and colon cancer. In a hypoxic environment, HIF-1 induces PDK3 expression by binding to the promoter of PDK3, resulting in conversion from mitochondrial respiration to glycolysis for energy production. Induction of hypoxia-mediated PDK3 or forced PDK3 overexpression can inhibit cell apoptosis and increase resistance to cisplatin or paclitaxel in colorectal cancer. Finally, fatty acid synthase (FASN) has been linked to acquired docetaxel/trastuzumab/Adriamycin resistance in breast cancer and intrinsic gemcitabine and radiation resistance in PDAC [74]. In pancreatic tumors, a previous study demonstrated that overexpression of FASN can lead to resistance to chemo- or radiotherapy. FASN is usually overexpressed in PDAC cells. Inhibition of fatty acid synthesis by an siRNA or FASN inhibitor reduces gemcitabine resistance, while ectopic overexpression of FASN contributes to intrinsic resistance to gemcitabine and radiation. The radiation-induced ceramide production leads to a reduction of caspase-8 and FASN-induced radiation resistance. The precise mechanism of FASN-induced gemcitabine resistance needs to be elucidated. In gastric cancer glutaminolysis, FASN was associated to cisplatin resistance, which is mediated by activation of mammalian rapamycin complex 1 (mTORC1) signaling [75]. Notably, in vitro and in vivo experiments showed that AZD8055 and erlotinib combined therapy significantly inhibited the mTORC1/C2 signaling pathway, EGFR/AKT feedback activation, and cell growth, as well as suppressed the progression of PDAC in a xenograft model. Therefore, new combinations of agents targeting these pathways might be necessary for overcoming chemoresistance caused by metabolic aberrations.

4 Critical Pathways and Targets in Cancer Metabolism

4.1 HIF-1

Adequate and continuous oxygen as a terminal electron acceptor is necessary for human mitochondrial respiratory chains that generate ATP to power most biochemical reactions. The intracellular O_2 concentration is tightly regulated. However, in cancer cells, dysfunctions in the regulatory pathways, e.g., HIF-1, are common. The regulator HIF-1 is indispensable for cell proliferation and survival, as well as glucose and iron metabolism. HIF-1 stability, subcellular localization, and transcriptional activity are closely related to oxygen levels [76]. Hypoxia leads to changes in the redox state of mitochondria, which directly increases the production and excretion of lactic acid. In hypoxia, HIF-1 induced p21 upregulation to reduce or stop proliferation, but in some cancers, mTOR or Notch pathway activation can maintain cell proliferation. With hypoxia, in PDAC, the expression of HIF-1 is increased, just as the expression of glucose metabolic enzymes PDK1, LDH-A, and PKM2. Accordingly, knockdown of HIF-1 under hypoxic conditions inhibited the production of lactate and the expression of PDK1, LDH-A, and PKM2. Knockdown of HIF-1 under hypoxia repressed the growth of the pancreatic cells BxPC-3 along with induction of apoptosis [77].

HIF-1 stability and activity in hypoxic conditions can be upregulated by almost all glycolytic genes and monocarboxylic acid transporters that produce lactic acid and even enhanced reprogramming of glycolytic metabolism. HIF-1 has also been shown to upregulate the expression of genes encoding the glucose transporters Glut1 and Glut3, glycolytic enzymes such as the hexokinases HK1 aldolase A and C, and GAPDH. HIF-1 could induce adaptation responses such as angiogenesis and anaerobic metabolism to promote cell survival. Consistently, constitutive expression of HIF-1 confers apoptosis resistance in PDAC cells. In the PDAC cells PCI-35, with constitutive HIF-1 expression, also Glut1 and aldolase A mRNAs were more abundantly expressed, thereby facilitating increased anaerobic metabolism and apoptosis resistance under conditions of hypoxia and glucose deprivation. Another key enzyme upregulated by hypoxia-induced HIF-1 activity, and altering pyruvate metabolism, is PDK1. HIF-1 mediates pyruvate dehydrogenase complex inactivation, causing subsequent loss of pyruvate oxidation [78]. In hypoxia conditions, the inhibition of the pyruvate dehydrogenase complex seems to be a protective mechanism, because it has recently been shown that activation of this enzyme complex by oncogenes is a key driver of oncogene-induced senescence through increased oxygen consumption and redox stress. It is interesting to note that inhibition of glycogen metabolism as well as one or more specific glucose metabolic pathways also induces cancer senescence. The importance of glucose metabolism is further illustrated by the fact that inhibition of PDK1 expression impairs cell growth and increases oxygen consumption and cell death under hypoxia in human cancer cell lines.

4.2 LDH

LDH catalyzes the reversible transformation of pyruvate to lactate under anaerobic conditions, coupled with the oxidation of NADH to NAD^+. LDH is a tetrameric enzyme consisting of two types of subunits designated M (LDH-A gene product) and H (LDH-B gene product). There are five different LDH isozymes in human cells, as a result of different hybrid forms of H and/or M subunits, LDH-1 (H4), LDH-2 (MH3), LDH-3 (M2H2), LDH-4 (M3H), and LDH-5 (M4), with LDH-A identified as a target of both c-Myc and HIF-1. LDH plays an essential role by catalyzing the anaerobic glycolysis through the regulation of glycolysis; usually, its upregulation is in favor of the efficiency of anaerobic glycolysis in tumor cells and allows ATP production in absence of oxygen. Of all LDH enzyme forms, LDH-5 is the predominant isoform found in skeletal muscle and other highly glycolytic tissues, including tumor tissues, and has the highest efficiency to catalyze the conversion of pyruvate to lactate. LDH-5 is mainly existed in the cytoplasm, where it participates in glucose metabolism. Several studies have illustrated the prognostic relevance of LDH in different tumor types including PDAC [79]. In tissue and xenograft studies, it was demonstrated that inhibition of LDH-A activity due to lysine 5 acetylation is reduced in human PDAC, thereby underlining its role in PDAC initiation and as a potential new target.

The potential oncogenic activity of LDH-A has also been reported in esophageal carcinoma and gastric cancer. Of further notice, the HIF-1-containing function must bind the two conserved hypoxia response elements (HREs) of the site, which strongly indicates the oxygen-dependent regulation of LDH-5 activity. The relevance of LDH is further supported by elevated expression levels of LDH-A observed in PDAC samples compared with the matched normal tissues and by the fact that LDH-A promoted the growth of the PDAC, both in vitro and in vivo. These results encourage further LDH-A-directed therapeutic interventions for PDAC [80].

4.3 mTOR

Mammalian target of rapamycin (mTOR) is a serine/threonine kinase, frequently activated in human cancers, including PDAC. Whether or not the mTOR gene acquires oncogenic properties through somatic mutations has remained unclear. However, KRAS signaling, including PI3K/Akt, links ligation of growth factor receptors to the phosphorylation and activation of mTOR [81]. mTOR engages a signaling program downstream from nutrient availability to stimulate metabolism leading to cell cycle progression. mTOR exists as two complexes: mTORC1, being sensitive to rapamycin inhibition, and mTORC2, being largely rapamycin inhibition insensitive. mTORC1 enhances the translation of mRNAs by interacting with the helper protein Raptor to phosphorylate S6 kinase 1, which includes ribosomal proteins, elongation factors, and insulin growth factor 2. mTOR signaling defects can cause both metabolic disorders and cancer, which indicates that mTOR is critical in cancer development and metabolism. The notion is supported by the observation that metformin inhibits mTORC1signaling by activating AMPK and REDD1 and a Rag GTPase-sensitive mechanism, in addition to suppressing cancer. It has been proposed that mTORC1 controls cell proliferation exclusively via 4E-BP while it regulates cell growth via S6K [82].

A recent study has shown that multi-dynamic mTORC1 signaling causes LTsc1KO mice to exhibit metabolic abnormalities, including defects in glucose and lipid homeostasis, and subsequently develop hepatocarcinoma. Glutaminolysis constitutes another mTOR link between metabolism and cancer. Highly proliferating cancer cells are typically glutamine-dependent, and tumor growth is associated with the activity of glutaminase (GLS), the enzyme that catalyzes the first step of glutaminolysis. Lastly, it has been shown that glutaminolysis can also activate mTORC1, thus promoting cell growth and inhibiting autophagy. Of note, mTOR activity also impacts expression of HIF-1, probably through the activation of S6K. As a consequence, inhibition of mTOR by rapamycin also suppresses HIF-1 expression [83]. Remarkably, several preclinical data have demonstrated that inhibition of mTOR in specific KRAS-dependent PDAC genetic subtypes leads to inhibition of tumorigenesis in vitro and in vivo. However, phase II trials of anti-mTOR regimens have not shown positive results [84]. Coordinated inhibition of mTOR and other steps along the mTOR signaling pathway, including critical factors in tumor metabolism, may lead to better responses in overcoming tumor chemoresistance.

5 Conclusion and Perspective

The mitochondria have occupied a central position in cancer biology research due to their essential roles in cancer initiation, growth, survival, metastasis, relapse, and acquired drug resistance. Mirroring the heterogeneity of cancer, mitochondria have diversity in their structure, DNA, energy metabolism, and ion channels, depending on the spatiotemporal conditions and types of cancer. With the changes in mitochondrial function, there are associated changes of mitochondria-dependent metabolites which are clinical biomarkers for monitoring and predicting drug resistance and efficacy [85–87]. Several hundreds of meaningful mitochondrial targets have been considered for cancer therapies at the same time, and there is no single master key target for the treatment of different types of cancers. Thus, for the successful development of mitochondria-targeting cancer treatments, we need to establish three main axes: (1) personalized diagnostic tools to determine the genetic and biological conditions of the cancer and mitochondria, e.g., mutation of nuclear and mtDNA, mass of mtDNA and mitochondria, quality of mitochondrial OXPHOS, levels of mitochondrial metabolites and ROS, and $\Delta\Psi m$; (2) development and optimization of novel therapeutic tools, e.g., drug repositioning, inhibitors/activators of specific pathways, genetic engineering techniques, and mitochondria-targeting anti- or prooxidants, overcoming drug resistance; and (3) cancer mitochondria-targeting delivery systems, e.g., targeting peptides, lipophilic cations, nanoparticles, nanovesicles, and nano-liposomes.

Acknowledgments The work was supported by the Dalian University of Technology Distinguished Professor Grant (Gary Guishan Xiao), the National Natural Science Foundation of China (81770846, 81642006, 81272430), Agi Hirshberg international pancreatic disease center funds (H2015PC01), talent introduction funds of Dalian University of Technology (852004), and major research projects of Dalian University of Technology (DUT17ZD308).

References

1. Taylor RW, Turnbull DM. Mitochondrial DNA mutations in human disease. Nat Rev Genet. 2005;6(5):389–402. [PubMed: 15861210]
2. Falk MJ, Sondheimer N. Mitochondrial genetic diseases. Curr Opin Pediatr. 2010;22(6):711–6. [PubMed: 21045694]
3. Mullen AR, Wheaton WW, Jin ES, Chen PH, Sullivan LB, Cheng T, Yang Y, Linehan WM, Chandel NS, DeBerardinis RJ. Suppression of human tumor cell proliferation through mitochondrial targeting. FASEB J. 2002;16(9):1010–6. [PubMed: 12087062]
4. Mullen AR, Wheaton WW, Jin ES, Chen PH, Sullivan LB, Cheng T, et al. Reductive carboxylation supports growth in tumour cells with defective mitochondria. Nature. 2011;481(7381):385–8. [PubMed: 22101431]
5. Tan AS, Baty JW, Dong LF, Bezawork-Geleta A, Endaya B, Goodwin J, Bajzikova M, Kovarova J, Peterka M, Yan B, Bezawork-Geleta A, Endaya B, Goodwin J, et al. Mitochondrial genome acquisition restores respiratory function and tumori-genic potential of cancer cells without mitochondrial DNA. Cell Metab. 2015;21(1):81–94. [PubMed: 25565207]
6. Bao L, Zhang Y, Wang J, Wang H, Dong N, Su X, Xu M, Wang X. Variations of chromosome 2 gene expressions among patients with lung cancer or non-cancer. Cell Biol Toxicol. 2016;32(5):419–35. [PMID: 27301951]
7. Possemato R, Marks KM, Shaul YD, Pacold ME, Kim D, Birsoy K, Sethumadhavan S, Woo HK, Jang HG, Jha AK, et al. Functional genomics reveal that the serine synthesis pathway is essential in breast cancer. Nature. 2011;476(7360):346–50. [PubMed: 21760589]

8. Sellers K, Fox MP, Bousamra M 2nd, Slone SP, Higashi RM, Miller DM, Wang Y, Yan J, Yuneva MO, Deshpande R, et al. Pyruvate carboxylase is critical for non-small-cell lung cancer proliferation. J Clin Invest. 2015;125(2):687–98. [PubMed: 25607840]
9. Shi L, Wang LY, Wang BB, Cretoiu SM, Wang Q, Wang XD, Chen CS. Regulatory mechanism of betacellulin in CXCL8 production from lung cancer cells through EGFR-PI3K pathway. J Transl Med. 2014;12:70. [PMID: 24629040]
10. Pelicano H, Martin DS, RH X, Huang P. Glycolysis inhibition for anticancer treatment. Oncogene. 2006;25(34):4633–46. [PubMed: 16892078]
11. Levine AJ, Puzio-Kuter AM. The control of the metabolic switch in cancers by oncogenes and tumor suppressor genes. Science. 2010;330(6009):1340–4. [PubMed: 21127244]
12. Bonuccelli G, Tsirigos A, Whitaker-Menezes D, Pavlides S, Pestell RG, Chiavarina B, Frank PG, Flomenberg N, Howell A, Martinez-Outschoorn UE, et al. Ketones and lactate "fuel" tumor growth and metastasis: evidence that epithelial cancer cells use oxidative mitochondrial metabolism. Cell Cycle. 2010;9(17):3506–14. [PubMed: 20818174]
13. Xu RH, Pelicano H, Zhou Y, Carew JS, Feng L, Bhalla KN, Keating MJ, Huang P. Inhibition of glycolysis in cancer cells: a novel strategy to overcome drug resistance associated with mitochondrial respiratory defect and hypoxia. Cancer Res. 2005;65(2):613–21. [PubMed: 15695406]
14. Thomas PD, Kahn M. Kat3 coactivators in somatic stem cells and cancer stem cells: biological roles, evolution, and pharmacologic manipulation. Cell Biol Toxicol. 2016;32(1):61–81. [PMID: 27008332]
15. DeBerardinis RJ, Chandel NS. Fundamentals of cancer metabolism. Sci Adv. 2016;2(5):e1600200. [PubMed: 27386546]
16. Seo JB, Jung SR, Hille B, Koh DS, Extracellular ATP. protects pancreatic duct epithelial cells from alcohol-induced damage through P2Y1 receptor-cAMP signal pathway. Cell Biol Toxicol. 2016;32(3):229–47. [PMID: 27197531]
17. Hardie DG, Schaffer BE, Brunet A. AMPK: an energy-sensing pathway with multiple inputs and outputs. Trends Cell Biol. 2016;26(3):190–201. [PubMed: 26616193]
18. DeBerardinis RJ, Sayed N, Ditsworth D, Thompson CB. Brick by brick: metabolism and tumor cell growth. Curr Opin Genet Dev. 2008;18:54–61. [PubMed: 18387799]
19. Cairns RA, Harris IS, Mak TW. Regulation of cancer cell metabolism. Nat Rev Cancer. 2011;11:85–95. [PubMed: 21258394]
20. Warburg O, Posener K, Negelein E, Über den Stoffwechsel d. Karzinomzellen. Biochem Z. 1924;152:309–44.
21. Warburg O. On the origin of cancer cells. Science (New York, NY). 1956;123:309–14. [PubMed: 13298683]
22. Semenza GL, Artemov D, Bedi A, Bhujwalla Z, Chiles K, Feldser D, Laughner E, Ravi R, Simons J, Taghavi P, et al. 'The metabolism of tumours': 70 years later. Novartis Found Symp. 2001;240:251–60. discussion 260- 254. [PubMed: 11727934]
23. Gogvadze V, Zhivotovsky B, Orrenius S. The Warburg effect and mitochondrial stability in cancer cells. Mol Asp Med. 2010;31:60–74. [PubMed: 19995572]
24. Kroemer G, Pouyssegur J. Tumor cell metabolism: cancer's Achilles' heel. Cancer Cell. 2008;13:472–82. [PubMed: 18538731]
25. Salway JG. Metabolism at a glance. Oxford: Blackwell Science; 2000.
26. Vander Heiden MG, Cantley LC, Thompson CB. Understanding the Warburg effect: the metabolic requirements of cell proliferation. Science (New York, NY). 2009;324:1029–33. [PubMed: 19460998]
27. Som P, Atkins HL, Bandoypadhyay D, Fowler JS, MacGregor RR, Matsui K, Oster ZH, Sacker DF, Shiue CY, Turner H, et al. A fluorinated glucose analog, 2-fluoro-2-deoxy-D-glucose (F- 18): nontoxic tracer for rapid tumor detection. J Nucl Med. 1980;21:670–5. [PubMed: 7391842]

28. Courtnay R, Ngo DC, Malik N, Ververis K, Tortorella SM, Karagiannis TC. Cancer metabolism and the Warburg effect: the role of HIF-1 and PI3K. Mol Biol Rep. 2015;42:841–51. [PubMed: 25689954]
29. Bayley JP, Devilee P. The Warburg effect in 2012. Curr Opin Oncol. 2012;24:62–7. [PubMed: 22123234]
30. Yeung SJ, Pan J, Lee MH. Roles of p53, MYC and HIF- 1 in regulating glycolysis – the seventh hallmark of cancer. Cell Mol Life Sci. 2008;65:3981–99. [PubMed: 18766298]
31. Li J, Zhu S, Tong J, Hao H, Yang J, Liu Z, Wang Y. Suppression of lactate dehydrogenase a compromises tumor progression by downregulation of the Warburg effect in glioblastoma. Neuroreport. 2016;27:110–5. [PubMed: 26694942]
32. Liberti MV, Locasale JW. The Warburg effect: how does it benefit cancer cells? Trends Biochem Sci. 2016;41:211–8. [PubMed: 26778478]
33. Yamamoto T, Seino Y, Fukumoto H, Koh G, Yano H, Inagaki N, Yamada Y, Inoue K, Manabe T, Imura H. Over-expression of facilitative glucose transporter genes in human cancer. Biochem Biophys Res Commun. 1990;170:223–30. [PubMed: 2372287]
34. Altenberg B, Greulich KO. Genes of glycolysis are ubiquitously overexpressed in 24 cancer classes. Genomics. 2004;84:1014–20. [PubMed: 15533718]
35. Levine AJ, Puzio-Kuter AM. The control of the metabolic switch in cancers by oncogenes and tumor suppressor genes. Science (New York, NY). 2010;330:1340–4. [PubMed: 21127244]
36. Cerella C, Dicato M, Diederich M. Modulatory roles of glycolytic enzymes in cell death. Biochem Pharmacol. 2014;92:22–30. [PubMed: 25034412]
37. Lippai M, Szatmári Z. Autophagy-from molecular mechanisms to clinical relevance. Cell Biol Toxicol. 2017;33(2):145–68. [PMID: 27957648]
38. Pastorino JG, Shulga N, Hoek JB. Mitochondrial binding of hexokinase II inhibits Bax-induced cytochrome c release and apoptosis. J Biol Chem. 2002;277:7610–8. [PubMed: 11751859]
39. Patra KC, Hay N. The pentose phosphate pathway and cancer. Trends Biochem Sci. 2014;39:347–54. [PubMed: 25037503]
40. Boros LG, Puigjaner J, Cascante M, Lee WN, Brandes JL, Bassilian S, Yusuf FI, Williams RD, Muscarella P, Melvin WS, et al. Oxythiamine and dehydroepiandrosterone inhibit the nonoxidative synthesis of ribose and tumor cell proliferation. Cancer Res. 1997;57:4242–8. [PubMed: 9331084]
41. Tudzarova S, Colombo SL, Stoeber K, Carcamo S, Williams GH, Moncada S. Two ubiquitin ligases, APC/C-Cdh1 and SKP1-CUL1-F (SCF)-beta-TrCP, sequentially regulate glycolysis during the cell cycle. Proc Natl Acad Sci U S A. 2011;108:5278–83. [PubMed: 21402913]
42. Santos CR, Schulze A. Lipid metabolism in cancer. FEBS J. 2012;279:2610–23. [PubMed: 22621751]
43. Porstmann T, Griffiths B, Chung YL, Delpuech O, Griffiths JR, Downward J, Schulze A. PKB/Akt induces transcription of enzymes involved in cholesterol and fatty acid biosynthesis via activation of SREBP. Oncogene. 2005;24:6465–81. [PubMed: 16007182]
44. Yuan HX, Xiong Y, Guan KL. Nutrient sensing, metabolism, and cell growth control. Mol Cell. 2013;49:379–87. [PubMed: 23395268]
45. Tsun ZY, Possemato R. Amino acid management in cancer. Semin Cell Dev Biol. 2015;43:22–32. [PubMed: 26277542]
46. P G, Tchernyshyov I, Chang TC, Lee YS, Kita K, Ochi T, Zeller KI, De Marzo AM, Van Eyk JE, Mendell JT, et al. c-Myc suppression of miR-23a/b enhances mitochondrial glutaminase expression and glutamine metabolism. Nature. 2009;458:762–5. [PubMed: 19219026]
47. Kim J, Lee JH, Iyer VR. Global identification of Myc target genes reveals its direct role in mitochondrial biogenesis and its E-box usage in vivo. PLoS One. 2008;3:e1798. [PubMed: 18335064]
48. Zhu M, Wang N, Tsao SW, Yuen MF, Feng Y, Wan TS, Man K. Up-regulation of microRNAs, miR21 and miR23a in human liver cancer cells treated with Coptidis rhizoma aqueous extract. Exp Ther Med. 2011;2:27–32. [PubMed: 22977465]

49. Nicklin P, Bergman P, Zhang B, Triantafellow E, Wang H, Nyfeler B, Yang H, Hild M, Kung C, Wilson C, Myer VE, MacKeigan JP, Porter JA, et al. Bidirectional transport of amino acids regulates mTOR and autophagy. Cell. 2009;136:521–34. [PubMed: 19203585]
50. Locasale JW. Serine, glycine and one-carbon units: cancer metabolism in full circle. Nat Rev Cancer. 2013;13:572–83. [PubMed: 23822983]
51. di Salvo ML, Contestabile R, Paiardini A, Maras B. Glycine consumption and mitochondrial serine hydroxymethyltransferase in cancer cells: the heme connection. Med Hypotheses. 2013;80:633–6. [PubMed: 23474074]
52. Phang JM, Liu W, Hancock CN, Fischer JW. Proline metabolism and cancer: emerging links to glutamine and collagen. Curr Opin Clin Nutr Metab Care. 2015;18:71–7. [PubMed: 25474014]
53. Liu W, Phang JM. Proline dehydrogenase (oxidase) in cancer. Biofactors. 2012;38:398–406. [PubMed: 22886911]
54. Daye D, Wellen KE. Metabolic reprogramming in cancer: unraveling the role of glutamine in tumorigenesis. Semin Cell Dev Biol. 2012;23:362–9. [PubMed: 22349059]
55. Wise DR, Thompson CB. Glutamine addiction: a new therapeutic target in cancer. Trends Biochem Sci. 2010;35:427–33. [PubMed: 20570523]
56. MO Y, Fan TW, Allen TD, Higashi RM, Ferraris DV, Tsukamoto T, Matés JM, Alonso FJ, Wang C, Seo Y, et al. The metabolic profile of tumors depends on both the responsible genetic lesion and tissue type. Cell Metab. 2012;15:157–70. [PubMed: 22326218]
57. Iacobazzi V, Infantino V. Citrate—new functions for an old metabolite. Biol Chem. 2014;395:387–99. [PubMed: 24445237]
58. Boulahbel H, Duran RV, Gottlieb E. Prolyl hydroxylases as regulators of cell metabolism. Biochem Soc Trans. 2009;37:291–4. [PubMed: 19143649]
59. DeBerardinis RJ, Mancuso A, Daikhin E, Nissim I, Yudkoff M, Wehrli S, Thompson CB. Beyond aerobic glycolysis: transformed cells can engage in glutamine metabolism that exceeds the requirement for protein and nucleotide synthesis. Proc Natl Acad Sci U S A. 2007;104:19345–50. [PubMed: 18032601]
60. XL Z, Guppy M. Cancer metabolism: facts, fantasy, and fiction. Biochem Biophys Res Commun. 2004;313:459–65. [PubMed: 14697210]
61. Yang C, Sudderth J, Dang T, Bachoo RM, McDonald JG, DeBerardinis RJ. Glioblastoma cells require glutamate dehydrogenase to survive impairments of glucose metabolism or Akt signaling. Cancer Res. 2009;69:7986–93. [PubMed: 19826036]
62. Metallo CM, Gameiro PA, Bell EL, Mattaini KR, Yang J, Hiller K, Jewell CM, Johnson ZR, Irvine DJ, Guarente L, et al. Reductive glutamine metabolism by IDH1 mediates lipogenesis under hypoxia. Nature. 2012;481:380–4. [PubMed: 22101433]
63. Wallace DC. Mitochondria and cancer. Nat Rev Cancer. 2012;12:685–98. [PubMed: 23001348]
64. Acin-Perez R, Enriquez JA. The function of the respiratory supercomplexes: the plasticity model. Biochim Biophys Acta. 2014;1837:444–50. [PubMed: 24368156]
65. Xu W, Yang H, Liu Y, Yang Y, Wang P, Kim SH, Ito S, Yang C, Wang P, Xiao MT, et al. Oncometabolite 2-hydroxyglutarate is a competitive inhibitor of alpha-ketoglutarate-dependent dioxygenases. Cancer Cell. 2011;19:17–30. [PubMed: 21251613]
66. Zhao S, Lin Y, Xu W, Jiang W, Zha Z, Wang P, Yu W, Li Z, Gong L, Peng Y, et al. Glioma-derived mutations in IDH1 dominantly inhibit IDH1 catalytic activity and induce HIF-1alpha. Science (New York, NY). 2009;324:261–5. [PubMed: 19359588]
67. Losman JA, Kaelin WG Jr. What a difference a hydroxyl makes: mutant IDH, (R)-2-hydroxyglutarate, and cancer. Genes Dev. 2013;27:836–52. [PubMed: 23630074]
68. Rahman M, Hasan MR. Cancer metabolism and drug resistance. Metabolites. 2015;5:571–600. [PubMed: 26437434]
69. Anderson M, Marayati R, Moffitt R, Yeh JJ. Hexokinase 2 promotes tumor growth and metastasis by regulating lactate production in pancreatic cancer. Oncotarget. 2016;2:3762–3. [PubMed: 27259263]
70. Colell A, Ricci JE, Tait S, Milasta S, Maurer U, Bouchier-Hayes L, Fitzgerald P, Guio-Carrion A, Waterhouse NJ, Li CW, et al. GAPDH and autophagy preserve survival after apoptotic

cytochrome c release in the absence of caspase activation. Cell. 2007;129:983–97. [PubMed: 17540177]

71. Lee DC, Sohn HA, Park ZY, Oh S, Kang YK, Lee KM, Kang M, Jang YJ, Yang SJ, Hong YK, et al. Article a lactate-induced response to hypoxia article alactate-Induced response to hypoxia. Cell. 2015;161:595–609. [PubMed: 25892225]
72. Zhou M, Zhao Y, Ding Y, Liu H, Liu Z, Fodstad O, Riker AI, Kamarajugadda S, Lu J, Owen LB, et al. Warburg effect in chemosensitivity: targeting lactate dehydrogenase–A re-sensitizes Taxol-resistant cancer cells to Taxol. Mol Cancer. 2010;9:1–12. [PubMed: 20144215]
73. Nakano Y, Tanno S, Koizumi K, Nishikawa T, Nakamura K, Minoguchi M, Izawa T, Mizukami Y, et al. Gemcitabine chemo resistance and molecular markers associated with gemcitabine transport and metabolism in human pancreatic cancer cells. Br J Cancer. 2007;96:457–63. [PubMed: 17224927]
74. Targeting cellular metabolism to improve cancer therapeutics. Cell Death Dis. 2012;71:4585–97.
75. Wang JB, Erickson JW, Fuji R, Ramachandran S, Gao P, Dinavahi R, Wilson KF, Ambrosio AL, Dias SM, Dang CV, et al. Targeting mitochondrial glutaminase activity inhibits oncogenic transformation. Cancer Cell. 2011;18:207–19. [PubMed: 20832749]
76. Greijer AE, van der Wall E. The role of hypoxia inducible factor 1 (HIF-1) in hypoxia induced apoptosis. J Clin Pathol. 2004;57:1009–14. [PubMed: 15452150]
77. He G, Jiang Y, Zhang B, Wu G. The effect of HIF-1 on glucose metabolism, growth and apoptosis of pancreatic cancerous cells. Asia Pac J Clin Nutr. 2014;23:174–80. [PubMed: 24561986]
78. Golias T, Papandreou I, Sun R, Kumar B, Brown NV, Swanson BJ, Pai R, Jaitin D, Le QT, Teknos TN, et al. Hypoxic repression of pyruvate dehydrogenase activity is necessary for metabolic reprogramming and growth of model tumours. Sci Rep. 2016;6:311–21. [PubMed: 27498883]
79. Haas M, Heinemann V, Kullmann F, Laubender RP, Klose C, Bruns CJ, Holdenrieder S, Modest DP, Schulz C, Boeck S. Prognostic value of CA 19-9, CEA, CRP, LDH and bilirubin levels in locally advanced and metastatic pancreatic cancer: results from a multicenter, pooled analysis of patients receiving palliative chemotherapy. J Cancer Res Clin Oncol. 2013;139:681–9. [PubMed: 23315099]
80. Fantin VR, St-Pierre J, Leder P. Attenuation of LDH-A expression uncovers a link between glycolysis, mitochondrial physiology, and tumor maintenance. Cancer Cell. 2006;9:425–34. [PubMed: 16766262]
81. Faivre S, Kroemer G, Raymond E. Current development of mTOR inhibitors as anticancer agents. Nature. 2006;5:671–88. [PubMed: 16883305]
82. Dowling RJ, Topisirovic I, Alain T, Bidinosti M, Fonseca BD, Petroulakis E, Wang X, Larsson O, Selvaraj A, Liu Y, et al. mTORC1-mediated cell proliferation, but not cell growth, controlled by the 4E-BPs. Science. 2010;328:1172–6. [PubMed: 20508131]
83. Thomas GV, Tran C, Mellinghoff IK, Welsbie DS, Chan E, Fueger B, Czernin J, Sawyers CL. Hypoxia-inducible factor determines sensitivity to inhibitors of mTOR in kidney cancer. Nat Med. 2006;12:122–7. [PubMed: 16341243]
84. Iriana S, Ahmed S, Gong J, Annamalai AA, Tuli R, Hendifar AE. Targeting mTOR in pancreatic ductal adenocarcinoma. Front Oncol. 2016;6:99. [PubMed: 27200288]
85. Wang X. New biomarkers and therapeutics can be discovered during COPD-lung cancer transition. Cell Biol Toxicol. 2016;32(5):359–61. [PMID: 27405768]
86. Gu J, Wang X. New future of cell biology and toxicology: thinking deeper. Cell Biol Toxicol. 2016;32(1):1–3. [PMID: 26874518]
87. Wang X. CBT profiles of cabozantinib approved for advanced renal cell carcinomas. Cell Biol Toxicol. 2016;32(4):259–61. [PMID: 27383755]

Yue Pan, An assistant professor at School of Pharmaceutical Science and Technology, Dalian University of Technology. A member of Professor Gary Guishan Xiao's team. She got a Ph.D. degree from Jilin University and a bachelor's degree from Sichuan University. Her previous studies are focused on the effect of berberine in reversing drug resistance and its mechanism, the microenvironment in cancer drug resistance and nanoparticles in cancer early diagnosis and therapy. Her current study is focused on cancer metabolism and the relationship between metabolism and cancer drug resistance.

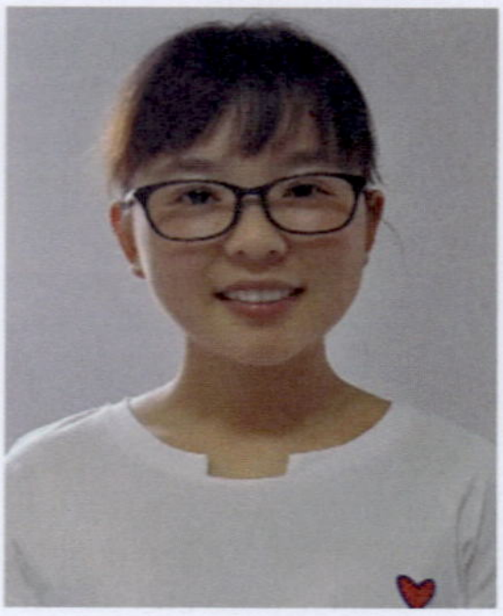

Min Cao, A master of School of Pharmaceutical Science and Technology, Dalian University of Technology. She got a bachelor's degree from Hebei Medical University in 2015, and is engaged in miRNAs as biomarkers for early diagnosis of cancer research. Her previous studies were focused on early diagnosis of oral cancer, through exploring relationship between miRNAs and oral cancer, and establishing the technology platform for tumor marker miRNAs detection and therapy. Her current study is focused on oral cancer metabolism and tumor metastasis.

Jianzhou Liu, A doctor of School of Pharmaceutical Science and Technology, Dalian University of Technology. He got a bachelor's degree from Dalian Ocean University in 2012, and a master's degree from Shanxi University in 2015. His previous studies are focused on the effects of forsythiaside A in inflammatory cells and its mechanism. His current study is focused on cancer metabolism and the relationship between metabolism and cancer drug resistance.

Qing Yang, A master of School of Pharmaceutical Science and Technology, Dalian University of Technology. She got a bachelor's degree from Yanshan University in 2016. She majored in miRNAs as biomarkers for early diagnosis of cancer. Her previous studies were focused on early diagnosis of oral cancer, through exploring relationship between miRNAs and protein pathways in oral cancer. Her current study is focused on oral cancer proliferation and early diagnosis.

Xiaoyu Miao, A master of School of Pharmaceutical Science and Technology, Dalian University of Technology. He got a bachelor's degree from Dalian University of Technology in 2016. He is focused on exploring the role of miRNAs in cancer metastasis. His previous research was associated with multidrug resistance in pancreatic cancer, through exploring the relationship between miRNAs and protein pathways in pancreatic cancer. His current study is related to miRNAs in exosomes.

Vay Liang W. Go, is currently the Co-Director of the UCLA Agi Hirshberg Center for Pancreatic Diseases and the Editor-in-Chief of Pancreas. His current research focuses on phytonutrient mechanism of action, utilizing molecular biology and metabolomic technologies. Dr. Go joined the UCLA Department of Medicine in 1988 as the executive chair of the Department of Medicine after serving as the director of the Division of Digestive Diseases and Nutrition at the NIH-National Institute of Diabetes, Digestive and Kidney Diseases (1985–1988). In 1996, he co-founded the UCLA Center for Human Nutrition with Dr. David Heber. Dr. Go was appointed jointly by the Secretaries of Department of Health and Human Services and Department of Agriculture to the US 2005 Dietary Guidelines Advisory Committee. He has authored and co-authored over 500 peer-review articles, book chapters and editorials. He co-founded the American Pancreatic Association in 1969 and the society's Lifetime Achievement Award is named after him. In 2007, he was the American Gastroenterological Association Foundation's Mentor Research Scholar Award Program honoree for having trained over 70 fellows and junior faculty, some of whom have now provided national and international leadership in gastroenterology, pancreatology and nutrition. He was elected to the Board of Directors of the North American Neuroendocrine Tumor Society and served from 2006 to 2009 and currently is a member of their Executive Committee. Over the course of his career he has received numerous honors and recognition awards including the 2013 University of Santo Tomas Medical Alumni Association Most Outstanding Academic Achievement Award and the Mayo Foundation Distinguished Alumni Award in 2002.

Paul W. N. Lee, A Professor of Pediatrics at UCLA and former Division Chief of Pediatric Endocrinology, the Director of the Stable Isotope Research Laboratory at Harbor-UCLA Medical Center and Sidmap. Dr. Lee is a pioneer in developing the use of the least squares approach to the quantitation of stable isotopes in mass spectrometry. In collaboration with Dr. Joseph Katz who was a pioneer in the application of tracers in metabolic studies, Dr. Lee published extensively on the use of mass and positional mass isotopomer to trace metabolic pathways, which has evolved into the field of metabolomics as we know it today. Dr. Lee has received numerous research and fellowship training grants and has published many seminal articles on the use of stable isotopes in metabolic research. The method was later adopted for mass isotopomer distribution analysis. Dr. Lee has been the Director of Stable Isotope Research Laboratory since 1986. His laboratory currently utilizes Agilent GC/MS, ABI5000 LC/MS/MS and LCQdeca for metabolite analysis. Through his research, Dr. Lee has gained unique understanding of cellular metabolism especially in the area of determining metabolic phenotypes in cells. Dr. Lee continues to serve on many committees within the Harbor-UCLA Medical Center community. He is Board Certified by the American Board of Pediatrics and regarded as an expert in the field of pediatric endocrinology and diabetes.

Gary Guishan Xiao, A distinguished Professor from School of Pharmaceutical Science and Technology, Dalian University of Technology. He earned his Ph.D. degree in Molecular Computational Biology, focusing on structural modeling of ligand-enzyme interaction and prediction of protein refolding pathways during protein translational process. He had his post-doctoral research training at Baylor College of Medicine and UCLA, focusing on pharmacokinetics and biochemistry of non-steroidal inflammatory drugs such as aspirin derivatives and their targets, human prostaglandin H synthases H1 and H2, and cell-cycle regulation, especially the role of c-Myc in cardiac termination differentiation. His research at Harbor-UCLA Medical Center and City of Hope National Cancer Center were highly focused on genomic and proteomic interpretations of diet-induced obesity/diabetes and its complications such cancers and osteoporosis. He joined Osteoporosis Research Center at the Creighton University Medical Center in 2007. His research focuses on functional genomics and proteomics analyses of complex disease including osteoporosis and cancer, especially breast cancer, pancreatic cancer, and prostate cancer. With the established new functional genomics and proteomics platform, he has established several projects covering function of microRNA in estrogen signaling, functional proteomics, focusing on the role of protein posttranslational modifications in cancer and osteoporosis, and proteomics technology development. He serves as a regular reviewer or as an ad hoc reviewer for several medical journals and different funding agencies, such as NIH, and he is a Member of several journal editorial boards.

Chapter 12
Mitochondrial Lon Protease and Cancer

Bin Lu

Abstract ATP-dependent Lon protease of mitochondrial matrix is encoded by nuclear DNA and highly evolutionarily conserved throughout all organisms, which is involved in the quality control of proteins by selective degradation of misfolded, oxidized, and short-lived regulatory proteins within mitochondrial matrix, maintenance of mitochondrial genome (mtDNA), and folding of mitochondria proteins. Various stimuli such as hypoxia and oxidative and ER stress lead to upregulation of Lon expression. Inhibition of protease activity or downregulation of Lon promotes cancer cell death and enhances sensitivity of cancer cells to anticancer drugs through metabolic reprogramming, thus reducing the viability of cancer cell in tumor microenvironment and epithelial to mesenchymal transition (EMT). Moreover, mitochondrial ATP-dependent Lon protease may serve as a potential biomarker for cancer diagnosis and novel target for the development of anticancer drugs and for predicting of the efficiency and effectiveness of chemotherapy of a variety of cancers.

Keywords Mitochondria • Lon • Cancer • Protein quality control • Protein degradation • Anticancer drug

1 Introduction

Most of mitochondrial proteins are encoded by nuclear DNA (nDNA) and synthesized in the cytosol as a precursor form containing a mitochondrial targeting sequence (MTS), then transported into mitochondria to their final destination. After the precursor mitochondrial proteins reach their destination, the MTS is cleaved, and the polypeptides must be folded and assembled into the active form of proteins. However, under pathological conditions, the mitochondrial proteins are continuously damaged by physical stimuli and chemical agents, such as heat, electric or

B. Lu (✉)
Institute of Biophysics, Attardi Institute of Mitochondrial Biomedicine and Zhejiang Provincial Key Laboratory of Medical Genetics, School of Life Sciences, Wenzhou Medical University, Wenzhou, Zhejiang, China
e-mail: lubmito@wmu.edu.cn

H. Sun, X. Wang (eds.), *Mitochondrial DNA and Diseases*, Advances in Experimental Medicine and Biology 1038,
https://doi.org/10.1007/978-981-10-6674-0_12

magnetic fields, reactive oxygen species (ROS), toxin, and toxic compounds [1–5]. All the stimuli may cause the misfolding or aggregation of protein and formation of adducts and ultimately lead to cellular protein dysfunction. These abnormal proteins are either repaired by molecular chaperone or removed by specific proteases through selective degradation within mitochondria as well as on the outer membrane of mitochondria by ubiquitin/proteasome system [1, 2, 6–8]. The defective protein quality control system may lead to accumulation of damaged proteins within the mitochondria and caused mitochondrial dysfunction [1, 5, 9].

In the mitochondrial matrix, there are three proteases which belong to the superfamily of ATPase associated with diverse cellular activities (AAA+) including m-AAA, ClpXP, and Lon [1, 10, 11]. Lon protease was first identified in *Escherichia coli* (*E. coli*) more than 30 years ago [12], and the *bacterial Lon* gene was cloned in 1984. Until in 1993, the human homologue of bacterial Lon (also named LONP1) was identified [3, 13], which is encoded by nDNA and transported into mitochondrial matrix after translated in ribosome within cytosol [1, 2]. Human mitochondrial Lon protease is ubiquitously expressed in all the tissues. Although Lon protease is highly conserved throughout evolution, limited attention has been paid to the functions of Lon protease of normal mitochondria physiology, as well as to its possible roles in the pathogenesis of various human diseases. In the recent years, Lon has been emerged as an essential and fundamental component of mitochondrial quality control system and as an important stress sensor in the mitochondrial alterations that could be observed in a number of chronic diseases and human cancers [1–3]. In this chapter, the recent discoveries concerning functions of mitochondrial ATP-dependent Lon protease are discussed and especially emphasized on the emerging role of Lon protease in human cancer.

2 Evolution and Functions of Mitochondrial Lon Protease

The bacteria Lon was also known as La protease, which was derived from the phenotype of a mutant *E. coli* lacking the *lon* gene. Upon the UV irradiation, this mutant *E. coli* inclines to grow in the longer forms compared with their wild-type counterparts [14–17]. Lon protease is highly conserved, and its homologues were identified in almost all the organisms, from *archaea* to mammalian mitochondria as well as in peroxisomes [14, 15, 18]. In *E. coli*, Lon protease degrades almost 50% of abnormal proteins [17]. In addition, Lon protease is also required for regulating physiological events through degradation of related regulatory proteins which are essential for the proper function of bacteria [1, 6]. The *Lon* gene of *E. coli* encodes an ATP-dependent Lon protease with molecular weight of approximately 87 kDa, which has the activity of ATPase and is also involved in the heat-shock response [19, 20]. By degradation of abnormal proteins as well as certain short-lived regulatory proteins, the *E. coli* Lon plays critical roles in regulating gene expression. Furthermore, in other prokaryotic organisms such as *Bacillus subtilis*, Lon protease plays an important role in gene expression regulation through controlling the transcription factor levels [21].

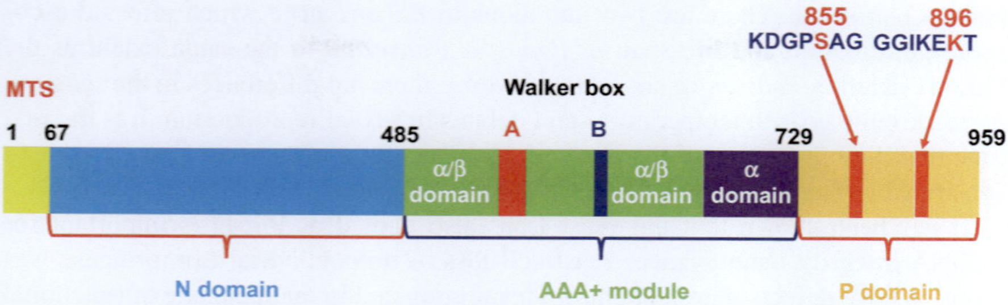

Fig. 12.1 Domain structure and functional motifs of human mitochondrial Lon protease. MTS, mitochondrial targeting sequence, of the Lon precursor is present at the N terminal and binds to the mitochondrial outer membrane translocation machinery, which mediates protein import into the matrix from cytosol and removed by proteolytic cleavage after importing in mitochondria. N domain, a highly variable amino-terminal domain proposed to function in substrate recognition and binding. AAA+ module, composed of the α/β subdomain, which contains the Walker Box A and B motifs for ATP-binding and hydrolysis, in addition to the small α subdomain. P domain, the protease domain at carboxyl terminal containing the serine (S) and lysine (K) residues forming the catalytic Ser-Lys dyad at the proteolytic active site

Lon protease is a soluble cytoplasmic protein in *eubacteria*; however, the Lon protease is bound to membrane in *archae*a [22–24]. The mitochondrial Lon protease localizes in the matrix of mitochondria [1, 2]. Lon-mediated protein degradation also plays a critical role in most pathogenic bacteria such as *Brucella abortus* and *Pseudomonas aeruginosa* in the virulence gene expression and promoting the infection to mammalian cells [25, 26]. The nDNA-encoded ATP-dependent mitochondrial Lon protease was first identified in humans [27] as well as in the yeast *Saccharomyces cerevisiae* [18, 22]. Mitochondrial Lon protease is translated in the ribosome as a precursor protein form that has a mitochondrial targeting sequence (MTS) at its N-terminal to lead the precursor form importing into mitochondria, and thus the precursor mitochondrial Lon protease was transported from the cytosol to mitochondrial matrix where it acts as a quality control protease after the cleavage of MTS to be the mature form of Lon and then refold into the functional protease (Fig. 12.1). Mitochondrial Lon protease exhibits most of the properties as bacterial Lon, which includes the binding capacity to the DNA and the crucial roles in the protein quality control system by removing abnormal proteins [28]. It is still unclear whether mitochondrial Lon protease has the same functions as bacterial Lon in regulating gene expression of certain proteins; however, it has been found that the mutated Lon may affect plant development [29].

PaLon1 gene of filamentous fungus *Podospora anserina* encodes mitochondrial Lon protease (named as PaLON1) that is important in the regulation of functions during fungal development. The absence of PaLON1 caused the reduction of its life span. PaLON1 was also important in some extreme growth conditions (such as high or low temperatures). PaLON1 is also essential for sexual reproduction and ascospore germination of *Podospora anserina*, which are the two regulated developmental stages. The defects caused by mutations of *PaLon1* may affect the N-domain of

PaLON1 protease. There are two mutations in *PaLon1 gene* which affected ascospore germination and life span of *Podospora anserina* to the same extent as the PaLon1 deletion *Podospora anserina*; however, there are differences in the sensitivity to extreme growth temperatures and defects in sexual reproduction. It is the first time that the importance of the N-terminal domain of a eukaryotic Lon protease in substrate binding was revealed by using three *Podospora anserina* mutations [30].

It has been shown that the yeast Lon (also named as Pim1) is important for mtDNA integrity maintenance. Yeast cell loss of mitochondrial Lon protease was found to have defects in mitochondrial respiration and in maintenance of functional mtDNA [27, 31]. The expression of *E. coli* Lon in yeast cells lacking Lon can partially rescue mtDNA integrity, which indicates that Lon protease exhibits remarkable functional conservation between distantly related organisms [32]. Lon selectively degrades abnormal proteins, such as misassembled, misfolded, oxidized, mutant proteins, and certain regulatory proteins with short half-lives. Lon protease is crucial for homeostasis of mitochondrial proteins and for survival of organisms from numerous stresses through the protein quality control machinery and regulation of metabolism in both bacteria and mitochondria [33].

3 Inhibitors of Lon Protease

Intracellular protein turnover and degradation of abnormal proteins are conducted by the ubiquitin-proteasome machinery within the cytosol as well as in the nucleus of eukaryotic cells. However, there are several ATP-dependent proteases including the Lon protease involved in the degradation of mitochondrial proteins to ensure the mitochondria function properly [1]. The upregulation of mitochondrial Lon protease expression is important for survival and proliferation of cancer cells by preventing mitochondrial proteotoxicity caused by oxidative, hypoxic, and ER stress. To date, there is no specific inhibitor of ATP-dependent mitochondrial Lon protease found. But several proteasome inhibitors including the widely used peptidyl aldehyde MG132 (Z-Leu-Leu-Leu-CHO) can spread into mitochondria and inhibit the degradation of the mitochondrial steroidogenic acute regulatory protein (StAR), which is an endogenous substrate of Lon protease, implying that the proteasome and Lon protease may have similarities to some extent in their proteolytic mechanism [34, 35].

Mitochondrial Lon protease is a major quality control protease in mitochondrial matrix which is responsible for the degradation of certain abnormal proteins including misfolded, misassembled, oxidized proteins or certain folded and short-lived regulator proteins as well as the maintenance of mtDNA integrity in cells. It has been well known that mitochondria produce the majority of cellular reactive oxygen species (ROS), and the excess ROS can cause oxidative damage to mtDNA and RNA, proteins, and lipids, therefore resulting in mitochondria dysfunction. Lon protease plays an essential role in the degradation of oxidized proteins in the mitochondrial matrix, such as oxidatively modified aconitase [36]. This degradation was blocked by PMSF, a serine protease inhibitor [36]. Moreover, Lon-mediated degradation of mitochondrial StAR was relied on ATP and could be inhibited by a number of proteasome inhibitors such as MG132 and *clasto*-lactacystin β-lactone (cLβL); however, epoxomicin, which belongs to a different class of proteasome inhibitors, has no effect to the activity of Lon protease.

In addition, 2-cyano-3,12-dioxooleana-1,9-dien-28-oic acid (CDDO) and its methyl ester derivatives (CDDO-Me) can directly and selectively inhibit the protease activity of Lon but have no effect on the 20S proteasome. The biotinylated CDDO can form a conjugate with Lon which results in the Lon protease activity being blocked. By using biotinylated CDDO as a probe, the inhibition mechanism of Lon protease by CDDO and its derivatives was elucidated. The electron microscopy (EM) data further demonstrated that Lon-knockdown or CDDO-treated lymphoma cells exhibit the accumulation of electron-dense inclusions within mitochondria matrix. Moreover, Lon knockdown causes lymphoma cell death, and CDDO-induced cell death of lymphoma cell is mediated through inactivating Lon protease, at least partially. All together, these results suggest that the inhibition of mitochondrial Lon protease activity may play an essential role in lymphoma cell death induced by CDDO or its derivatives, which further support that mitochondrial Lon protease is a novel anticancer drug target [37, 38].

It has been demonstrated that coumarinic derivatives are highly efficient small-molecule, non-peptidic, as well as versatile protease inhibitors. By employing the casein-FITC assay, certain coumarinic derivatives were found to inhibit the human Lon protease activity; however, these compounds have no effects to yeast proteasome [39]. Among these compounds, there are two small-molecule compounds, obtusilactone A (OA) and (−)-sesamin from C. kotoense, which were demonstrated to be a potential Lon protease inhibitor, through an enzyme-based screening system. Both obtusilactone A and (−)-sesamin could interact with Ser855 and Lys898 residues within the active site of Lon protease based on the molecular docking analysis [40]. Recently, we have identified a natural small-molecular compound from Chinese medicine chuanxiong, which specifically inhibits TFAM degradation by mitochondrial Lon protease [41]. Thus, the upregulation of Lon in response to various stresses may be critical for the survival and proliferation of cancer cells by efficiently preventing damaged proteins and regulatory proteins with short half-life accumulation in the mitochondria and therefore ensuring mitochondria function well.

4 Lon and Cancer

The ATP-dependent mitochondrial Lon protease is crucial in the quality control of mitochondrial function under various stress conditions; either up- or downregulation in the protein expression levels of *Lon* gene is associated with a number of human diseases [1]. However, the function of Lon protease in the tumorigenesis is still less understood. The rapid proliferation is one of the main characteristics of almost all cancer cells, which causes a hypoxic microenvironment in cancer cells. Thus, mitochondrial oxidative phosphorylation (OXPHOS) is inhibited due to less oxygen, and the cancer cells tend to rely on glycolysis to provide ATP for the survival and rapid division [42, 43]. In addition, the metabolic reprogramming from OXPHOS to glycolysis of glucose can provide the metabolic intermediates required for the biosynthesis of macromolecules and generation of precursors, reducing equivalents important for cellular biogenesis and antioxidant defenses, which are essential for cancer cell proliferation and growth [44].

Mitochondrial Lon protease mainly localizes in the matrix; however, it is also found in the nucleoids of mitochondria in the maintenance of mtDNA integrity [8]. Growing evidence indicates that mitochondrial Lon protease is involved in the metabolic reprogramming and in the tumorigenesis [6, 25, 45, 46]. Since mitochondrial respiratory chain enzyme complexes are the main source of intracellular ROS, ROS can cause mitochondrial dysfunction by damaging mtDNA, lipids, and proteins [25, 46]. For the cancer cells, the upregulation of Lon can ensure the mitochondrial protein homeostasis by effective removal of abnormal proteins such as oxidized, misfolded, misassembled, aggregated proteins, as well as short-lived regulatory proteins. Thus, the cancer cells can maintain mitochondrial function well, which further supports cancer cell growth.

Hypoxia microenvironment is a hallmark of solid tumor [47]; the Lon protease expression in cancer cells of solid tumor is increased by the hypoxia-inducible factor-1α (HIF1-α) [48]. Upregulation of Lon protease promotes the degradation of isoform 1 of cytochrome c oxidase subunit 4 (COX4-1), thus permitting the assembly of isoform 2 of cytochrome c oxidase subunit 4 (COX4-2), which optimizes enzyme activity and respiration efficiency under hypoxia and helps cancer cells to adapt to the hypoxic microenvironment [48].

In addition, high level of Lon protease is also responsible for the resistance of chemotherapeutics by highly efficient removal of oxidized or damaged proteins caused by various anticancer drugs [25, 46].

Overexpression of mitochondrial Lon protease has been found in a variety of human cancer, including malignant B cell lymphoma [25], cervical cancer [45], bladder cancer [46], non-small cell lung cancer [40, 49], and colon cancer tissues and cells [6]. High expression of Lon protease promotes cancer cell proliferation and invasion and the capability to migrate and form metastasis in nude mice [6]. Downregulation of Lon protease expression suppresses bladder cancer cell proliferation and enhances the chemosensitivity through increasing cell death and reducing cellular bioenergetics metabolism of bladder cancer cells [46]. In bladder cancer cells, downregulation of Lon protease reduced mitochondrial ROS generation and blocked the activation of c-Jun N-terminal kinase (JNK); therefore, cell proliferation of bladder cancer decreased. The UM-UC-3 bladder cells abolished Lon protease and exhibited increased sensitivity to anticancer drug doxorubicin through facilitating caspase-dependent cell death [46]. High expression of mitochondrial Lon protease in melanoma cells promotes experimental metastasis formation, whereas downregulation of Lon protease expression inhibits cancer cell proliferation and lung metastasis [6]. In high-grade gliomas, Lon is overexpressed and mediates hypoxic adaptation [50].

The Lon knockout mice ($Lon^{-/-}$) have embryonic lethal phenotype, which indicates that Lon is extremely critical for the development of mice. It has also been shown the Lon heterozygous mice ($Lon^{+/-}$) with half amount of Lon protease in all the tissues of wild-type mice are less sensitive to chemical carcinogens known to induce colorectal and skin cancer and are more resistant to development of cancers, and there are much less tumors formed than the wild-type mice ($Lon^{+/+}$). These results suggest that mitochondrial ATP-dependent Lon protease has a crucial role in the development of colon and skin tumors, and downregulation of Lon protease expression can protect against colorectal carcinoma and papilloma formation in mice [6].

5 Conclusion

In conclusion, the ATP-dependent mitochondrial Lon protease exhibits a major regulator of multiple mitochondrial functions including (but not limited to) the degradation of abnormal proteins, the maintenance of mtDNA integrity, and bioenergetics metabolism. Accumulating data clearly indicate that mitochondrial Lon protease is one of proteins within mitochondrial matrix higher expression of which can favor a variety of tumor growth; however its downregulation is associated with slow proliferation and metastasis. Lon protease expression level may serve as a potential valuable diagnostic and prognostic biomarker and therapeutic target for cancer patients, as well as in predicting the effectiveness of chemotherapy.

Acknowledgments Many thanks to Carolyn K. Suzuki and the members in my laboratory. This study has been supported by grants from the National Basic Research Program of China (973 Program, No. 2013CB531700), National Natural Science Foundation of China (No. 31070710, No. 31171345, No. 31570772, No. 31771543), and Zhejiang Qianjiang Talent Project B (No. 2010R10045) to BL.

Conflicting Declaration No.

References

1. Venkatesh S, Lee J, Singh K, Lee I, Suzuki CK. Multitasking in the mitochondrion by the ATP-dependent Lon protease. Biochim Biophys Acta. 2012;1823(1):56–66.
2. Pinti M, Gibellini L, Liu Y, Xu S, Lu B, Cossarizza A. Mitochondrial Lon protease at the crossroads of oxidative stress, ageing and cancer. Cell Mol Life Sci. 2015;72(24):4807–24.
3. Bao L, Diao H, Dong N, Su X, Wang B, Mo Q, Yu H, Wang X, Chen C. Histone deacetylase inhibitor induces cell apoptosis and cycle arrest in lung cancer cells via mitochondrial injury and p53 up-acetylation. Cell Biol Toxicol. 2016;32(6):469–82. PMID: 27423454
4. Liu T, Lu B, Lee I, Ondrovicova G, Kutejova E, Suzuki CK. DNA and RNA binding by the mitochondrial lon protease is regulated by nucleotide and protein substrate. J Biol Chem. 2004;279(14):13902–10.

5. Lu B, Lee J, Nie X, Li M, Morozov YI, Venkatesh S, Bogenhagen DF, Temiakov D, Suzuki CK. Phosphorylation of human TFAM in mitochondria impairs DNA binding and promotes degradation by the AAA+ Lon protease. Mol Cell. 2013;49(1):121–32.
6. Quiros PM, Espanol Y, Acin-Perez R, Rodriguez F, Barcena C, Watanabe K, Calvo E, Loureiro M, Fernandez-Garcia MS, Fueyo A, Vazquez J, Enriquez JA, Lopez-Otin C. ATPdependent Lon protease controls tumor bioenergetics by reprogramming mitochondrial activity. Cell Rep. 2014;8(2):542–56.
7. Lu B, Yadav S, Shah PG, Liu T, Tian B, Pukszta S, Villaluna N, Kutejova E, Newlon CS, Santos JH, Suzuki CK. Roles for the human ATP-dependent Lon protease in mitochondrial DNA maintenance. J Biol Chem. 2007;282(24):17363–74.
8. Vedi M, Sabina EP. Assessment of hepatoprotective and nephroprotective potential of withaferin A on bromobenzene-induced injury in Swiss albino mice: possible involvement of mitochondrial dysfunction and inflammation. Cell Biol Toxicol. 2016;32(5):373–90. PMID: 27250656
9. Bulteau AL, Szweda LI, Friguet B. Mitochondrial protein oxidation and degradation in response to oxidative stress and aging. Exp Gerontol. 2006;41(7):653–7.
10. Goard CA, Schimmer AD. Mitochondrial matrix proteases as novel therapeutic targets in malignancy. Oncogene. 2014;33(21):2690–9.
11. Bulteau AL, Bayot A. Mitochondrial proteases and cancer. Biochim Biophys Acta. 2011;1807(6):595–601.
12. Swamy KH, Goldberg AL. Escherichia coli contains eight soluble proteolytic activities, one being ATP dependent. Nature. 1981;292(5824):652–4.
13. Amerik A, Petukhova GV, Grigorenko VG, Lykov IP, Yarovoi SV, Lipkin VM, Gorbalenya AE. Cloning and sequence analysis of cDNA for a human homolog of eubacterial ATPdependent Lon proteases. FEBS Lett. 1994;340(1–2):25–8.
14. Howard-Flanders P, Simson E, Theriot L. The excision of thymine dimers from DNA, filament formation and sensitivity to ultraviolet light in Escherichia coli K-12. Mutat Res. 1964;106:219–26.
15. Seo JB, Jung SR, Hille B, Koh DS, Extracellular ATP. Protects pancreatic duct epithelial cells from alcohol-induced damage through P2Y1 receptor-cAMP signal pathway. Cell Biol Toxicol. 2016;32(3):229–47. PMID: 27197531
16. Charette MF, Henderson GW, Markovitz A. ATP hydrolysis-dependent protease activity of the lon (capR) protein of Escherichia coli K-12. Proc Natl Acad Sci U S A. 1981;78:4728–32.
17. Chung CH, Goldberg AL. The product of the lon (capR) gene in Escherichia coli is the ATP-dependent protease, protease La. Proc Natl Acad Sci U S A. 1981;78:4931–5.
18. Suzuki CK, Suda K, Wang N, Schatz G. Requirement for the yeast gene LON in intramitochondrial proteolysis and maintenance of respiration. Science. 1994;264(5156):273–6.
19. Park SC, Jia B, Yang JK, Van DL, Shao YG, Han SW, Jeon YJ, Chung CH, Cheong GW. Oligomeric structure of the ATP-dependent protease La (Lon) of *Escherichia coli*. Mol Cells. 2006;21(1):129–34.
20. Vieux EF, Wohlever ML, Chen JZ, Sauer RT, Baker TA. Distinct quaternary structures of the AAA? Lon protease control substrate degradation. Proc Natl Acad Sci U S A. 2013;110(22):E2002–8.
21. Schmidt R, Decatur AL, Rather PN, Moran CP Jr, Losick R. Bacillus subtilis lon protease prevents inappropriate transcription of genes under the control of the sporulation transcription factor sigma G. J Bacteriol. 1994;176(21):6528–37.
22. Van Dyck L, Langer T. ATP-dependent proteases controlling mitochondrial function in the yeast Saccharomyces cerevisiae. Cell Mol Life Sci. 1999;56(9–10):825–42.
23. Opperman CM, Sishi BJ. Tumor necrosis factor alpha stimulates p62 accumulation and enhances proteasome activity independently of ROS. Cell Biol Toxicol. 2015;31(2):83–94. PMID: 25761618
24. Gottesman S, Wickner S, Maurizi MR. Protein quality control: triage by chaperones and proteases. Genes Dev. 1997;11(7):815–23.

25. Ganta KK, Mandal A, Chaubey B. Depolarization of mitochondrial membrane potential is the initial event in non-nucleoside reverse transcriptase inhibitor efavirenz induced cytotoxicity. Cell Biol Toxicol. 2017;33(1):69–82. PMID: 27639578
26. Robertson GT, Kovach ME, Allen CA, Ficht TA, Roop RM 2nd. The *Brucella abortus* Lon functions as a generalized stress response protease and is required for wild-type virulence in BALB/c mice. Mol Microbiol. 2000;35(3):577–88.
27. Wang N, Gottesman S, Willingham MC, Gottesman MM, Maurizi MR. A human mitochondrial ATP-dependent protease that is highly homologous to bacterial Lon protease. Proc Natl Acad Sci U S A. 1993;90:11247–51.
28. GK F, Smith MJ, Markovitz DM. Bacterial protease Lon is a site-specific DNA-binding protein. J Biol Chem. 1997;272(1):534–8.
29. Rigas S, Daras G, Laxa M, Marathias N, Fasseas C, Sweetlove LJ, Hatzopoulos P. Role of Lon1 protease in post-germinative growth and maintenance of mitochondrial function in Arabidopsis thaliana. New Phytol. 2009;181(3):588–600.
30. Adam C, Picard M, Dequard-Chablat M, Sellem CH, Hermann-Le Denmat S, Contamine V. Biological roles of the Podospora anserina mitochondrial Lon protease and the importance of its N-domain. PLoS One. 2012;7(5):e38138.
31. Coleman JL, Katona LI, Kuhlow C, Toledo A, Okan NA, Tokarz R, Benach JL. Evidence that two ATP-dependent (Lon) proteases in Borrelia burgdorferi serve differentfunctions. PLoS Pathog. 2009;5(11):e1000676.
32. Teichmann U, van Dyck L, Guiard B, Fischer H, Glockshuber R, Neupert W, Langer T. Substitution of PIM1 protease in mitochondria by Escherichia coli Lon protease. J Biol Chem. 1996;271(17):10137–42.
33. Jonas K, Liu J, Chien P, Laub MT. Proteotoxic stress induces a cell-cycle arrest by stimulating Lon to degrade the replication initiator DnaA. Cell. 2013;154:623–36.
34. Granot Z, Kobiler O, Melamed-Book N, Eimerl S, Bahat A, Lu B, Braun S, Maurizi MR, Suzuki CK, Oppenheim AB, Orly J. Turnover of mitochondrial steroidogenic acute regulatory (StAR) protein by Lon protease: the unexpected effect of proteasome inhibitors. Mol Endocrinol. 2007;21(9):2164–77.
35. Granot Z, Geiss-Friedlander R, Melamed-Book N, Eimerl S, Timberg R, Weiss AM, Hales KH, Hales DB, Stocco DM, Orly J. Proteolysis of normal and mutated steroidogenic acute regulatory proteins in the mitochondria: the fate of unwanted proteins. Mol Endocrinol. 2003;17(12):2461–76.
36. Bota DA, Davies KJ. Lon protease preferentially degrades oxidized mitochondrial aconitase by an ATP-stimulated mechanism. Nat Cell Biol. 2002;4(9):674–80.
37. Bernstein SH, Venkatesh S, Li M, Lee J, Lu B, Hilchey SP, Morse KM, Metcalfe HM, Skalska J, Andreeff M, Brookes PS, Suzuki CK. The mitochondrial ATP-dependent Lon protease: a novel target in lymphoma death mediated by the synthetic triterpenoid CDDO and its derivatives. Blood. 2012;119(14):3321–9.
38. Gibellini L, Pinti M, Bartolomeo R, De Biasi S, Cormio A, Musicco C, Carnevale G, Pecorini S, Nasi M, De Pol A, Cossarizza A. Inhibition of Lon protease by triterpenoids alters mitochondria and is associated to cell death in human cancer cells. Oncotarget. 2015;6(28):25466–83.
39. Bayot A, Basse N, Lee I, Gareil M, Pirotte B, Bulteau AL, Friguet B, Reboud-Ravaux M. Towards the control of intracellular protein turnover: mitochondrial Lon protease inhibitors versus proteasome inhibitors. Biochimie. 2008;90(2):260–9.
40. Wang HM, Cheng KC, Lin CJ, Hsu SW, Fang WC, Hsu TF, Chiu CC, Chang HW, Hsu CH, Lee AY. Obtusilactone A and (−)-sesamin induce apoptosis in human lung cancer cells by inhibiting mitochondrial Lon protease and activating DNA damage checkpoints. Cancer Sci. 2010;101(12):2612–20.
41. Lan L, Guo M, Ai Y, Chen F, Zhang Y, Xia L, Huang D, Niu L, Zheng Y, Suzuki CK, Zhang Y, Liu Y, Lu B. Tetramethylpyrazine blocks TFAM degradation and up-regulates mitochondrial DNA copy number by interacting with TFAM. Biosci Rep. 2017;37(3):pii: BSR20170319. https://doi.org/10.1042/BSR20170319.

42. Vander Heiden MG. Targeting cancer metabolism: a therapeutic window opens. Nat Rev Drug Discov. 2011;10(9):671–84. https://doi.org/10.1038/nrd3504.
43. Cairns RA, Harris IS, Mak TW. Regulation of cancer cell metabolism. Nat Rev Cancer. 2011;11(2):85–95. https://doi.org/10.1038/nrc2981.
44. Zhu LZ, Hou YJ, Zhao M, Yang MF, XT F, Sun JY, XY F, Shao LR, Zhang HF, Fan CD, Gao HL, Sun BL. Caudatin induces caspase-dependent apoptosis in human glioma cells with involvement of mitochondrial dysfunction and reactive oxygen species generation. Cell Biol Toxicol. 2016;32(4):333–45. PMID: 27184666
45. Nie X, Li M, Lu B, Zhang Y, Lan L, Chen L, Lu J. Downregulating overexpressed human Lon in cervical cancer suppresses cell proliferation and bioenergetics. PLoS One. 2013;8(11):e81084.
46. Liu Y, Lan L, Huang K, Wang R, Xu C, Shi Y, Wu X, Wu Z, Zhang J, Chen L, Wang L, Yu X, Zhu H, Lu B. Inhibition of Lon blocks cell proliferation, enhances chemosensitivity by promoting apoptosis and decreases cellular bioenergetics of bladder cancer: potential roles of Lon as a prognostic marker and therapeutic target in baldder cancer. Oncotarget. 2014;5(22):11209–24.
47. Hanahan D, Weinberg RA. Hallmarks of cancer: the next generation. Cell. 2011;144(5):646–74.
48. Fukuda R, Zhang H, Kim JW, Shimoda L, Dang CV, Semenza GL. HIF-1 regulates cytochrome oxidase subunits to optimize efficiency of respiration in hypoxic cells. Cell. 2007;129(1):111–22.
49. Cheng CW, Kuo CY, Fan CC, Fang WC, Jiang SS, Lo YK, Wang TY, Kao MC, Lee AY. Overexpression of Lon contributes to survival and aggressive phenotype of cancer cells through mitochondrial complex I-mediated generation of reactive oxygen species. Cell Death Dis. 2013;4:e681.
50. Di K, Lomeli N, Wood SD, Vanderwal CD, Bota DA. Mitochondrial Lon is over-expressed in high-grade gliomas, and mediates hypoxic adaptation: potential role of Lon as a therapeutic target in glioma. Oncotarget. 2016;7(47):77457–67.

Bin Lu, Ph.D., Professor of Biochemistry, Vice-director of Zhejiang Key Laboratory of Medical Genetics, Attardi Institute of Mitochondrial Biomedicine, School of Life Sciences, Wenzhou Medical University, Wenzhou, China. His main research is focused on the molecular mechanism of mitochondrial protein quality control system and diseases. More specifically, he focused on his research efforts on the mitochondrial ATP-dependent matrix protease Lon and ClpXP, as well as the mitochondrial DnaJ-like co-chaperone Tid1. He is the author of more than 80 scientific publications including Molecular Cell, Blood, Journal of Biological Chemistry, etc.

Chapter 13
Regulatory Roles of Mitochondrial Ribosome in Lung Diseases and Single Cell Biology

Linlin Zhang, William Wang, Bijun Zhu, and Xiangdong Wang

Abstract The mitochondria have the most vital processes in eukaryotic cells to produce ATP composed of polypeptides that are produced via ribosomes, as oxidative phosphorylation. Initially, studies regarding human mitochondrial ribosomes were performed in the model system, bovine mitochondrial ribosome, to investigate how ribosomes are biosynthesized and evolved as well as what their structure and function are. Advances in X-ray crystallography have led to dramatic progresses in structural studies of the ribosome. In recent years, there has been a growing interest in the properties of the mitochondrial ribosome. Although one of its main functions is the production of ATP, it was also linked to multiple diseases. A key area that remains unexplored and requires investigation and exploration is how mitochondrial ribosomal RNA (mt-rRNA) variations can affect the mitochondrial ribosomes in developing disease. This review summarizes the structure, elements, functions, and regulatory roles in associated diseases. With the continuous development of technology, studies on the mechanism of mitochondrial ribosome related diseases are crucial, in order to identify methods of prevention and treatment of these disorders.

Keywords Mitochondria • Ribosome • Element • Function • Structure • Diseases • Single cell

---

L. Zhang • W. Wang • B. Zhu • X. Wang (✉)
Zhongshan Hospital Institute of Clinical Science, Fudan University, Shanghai Medical College, Shanghai, China
e-mail: xiangdong.wang@clintransmed.org

H. Sun, X. Wang (eds.), *Mitochondrial DNA and Diseases*, Advances in Experimental Medicine and Biology 1038,
https://doi.org/10.1007/978-981-10-6674-0_13

1 Introduction

Mitochondria are cellular organelles that produce adenosine triphosphate (ATP) and convert different energy types within eukaryotic cells. In addition to their function in energy metabolism, they play an important part in a diverse number of cellular processes, such as apoptosis, disease, or aging [1–3]. The structures of different complexes of mitochondrial ribosomes (mitoribosomes) have been increasingly understood with the development of cryo-EM field. Crucially we have also been able to gain a better understanding of insertion and extension sequences within translation factors as well as how protein synthesis functions within the mitochondria at a molecular level [4]. In addition, electron microscopy developments have revealed the structure of yeast mitochondrial large subunit (mt-LSU) and both mammalian mt-LSU and mitochondrial small subunit (mt-SSU) at a near-atomic resolution. Novel data on the structure of the functionally important ribosomal domain, L12/P stalk, of the large ribosomal subunit has also been discovered. It is the most mobile site of the ribosome and has been found in ribosomes of all living cells; furthermore, it is involved in interactions between ribosomes and translation factors [5]. Our aim is to conclude the biology, function, metabolism, and associated diseases of mitochondrial ribosome. Through systematic carding, we now possess a general understanding of mitochondrial ribosome. However, it is essential to explore its finer structure and function in order to discover how we can apply this knowledge for clinical purposes.

2 Biology of Mitochondrial Ribosome

2.1 Elements

Human cells contain two sets of genomes and two sets of protein synthesis systems, illustrated in Fig. 13.1. The ribosomes exist in both bacterial and eukaryotic cells and can be located within the cytoplasm; however, they are also present within mitochondria and chloroplasts from bacterial ancestry [6]. A key role of the ribosome is to catalyze protein synthesis. Ribosomal RNAs (rRNAs) and proteins assemble functional ribosomes which are macromolecular ribonucleoprotein nanomachines. The human mitochondrial genome codes for 13 proteins that participate in oxidative phosphorylation (OXPHOS). These proteins are translated by a dedicated set of ribosomes (mitoribosomes). Instead of having a sedimentation coefficient of 70S or 80S, which is what one would predict for a prokaryotic ribosome or a eukaryotic ribosome, respectively, the mammalian mitochondrial ribosomes have a coefficient of 55S [7] and are composed of two main subunits, a large subunit (39S) and a small subunit (28S) subunit. These subunits contain a 16S rRNA and 12S rRNA, respectively, and no 5S rRNA [8]. Within the small subunit, the 12S RNA in the human mitochondrial ribosomes has a length of 954 nucleotides (323 kDa), which is 40% shorter than the 16S RNA of a bacteria. In addition, the 16S RNA in the large subunit of human mitochondrial ribosomes is 1558 nucleotides long (528 kDa), half the length of a bacterial 23S RNA (Table 13.1) [6]. Human mitoribosomes have a higher

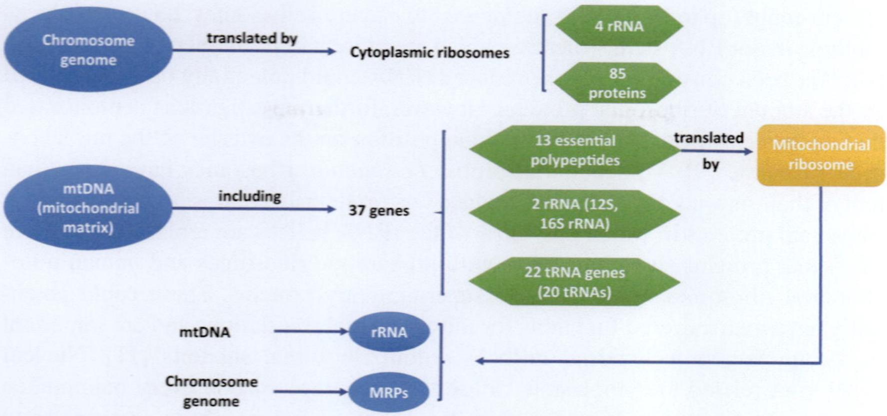

Fig. 13.1 Two sets of genomes and protein synthesis systems. Human cells contain two sets of genomes and two sets of protein synthesis systems. The first set is the chromosome genome which encoded protein is responsible for the translation of cytoplasmic ribosomes [63]. Second sets of genomic mtDNA are located in the mitochondrial matrix, including 37 genes, 13 protein genes associated with oxidative phosphorylation, 2 rRNA genes (12srRNA and 16srRNA), and 22 tRNA genes (20 kinds of tRNA). The protein encoded by mtDNA is responsible for the translation of mitochondrial ribosomes. The cytoplasmic ribosome consists of four rRNAs and 85 proteins, all of which are encoded by chromosome DNA. In contrast, the components of the mitochondrial ribosome are encoded by two genomes, the ribosomal RNA is encoded by mtDNA, and the MRPs are encoded by the genome, which is synthesized in the cytoplasmic ribosome and then is fed into the mitochondria

Table 13.1 Mammalian mitoribosomes have lower sedimentation coefficients (~55S) and, like all ribosomes, consist of two subunits (the 28S small subunit or SSU and the 39S large subunit or LSU)

	Prokaryote mt	Eukaryote mt	Mammalian mt	Human mt
Sedimentation coefficient	70S	80S	55S	55S
LSU	50S	60S	39S	39S
RNA	23S	28S	16S	16S
	5S	5S	$tRNA^{Phe}$	$tRNA^{Val}$
		5.8S		
SSU	30S	40S	28S	28S
RNA	16S	18S	12S	12S
RNA:Protein	2:1	1:2	1:2	1:2
Mass	2.5 MDa	3.9 ~ 4.5 MDa	2.7 MDa	nd

The two mitochondrially encoded rRNA species: 12S in the SSU and 16S in the LSU. RNA-protein ratio of 2:1 observed in bacteria. In comparison to the bacterial ribosome, the mammalian mitoribosome has lost a significant amount of RNA, which has been replaced by mito-specific proteins giving rise to a 1:2 RNA-protein ratio. Compared to mammalian mitoribosome, human mitoribosome also has difference protein count (up to 80 proteins) compared to bacterial ribosomes; however, it lacks multiple major RNA stem structures which are present in bacterial ribosomes. It has however been considered that the reduced rRNA could potentially be compensated by the amount of ribosomal proteins. However, further investigations demonstrated that the proteins instead occupy a unique position on the exterior of the mitochondrial ribosome [9]. Mitochondrial protist *Leishmania* ribosomes have more than half of their missing rRNA helices replaced by proteins in contrast to mitochondrial ribosomal proteins in which only 20% of the rRNA helices are replaced [10]. Core ribosomal proteins share homology with prokaryotic ribosomes and human mitochondrial ribosomes which have 35 supernumerary proteins. These could potentially have undiscovered functions for mitochondrial translations and are important in recognizing mitochondrial mRNAs within ribosomal subunits [11]. Nuclear genes (not related to cytoplasmic ribosomal encoding) encode every mammalian mitochondrial ribosomal proteins (MRPs). These evolve rapidly in comparison to ones that encode cytoplasmic ribosomal proteins [12]. Recently, a higher-resolution cryo-EM provided a significantly more detailed insight into the structure of the 55S architecture. The core of the 55S mitoribosome is highly porous in comparison to other ribosomes; the proteinaceous shell which surrounds it also has a more intricate complex compared to an ancestral bacterial ribosome [13].

2.2 *Structure*

2.2.1 Mitoribosomal Small Subunit Structure

Decoding of mRNA is done by a 28S mt-SSU from the human mitoribosome [14]. In contrast, the larger subunit, 39S mt-LSU, is responsible for peptidyl transferase activity [2]. The mammalian mt-SSU is estimated to be composed of a 12S mt-rRNA and 30 distinct types of proteins, of which 15 porcine and 14 human proteins are mt specific. An elongated subunit is observed due to an increase within its protein content in the mitoribosome in contrast to the structure of the bacterial subunit [15]. A similar trend is observed in the 28S small subunit from the mitoribosome, which is significantly more elongated in a mammalian when contrasted to bacteria. The head, body, and platform regions of a small subunit are in a solvent within the interface on a subunit, which is connected to the enlarged mitoribosomal head of the small subunit via an obliquely oriented channel. Of these structures, the platform is conserved to maintain its structural integrity, and additional protein mass can be found within the head and body region of the mitoribosomal small subunit [16].

2.2.2 Inter-Subunit Bridges

Fifteen inter-subunit bridges are important in binding together the two subunits. Six of these (B2a–B2c, B3, B5, and B7a) have a similar structural and chemical composition when compared to bacterial ribosomes [17]. The remaining nine bridges contain proteins and differ from the bacterial counterparts in either spatial position or chemical. Four (B1a–B1d) of the nine bridges correspond to the bacterial B1 bridge group. The group represents where the bridges are located, and the following nomenclature is for the bacterial ribosome [18]. The four other bridges following the nomenclature are denoted as B2d, B2e, B7b, and B7c. B9 is responsible for connecting the lower regions of both large subunit and small subunit and has no bacterial equivalent. To summarize, the mitoribosomal subunits are connected via different types of bridges; five RNA-RNA bridges, two RNA-protein bridges, seven protein-protein bridges, and one bridge that is formed from both RNA and secondary protein components from each subunit [16].

2.2.3 Structure of the Mitoribosomal Large Subunit

The bacterial equivalent of 39S large subunit of the mitoribosome is comparatively smaller compared to the mammalian unit; this also is true for features of the small subunit. Additionally, three key features of the large subunit, L7/L12 stalk counterpart, and L1 stalk are also larger when compared to its bacterial counterpart. A study [19] comparing the secondary structures of the mitoribosomal large subunit and the rRNA of the bacterial large subunit, 16S rRNA and 23S rRNA, respectively [19], discovered that there are bacterial rRNA segments that are not present within the mitoribosome. Using a cryo-EM map, one important observation was that domain I, one of the six secondary structure domains within the bacterial 23S RNA, is not present within the mitoribosome. Additionally, the fragment of domain III is also absent in the mitoribosome. The cryo-EM map also revealed that several missing sections of domain III were filled by proteins, whereas the domain I areas were not replaced by anything [16]. The large subunit harbors the peptidyl transferase center, which is composed of rRNA. This center forms the active site responsible for peptide bond formation. It also contains the ribosomal exit tunnel for nascent polypeptides [14].

2.2.4 The Sites: A, P, and E Within the Mitoribosome

The aminoacyl site (A site), peptidyl site (P site), and transfer RNA-exit site (E site) of the mitoribosome are easily identified, a common trait shared by many ribosomes. Each site has an important and distinctive role. The aminoacyl site main function is to both recognize aminoacyl-tRNA and to detect how the aminoacyl moiety is positioned for the peptidyl transferase reaction [1]. The mt-SSU helps in producing residues which play a vital role in decoding. In addition, the mt-LSU is important in facilitating the contact between mt-transfer RNA (mt-tRNA) and 16S mt-rRNA. As a consequence of the evolution of the rRNA, the A-site finger structure is commonly lost. This loss has been theorized to be related to how characteristics of the mammalian mt-tRNA do not form a cloverleaf structure [15]. The peptidyl site has multiple functions, one is to bind to the initiating tRNA when protein synthesis is initialized in order to maintain its accurate translational reading frame and to prevent loss of the nascent polypeptide chain [1]. Since the peptidyl site finger of the mitochondrial large subunit interacts with the T-loop of the mt-RNA, this site is relatively preserved, and this interaction is more powerful compared to bacterial ribosome. In addition, due to reduced size of the loop compared to the bacterial tRNA, this finger structure is crucial in order to maintain the orientation of the RNA species when the peptide bonds start to form. Lastly, whether the E site is present is widely debated among scientists; this is because contact points between bacterial ribosomes and tRNA are reduced or lost due to a shortened mt-rRNA [15]. The transfer RNA-exit site does however play a key role in stabilizing the mRNA reading frame in both cytoplasmic and bacterial ribosomes. This is because deacylated tRNA and mRNA codon from the peptidyl site is formed from elongation factor GTP (or EFG1mt in mammalian mitochondrion)-dependent tRNA translocation reaction that both interact with the site [4]. However, recent cryo-EM studies have shown that the E site is present due to the existence of a modified binding pocket [2].

3 Function

3.1 Cellular Powerhouse

The mitochondria act as a cellular powerhouse for the cell, and it also contains an important piece of circular chromosome (mtDNA). This mtDNA is crucial as it is involved in the coding of the polypeptide structures of the mitochondrial oxidative phosphorylation chain which is responsible for many of the cells' ability to produce ATP. Human mtDNA codes for 13 essential polypeptide components of the mitochondrial oxidative phosphorylation chain, which generates most of the cells' ATP. It does not only code for 13 polypeptides, but it also is involved in the coding of the producers of these polypeptides, 2 rRNAs and 22 tRNAs. The RNA mitoribosomal components are encoded by a gene called mt-rRNA [2]. Oxidative phosphorylation chain is responsible for the production of ATP by generating a proton gradient across the inner membrane of the mitochondria which is used by the mammalian cells [20]. The basic process of protein synthesis is in Fig. 13.2.

3.2 Apoptosis

The cell's Pandora's box, the mitochondrion, does not only provide organisms with ATP, but it also contains potentially harmful proteins that it keeps stored away. Activation of these harmful proteins sets in motion programmed cell death (apoptosis) pathways that result in the demise of the cell. In many of these pathways, permeabilization of mitochondrial membranes is a critical event that results in the release (from the mitochondrial intermembrane space) of various molecules that are crucial for apoptosis [21]. The mitoribosomes were identified to play an important

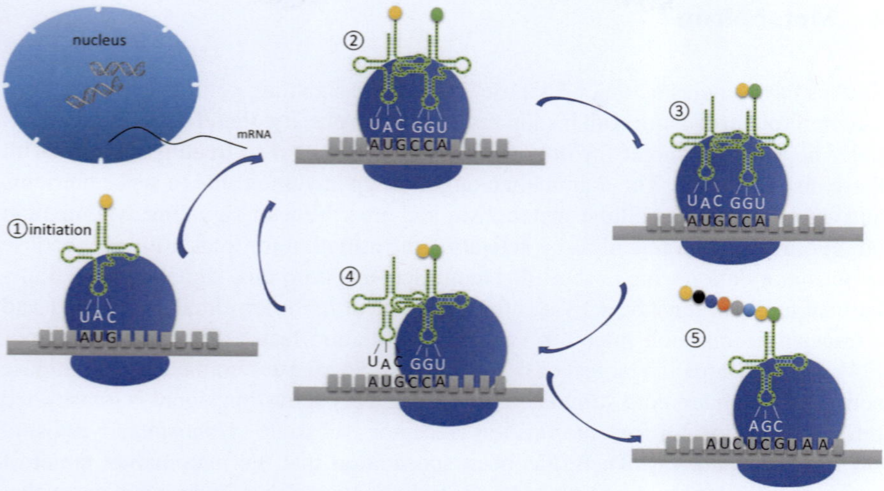

Fig. 13.2 Protein synthesis. ①After the synthesis of mt-mRNA in the nucleus, the cytosol then enters through the nuclear pore and connected with mitoribosome. mt-tRNAMet enter site 1 via complementary base (AUG) pairing. ②According to the codon sequence, the corresponding tRNA of amino acid enters the site 2. ③Transpeptidase catalytic methionine in site 1 and amino acid in site 2 tRNA reacting the dehydration condensation. ④The tRNA then leaves that amino acid behind and moves to the next site on the ribosome, along with a synchronous movement of the mRNA to bring the next codon and its associated amino acid into position for bond formation and repeat the three steps until a stop codon (UAA, UAG, UGA) reaches the A site

role in cell apoptosis; this was proven by the presence of two important proteins: death-associated protein 3 and the programmed cell death protein 9 (PDCD9) [16]. These two pro-apoptotic proteins, death-associated protein 3 and PDCD9, were investigated using a proteolytic digestion of the whole mitochondrial ribosomal subunits. The peptides were then analyzed using a liquid chromatography-tandem mass spectrometry which revealed that they were both part of the mitoribosome. Death-associated protein 3 was shown to have motif characteristic of guanine nucleotide-binding proteins and is most likely the protein that is responsible for nucleotide-binding activity in mammalian mitoribosomes [5]. However questions remain about the death-associated protein 3, such as how its position could affect apoptosis. Although death-associated protein 3 is crucial, other mitoribosomes that are involved in apoptosis should not be neglected. The mitochondrial ribosomal protein, MRPS30, has also been shown to affect the apoptosis process and require further investigation. To clarify, MRPS30 and PDCD9 are the same protein [22].

4 Metabolism

Studies have suggested that S6 kinase 1 and other signaling components may have a central role in causing both insulin resistance and obesity; therefore it could potentially be a crucial target for drug development in regard to treating patients with these disorders [23]. The mammalian target of rapamycin is able to sense nutrients and able to integrate cellular metabolism and growth factor signaling. Mammalian target of rapamycin functions by activating the initiation for translation in response to phosphorylation which occurs due to nutrient and hormonal signaling. Two translational modular must be activated for this to occur: p70 ribosomal S6 kinase 1 and a translation initiation inhibitor, eukaryotic initiation factor 4E-binding protein 1 [24]. Both mammalian targets of rapamycin and S6 kinase 1 contain Ser/Thr kinase activities which are both stimulated by insulin. This activation signal is transmitted through the insulin receptor/insulin receptor substrate 1/phosphatidylinositol 3-kinase/Akt pathway. There has been speculation that the mammalian target of rapamycin/S6 kinase 1pathway acts as a physiological feedback mechanism that can negatively modulate the ability of insulin to transmit signals to phosphatidylinositol 3-kinase via insulin receptor substrate 1 [25]. Sun and his fellow colleagues [26] noted an increase in serine kinase activity in the liver and muscle of obese insulin-resistant rodents. This results in insulin receptor substrate 1, an important molecular mechanism for insulin resistance in glucose metabolism to be inhibited due to inhibition of serine phosphorylation [27].

Nonalcoholic fatty liver disease is closely associated with metabolic syndromes and insulin resistance. These diseases may represent the hepatic component of metabolic syndrome. Pharmacological treatments of fatty liver disease have identified two kinases involved in the process of insulin sensitization: adenosine monophosphate-activated protein kinase (AMPK) and p70 ribosomal S6 kinase 1. The critical role of S6 kinase 1 was evidenced by the fact that the deletion of S6 kinase 1 gene can avoid the development of insulin resistance. Although S6 kinase 1 was targeted as a potential target for insulin resistance, the functional role of AMPK-S6 kinase 1 pathway in lipogenesis in the liver has yet to be explored [28–30]. Consequently, further studies are needed to better understand the pathology and treatment of the disease.

5 Mitochondrial Ribosome Diseases

The ribosome is a subcellular organelle composed of two different subunits [30]; each subunit contains various numbers of rRNAs and ribosomal proteins. Several ribosome-associated proteins (RAPs), which are not ribosomal proteins, are able to bind to the ribosome with the help of ribosomal proteins to compromise the ribosomal proteome. The ribosome acts as a central effector of protein synthesis in all organism kingdoms [31], and the rate of protein synthesis within the cell is determined by both the number of ribosomes per cell and the protein synthesis rate per ribosome. Several lines of evidence suggest that the ribosome may be involved in a range of pathological processes, such as cancer, genetic diseases, and viral infection [32].

5.1 Cancer

Cryo-EM development, and especially single-particle data collection through both image reconstruction and direct-electron detectors, has been a crucial factor in developing detailed functional and structural models of a ribosome. Both ribosome biogenesis and translation control are crucial cellular processes that are regulated at multiple levels. Certain tumor suppressors and proto-oncogenes might regulate malignant progression by altering the protein synthesis machinery [33].

Oncogenesis and tumor progression need to be studied to understand these factors. Additionally, processes such as oncogene activation and the inactivation of tumor suppressors are important factors to research [34]. Recently, new data has shown that the regulation and control of ribosome processing could potentially be related with how some oncogenes and tumor suppressors function. For example, c-myc, an oncogene, cannot only activate RNA pol I and III (involved in rRNA synthesis) but also RNA pol II (involved in synthesis of mRNA coding for ribosomal proteins) [35]. Additionally, when protein RB and TP53, both of which are important tumor suppressor proteins, lose their functions, an increase in ribosome biogenesis activity in the cancer cell can be observed. It has however been demonstrated that oncogenic activity of c-myc could potentially restore an IRES-dependent translational ability within the cell that is commonly deregulated by c-myc in this situation and result in the suppression of genetic control of the ribosome biogenesis [36, 37].

Furthermore, mutations in genes responsible for proteins that partake in ribosome biogenesis production have been discovered to affect cancer and human disease. Mutations in the gene that encodes ribosomal protein S19 have been identified in a syndrome that is characterized by increased susceptibility to cancer—Diamond-Blackfan anemia [38]. Mutations in K-ras or Ki-ras (KRAS) (15–25%), mutations and amplifications in epidermal growth factor receptor (EGFR) (10–50%) and MET (4–8%), and rearrangements in EML4-ALK (4–7%) are part of oncogenic changes within lung adenocarcinoma. By using confocal microscopy experiments on the site of rRNA synthesis, which is where epithelial cell transforming 2 (Ect2) is localized in the nucleolus, it was discovered to be associated with how Ect2 mRNA and 45S pre-ribosomal RNA express themselves. In a human mutant KRAS lung adenocarcinoma, it was noted that there is some ribosomal DNA (rDNA) transcription [39]. While the survival time for breast cancer patients can be difficult to estimate, it has been suggested that it is directly correlated to what the metastatic recurrence rate is. In a study, a Cox regression model was used to investigate which genes were associated with metastasis, both during and after the first 3 years of surgery (early and late type genes). Ribosome-related factors eukaryotic initiation factor (*EIF4B*), ribosomal protein L5 (*RPL5*), ribosomal protein L5 (*RPL3*) which are all late-type genes, and epsin (*EPN3*) which is a tumor angiogenesis modifier were found to be directly linked with what the metastasis-free survival (MFS) was in the later period as well as what the meta-analysis was in tamoxifen-treated breast cancer cohorts. In contrast to this result, only *EPN3* of the late-type genes was shown to survive consistently in more than one cohort in other cancer types. It was also associated to have a worse outcome in two non-small cell lung cancer cohorts [40].

The most common lung cancer type is non-small cell lung cancer (NSCLC) which also accounts for roughly 85–90% of all lung cancers. Accurate open reading frames of both (Rio) kinase 2 (RIOK2) and Nin one binding-1 protein are important as they relate to the assembly of ribosomes and also have elevated levels in malignant tumors. RIOK2 and RIOK1 are both members of the RIO family, and they both have crucial roles in synthesizing the 40S ribosomal subunit by promoting 20S pre-rRNA mature into 18S rRNA. The results of a study [41] demonstrated that RIOK2 and Nin one binding 1 were both at elevated levels in NSCLC cells and tissues. They also observed that tumor node metastasis was associated with expressed profiles and correlations could be drawn from the clinical stage to lymph node metastasis and differentiation.

5.2 *Inflammatory Cells*

5.2.1 Epithelial Cells

Commonly, the cell junction formation and epithelial barrier maintenance and regulation are dictated by the alveolar epithelium. When disruption occurs in type I and II epithelial cells, it can consequently result in increased transport of fluids and an increased permeability in the epithelial barrier [42]. There were two studies that investigated what the role of autophagy in hyperoxia-induced epithelial cell death is. They noted that hyperoxia-induced autophagosome formation in Beas-2B cells occurred, whereas light chain 3B-siRNA caused hyperoxia-induced cell death. However, an overexpression of light chain 3B would result in cytoprotection for cells [43].

5.2.2 Endothelial Cells

The vascular endothelial cells of the lung play a critical role in the semipermeable barrier between the vascular, interstitial, and alveolar spaces. Injury of vascular endothelial cells leads to increased capillary permeability [44]. There are very limited studies focusing on the role in lung microvascular endothelial cells in the context of acute lung injury. Only one related study has reported that mice with knockdown of autophagic protein light chain 3B display an exaggerated pulmonary hypertension during hypoxia [45].

5.2.3 Macrophages

Macrophages can commonly be found in the lung interstitium and alveoli and are available for recruitment into the lung if an inflammatory stimulus is detected. It has been demonstrated that the alveolar macrophages do not only have an important role in both initially responding to the stimuli but also in maintaining the pulmonary inflammation throughout its duration. In the inflammatory pathways, alveolar macrophages result in cell damage in acute lung injury, and a large variety of cytokines, such as tumor necrosis factor-α, interleukin-1β, interleukin 6, interleukin 8, and interferon-g, are secreted by macrophages which are involved in the initial stages in initiation of acute lung injury [46].

5.2.4 Neutrophils

Neutrophils, classified under leukocytes, are crucial in the development of pathology such as acute lung injury/ARDS. Excessive transepithelial leukocyte migration, with the majority being neutrophils, is classified as one of the cardinal features for acute lung injury. Neutrophils stimulated with lipopolysaccharide or TNF-α demonstrated an activation of nuclear factor kappa B (NF-kB), p38, and protein kinase B, and early alterations in neutrophil activation are associated with ventilator time and survival in acute lung injury patients [47]. Polymorphonuclear recruitment is induced when harmful mediators such as cytokines, chemokines, proteases, reactive oxygen species, and matrix metalloproteinases are released in reaction to neutrophil stimulation. Additionally, tumor necrosis factor-α, interleukin-1β, macrophage inflammatory protein 2, and interleukin 8 are also related to neutrophil secretion. Although crucial to the immune response these harmful mediators can cause severe consequences if produced at excess levels, examples of these are membrane damage and increased permeability of the epithelial barrier [48].

5.3 *Lung Injury*

Acute lung injury is defined by an elevation of pulmonary vascular permeability resulting in noncardiogenic pulmonary edema and ultimately hypoxemic respiratory deficiencies. Lipopolysaccharide is known to be able to cause pulmonary inflammation by secreting inflammatory mediators from microbial pathogens. This has also been experimentally proven where animal acute lung injury models were used to cause inflammation in the specimens [49]. Although the exact mechanism remains unknown of a renowned traditional Chinese medicine, Jie-Geng-Tang, it is known for its ability to "clear heat and relieve toxicity." One study demonstrates that Jie-Geng-Tang could mitigate the inflammation of acute lung injury. This study was performed using three different proteins (immunoregulation and anti-inflammation, ribosome, and muscle contraction, resp.). The study concluded that the three protein group functions were indistinguishable from the pharmacological actions of Jie-Geng-Tang [50]. Consequently, we can infer that ribosomal protein relates to acute lung injury.

6 Single Cell Biology

The most basic and fundamental unit of all living tissue is the cell. Importantly, each cell is heterogeneous, and the composition of cells varies greatly, even though they might look identical from a morphological and genetic perspective. Although miniscule, these changes are fundamentally crucial for organism in both how it functions and its health. In recent years scientists have started to integrate cell genomics with single-cell proteomics to investigate issues that remain unsolved in areas such as cancer, stem cell biology, neuroscience, developmental biology, and infectious disease [51–55]. The strength of this approach is its ability to be applied to any kind of tissue with the purpose to delineate cell types, define progenitors, lineage hierarchies, and identify lineage-specific regulatory factors [53].

Renato Dulbecco suggested that [54] for people to understand human cancer, it is crucial that we make efforts to understand the details of the cancer genome. The depth that we can investigate the cancer genome is closely linked to technological advancements. An example would be how gene expression profiling and sequencing of DNA and RNA have allowed us to investigate how transcriptomic and genomic heterogeneity functions in normal health and disease. However, for us to perform gene expression profiling, it was crucial to first develop whole-genome and whole-transcriptome amplification methods [52]. Another example could be single-cell sequencing technology which has recently been under the spotlight. This technology is able to analyze genetic polymorphisms of individual cells at the genome-wide level to target single cells that are isolated from tumor biopsies. This could consequently lead to a potential application to circulating tumor cells, which could result in easier prediction of how a tumor would progress or metastasize [51]. Additionally, multiple annealing and looping-based amplification cycles have a potential application for genome-wide amplification of single circulating tumor cells from lung cancer patients enabling easier diagnostic investigations for patients [55]. Single-cell RNA sequencing was measured [56], and then single-cell DNA sequencing was described 3 years later [57]. This description divided single-cell sequencing into multiple areas, such as sample collection, single-cell isolation, nucleotide sequence (DNA or RNA) amplification, and DNA sequencing and analysis.

Unfortunately, many problems remain to be solved. Due to its high cost, high technical requirement, and lack of computational analysis, it will require time before we will see single-cell sequencing technology applied to clinic [58]. Additionally, further investigation and experimentation in both clinic and laboratory are required in order to develop a more comprehensive understanding of how mutations in the trunk and branch in the cancer phylogenic tree are clinically significant [52].

7 Conclusion

The evolution process of ribosome is undergoing a long time. The discovery of the detailed ribosome structure via X-ray crystallography and cryo-electron microscopy has been an important discovery in furthering our understanding in the field of mitochondrial studies. These include how r-proteins function and operate in order to result in protein synthesis and also what function the multiple regions within the rRNA have [59]. From our discussion in this paper, a few conclusions can be established: firstly, ribosome biogenesis greatly contributes to how diseases progress and develop. Secondly, the increased ribosome production in dysfunctional tissue could be linked to the disease progression. However, there is still a need for further investigation regarding this topic. Although biogenesis contributes to disease progression, it is not yet clear what parts of the biogenesis and ribosome assembly are affected, whether it is only a part of this process or if the entire process is affected. Additionally, the full effect of an increase ribosome production in dysfunctional tissues is unknown [12]. Throughout the years, the progresses in ribosomal proteomics have been lengthy and slow due to the initial beliefs that the ribosome was only involved in protein translation. Mauro and Edelman proposal of the ribosome filter hypothesis was a crucial step in understanding how ribosomal proteomics was not as simple as previously believed [15]. Following their study, more scientists started to investigate this field and a few ribosome-specific regulatory capacity studies were conducted which resulted in an explosive interest in the field of ribosome proteomics [60]. On the other hand, the development of single-cell sequencing technology has led us to see the possibility of a cure for cancer.

8 Future Outlooks

Although the recent progress of high-resolution cryo-EM structure from mitoribosomes has been important step in studying the mammalian ribosome as mentioned above, it is important to note that we are only able to now investigate the more complex mechanisms [4]. Mitochondrial diseases affect at least 1 in 5000 of the population and produce diverse clinical phenotypes often presented as multi-systemic disorders [5]. It is vital to study the mechanism of abnormal mitochondrial ribosomal proteins act on the target sites. This information has the potential to play a key role in the treatment of various diseases caused by abnormal mitoribosomal proteins. It can however not be ignored that the mitoribosome is a common target for antibacterial therapeutics. Due to its complexity and varied sites, there are multiple options to consider for future investigations. Through studying the complexes formed by the ribosome and various antibiotics, scientists have been able to further their understanding of how the drug-binding sites on the ribosomes interact with drugs [61]. However, although there has been a relatively large amount of site studied [62], only a fraction of these have showed to be able to bind to antibiotics. What we can expect is that we will one day be able to fully unscramble the structure and functional sites of the mitoribosome, which would be invaluable for the treatment of ribosomal diseases. The advancement and development of single-cell sequencing technology in recent years have occurred in an extraordinary pace, and their use in different fields in the future cannot be underestimated. However, single-cell methods are not perfect and have known drawbacks and limitations, and thus continuous investigation and exploration are crucial to progress the field of ribosome proteomics.

Acknowledgments The work was supported by Zhongshan Distinguished Professor Grant (XDW), The National Nature Science Foundation of China (91230204, 81270099, 81320108001, 81270131, 81300010), The Shanghai Committee of Science and Technology (12JC1402200, 12431900207, 11410708600, 14431905100). Operation funding of Shanghai Institute of Clinical Bioinformatics, Ministry of Education for Academic Special Science and Research Foundation for PhD Education (20130071110043), and National Key Research and Development Program (2016YFC0902400, 2017YFSF090207).

References

1. Korostelev A, Noller HF. The ribosome in focus: new structures bring new insights. Trends Biochem Sci. 2007;32:434–41. [PubMed: 17764954]
2. Kakkar P, Mehrotra S, Viswanathan PN. Influence of antioxidants on the peroxidative swelling of mitochondria in vitro. Cell Biol Toxicol. 1998;14:313–21. [PubMed: 9808359]
3. Kikuchi S, Ninomiya T, Kohno T, Kojima T, Tatsumi H. Cobalt inhibits motility of axonal mitochondria and induces axonal degeneration in cultured dorsal root ganglion cells of rat. Cell Biol Toxicol. 2017. https://doi.org/10.1007/s10565-017-9402-0. [PubMed: 28656345]
4. Greber BJ, Ban N. Structure and function of the mitochondrial ribosome. Annu Rev Biochem. 2016;85:103–32. [PubMed: 27023846]
5. Kaushal PS, Sharma MR, Agrawal RK. The 55S mammalian mitochondrial ribosome and its tRNA-exit region. Biochimie. 2015;114:119–26. [PubMed: 25797916]
6. Cavdar KE, Ranasinghe A, Burkhart W, Blackburn K, Koc H, Moseley A, Spremulli LL. A new face on apoptosis: death-associated protein 3 and CDPD9 are mitochondrial ribosomal proteins. FEBS Lett. 2001;492:166–70. [PubMed: 11248257]
7. De SD, YT T, Amunts A, Fontanesi F, Barrientos A. Mitochondrial ribosome assembly in health and disease. Cell Cycle. 2015;14:2226–50. [PubMed: 26030272]
8. Pietromonaco SF, Denslow ND, O'Brien TW. Proteins of mammalian mitochondrial ribosomes. Biochimie. 1991;73:827–36. [PubMed: 1764527]
9. Brown A, Amunts A, Bai XC, Sugimoto Y, Edwards PC, Murshudov G, Scheres SH, Ramakrishnan V. Structure of the large ribosomal subunit from human mitochondria. Science. 2014;346:718–22. [PubMed: 25278503]
10. Rackham O, Filipovska A. Supernumerary proteins of mitochondrial ribosomes. Biochim Biophys Acta. 1840;2014:1227–32. [PubMed: 23958563]
11. Sharma MR, Booth TM, Simpson L, Maslov DA, Agrawal RK. Structure of a mitochondrial ribosome with minimal RNA. Proc Natl Acad Sci U S A. 2009;106:9637–42. [PubMed: 19497863]
12. Maayah ZH, Abdelhamid G, El-Kadi AO. Development of cellular hypertrophy by 8-hydroxyeicosatetraenoic acid in the human ventricular cardiomyocyte, RL-14 cell line, is implicated by MAPK and NF-κB. Cell Biol Toxicol. 2015;31:241–59. [PubMed: 26493311]

13. O'Brien TW. Properties of human mitochondrial ribosomes. IUBMB Life. 2003;55:505–13. [PubMed: 14658756]
14. van der Sluis EO, Bauerschmitt H, Becker T, Mielke T, Frauenfeld J, Berninghausen O, Neupert W, Herrmann JM, Beckmann R. Parallel structural evolution of mitochondrial ribosomes and OXPHOS complexes. Genome Biol Evol. 2015;7:1235–51. [PubMed: 25861818]
15. Kramer G, Boehringer D, Ban N, Bukau B. The ribosome as a platform for co-translational processing, folding and targeting of newly synthesized proteins. Nat Struct Mol Biol. 2009;16:589–97. [PubMed: 19491936]
16. Ganta KK, Mandal A, Chaubey B. Depolarization of mitochondrial membrane potential is the initial event in non-nucleoside reverse transcriptase inhibitor efavirenz induced cytotoxicity. Cell Biol Toxicol. 2017;33:69–82. [PubMed: 27639578]
17. Sharma MR, Koc EC, Datta PP, Booth TM, Spremulli LL, Agrawal RK. Structure of the mammalian mitochondrial ribosome reveals an expanded functional role for its component proteins. Cell. 2003;115:97–108. [PubMed: 14532006]
18. Yusupov MM, Yusupova GZ, Baucom A, Lieberman K, Earnest TN, Cate JH, Noller HF. Crystal structure of the ribosome at 5.5 A resolution. Science. 2001;292:883–96. [PubMed: 11283358]
19. Gabashvili IS, Agrawal RK, Spahn CM, Grassucci R, Svergun D, Frank J, Penczek P. Solution structure of the E. coli 70S ribosome at 11.5 A resolution. Cell. 2000;100:537–49. [PubMed: 10721991]
20. Mears JA, Cannone JJ, Stagg SM, Gutell RR, Agrawal RK, Harvey SC. Modeling a minimal ribosome based on comparative sequence analysis. J Mol Biol. 2002;321:215–34. [PubMed: 12144780]
21. Zerin T, Kim JS, Gil HW, Song HY, Hong SY. Effects of formaldehyde on mitochondrial dysfunction and apoptosis in SK-N-SH neuroblastoma cells. Cell Biol Toxicol. 2015;31:261–72. [PubMed: 26728267]
22. Zhu LZ, Hou YJ, Zhao M, Yang MF, XT F, Sun JY, XY F, Shao LR, Zhang HF, Fan CD. Caudatin induces caspase-dependent apoptosis in human glioma cells with involvement of mitochondrial dysfunction and reactive oxygen species generation. Cell Biol Toxicol. 2016;32:333–45. [PubMed: 27184666]
23. Fenton TR, Gout IT. Functions and regulation of the 70kDa ribosomal S6 kinases. Int J Biochem Cell Biol. 2011;43:47–59. [PubMed: 20932932]
24. Gómez-Sagasti MT, Becerril JM, Epelde L, Alkorta I, Garbisu C. Early gene expression in Pseudomonas fluorescents exposed to a polymetallic solution. Cell Biol Toxicol. 2015;31:39–81. [PubMed: 25754557]
25. Zhou Y, Zhang S, Dai C, Tang S, Yang X, Li D, Zhao K, Xiao X. Quinocetone triggered ER stress-induced autophagy via ATF6/DAPK1-modulated mAtg9a trafficking. Cell Biol Toxicol. 2016;32:141–52. [PubMed: 27085326]
26. Khamzina L, Veilleux A, Bergeron S, Marette A. Increased activation of the mammalian target of rapamycin pathway in liver and skeletal muscle of obese rats: possible involvement in obesity-1inked insulin resistance. Endocrinology. 2005;146:1473–81. [PubMed: 15604215]
27. Qiao LY, Zhande R, Jetton TL, Zhou G, Sun XJ. In vivo phosphorylation of insulin receptor substrate 1 at serine 789 by a novel serine kinase in insulin resistant rodents. J Biol Chem. 2002;277:26530–9. [PubMed: 12006586]
28. Tandon P, Gallo CA, Khatri S, Barger JF, Yepiskoposyan H, Plas DR. Requirement for ribosomal protein S6 kinase 1 to mediate glycolysis and apoptosis resistance induced by Pten deficiency. Proc Natl Acad Sci U S A. 2011;108:2361–5. [PubMed: 21262837]
29. Kaarniranta K, Tokarz P, Koskela A, Paterno J, Blasiak J. Autophagy regulates death of retinal pigment epithelium cells in age-related macular degeneration. Cell Biol Toxicol. 2017;33:113–28. [PubMed: 27900566]
30. Seo JB, Jung SR, Hille B, Koh DS, Extracellular ATP. Protects pancreatic duct epithelial cells from alcohol-induced damage through P2Y1 receptor-cAMP signal pathway. Cell Biol Toxicol. 2016;32:229–47. [PubMed: 27197531]

31. Hwahng SH, Ki SH, Bae EJ, Kim HE, Kim SG. Role of adenosine monophosphate-activated protein kinase-p70 ribosomal S6 kinase-l pathway in repression of liver X receptor-alpha-dependent lipogenie gene induction and hepatic steatosis by a novel class of dithiolethiones. Hepatology. 2009;49:1913–25. [PubMed: 19378344]
32. Zhang K, Zheng W, Zheng H, Wang C, Wang M, Li T, Wang X, Zhang L, Xiao S, Fei C, Xue F. Identification of oxidative stress and responsive genes of HepG2 cells exposed to quinocetone, and compared with its metabolites. Cell Biol Toxicol. 2014;30:313–29. [PubMed: 25223261]
33. Graifer D, Karpova G. Roles of ribosomal proteins in the functioning of translational machinery of eukaryotes. Biochimie. 2015;109:1–17. [PubMed: 25433209]
34. Teng T, Mercer CA, Hexley P, Thomas G, Fumagalli S. Loss of tumor suppressor RPL5/RPL11 does not induce cell cycle arrest but impedes proliferation due to reduced ribosome content and translation capacity. Mol Cell Biol. 2013;33:4660–71. [PubMed: 24061479]
35. Kim HJ, Maiti P, Barrientos A. Mitochondrial ribosomes in cancer. Semin Cancer Biol. 2017. https://doi.org/10.1016/j.semcancer.2017.04.004. [PubMed: 28445780]
36. Kim WH, Shen H, Jung DW, Williams DR. Some leopards can change their spots: potential repositioning of stem cell reprogramming compounds as anti-cancer agents. Cell Biol Toxicol. 2016;32:157–68. [PubMed: 27156576]
37. Dejure FR, Royla N, Herold S, Kalb J, Walz S, Ade CP, Mastrobuoni G, Vanselow JT, Schlosser A, Wolf E, Kempa S, Eilers M. The MYC mRNA 3′-UTR couple RNA polymerase II function to glutamine and ribonucleotide levels. EMBO J. 2017;36:1854–68. [PubMed: 28408437]
38. Huun J, Lønning PE, Knappskog S. Effects of concomitant inactivation of p53 and pRb on response to doxorubicin treatment in breast cancer cell lines. Cell Death Discov. 2017;3:17026. [PubMed: 28580174]
39. Boesewetter DE, Collier JL, Kim AM, Riley MR. Alterations of A549 lung cell gene expression in response to biochemical toxins. Cell Biol Toxicol. 2006;22:101–18. [PubMed: 16528450]
40. Draptchinskaia N, Gustavsson P, Andersson B, Pettersson M, Willig TN, Dianzani I, Ball S, Tchernia G, Klar J, Matsson H, et al. The gene encoding ribosomal protein S19 is mutated in Diamond-Blackfan anaemia. Nat Genet. 1999;21:169–75. [PubMed: 9988267]
41. Martin JB, Mariana C, Marcelo GK. Nuclear PKCi-ECT2-Rac1 and ribosome biogenesis: a novel Axis in lung tumorigenesis. Cancer Cell. 2017;31:167–9. [PubMed: 28196591]
42. Hellwig B, Madjar K, Edlund K, Marchan R, Cadenas C, Heimes AS, Almstedt K, Lebrecht A, Sicking I, Battista MJ, et al. Epsin family member 3 and ribosome-related genes are associated with late metastasis in estrogen receptor-positive breast cancer and long-term survival in non-small cell lung cancer using a genome-wide identification and validation strategy. PLoS One. 2016;11:e0167585. [PubMed: 27926932]
43. Minghetti M, Drieschner C, Bramaz N, Schug H, Schirmer K. A fish intestinal epithelial barrier model established from the rainbow trout (Oncorhynchus mykiss) cell line. RTgutGC Cell Biol Toxicol. 2017. https://doi.org/10.1007/s10565-017-9385-x. [PubMed: 28251411]
44. Matthay MA, Ware LB, Zimmerman GA. The acute respiratory distress syndrome. J Clin Invest. 2012;122:2731–40. [PubMed: 22850883]
45. Chen C, Shi L, Li Y, Wang X, Yang S. Disease-specific dynamic biomarkers selected by integrating inflammatory mediators with clinical informatics in ARDS patients with severe pneumonia. Cell Biol Toxicol. 2016;32:169–84. [PubMed: 27095254]
46. Stevens T. Functional and molecular heterogeneity of pulmonary endothelial cells. Proc Am Thorac Soc. 2011;8:453–7. [PubMed: 22052919]
47. Lee SJ, Smith A, Guo L, Alastalo TP, Li M, Sawada H, Liu X, Chen ZH, Ifedigbo E, Jin Y, et al. Autophagic protein LC3B confers resistance against hypoxia-induced pulmonary hypertension. Am J Respir Crit Care Med. 2011;183:649–58. [PubMed: 20889906]
48. Lippai M, Szatmári Z. Autophagy-from molecular mechanisms to clinical relevance. Cell Biol Toxicol. 2017;33:145–68. [PubMed: 27957648]

49. Nowak K, Ratajczak-Wrona W, Garley M, Jabłońska E. The effect of ethanol and N-nitrosodimethylamine on the iNOS-dependent NO production in human neutrophils. Role of NF-κB. Xenobiotica. 2017;30:1–8. [PubMed: 28608757]
50. Mondrinos MJ, Zhang T, Sun S, Kennedy PA, King DJ, Wolfson MR, Knight LC, Scalia R, Kilpatrick LE. Pulmonary endothelial protein kinase C-delta (PKCδ) regulates neutrophil migration in acute lung inflammation. Am J Pathol. 2014;184:200–13. [PubMed: 24211111]
51. Chu MP, Kriangkum J, Venner CP, Sandhu I, Hewitt J, Belch AR, Pilarski LM. Addressing heterogeneity of individual blood cancers: the need for single cell analysis. Cell Biol Toxicol. 2017;33:83–97. [PubMed: 27761761]
52. Wang W, Gao D, Wang X. Can single-cell RNA sequencing crack the mystery of cells? Cell Biol Toxicol. 2017. https://doi.org/10.1007/s10565-017-9404-y. [PubMed: 28733864]
53. Phiboonchaiyanan PP, Busaranon K, Ninsontia C, Chanvorachote P. Benzophenone-3 increases metastasis potential in lung cancer cells via epithelial to mesenchymal transition. Cell Biol Toxicol. 2017;33:251–61. [PubMed: 27796700]
54. Ellsworth DL, Blackburn HL, Shriver CD, Rabizadeh S, Soon-Shiong P, Ellsworth RE. Single-cell sequencing and tumorigenesis: improved understanding of tumor evolution and metastasis. Clin Transl Med. 2017;6:15. [PubMed: 28405930]
55. Niu F, Wang DC, JP L, Wu W, Wang XD. Potentials of single-cell biology in identification and validation of disease biomarkers. J Cell Mol Med. 2016;20:1789–95. [PubMed: 27113384]
56. Treutlein B, Brownfield DG, AR W, Neff NF, Mantalas GL, Espinoza FH, Desai TJ, Krasnow MA, Quake SR. Reconstructing lineage hierarchies of the distal lung epithelium using single cell RNA-seq. Nature. 2014;509:371–5. [PubMed: 24739965]
57. Qian M, Wang DC, Chen H, Cheng Y. Detection of single cell heterogeneity in cancer. Semin Cell Dev Biol. 2017;64:143–9. [PubMed: 27619166]
58. Ni X, Zhuo M, Su Z, Duan J, Gao Y, Wang Z, Zong C, Bai H, Chapman AR, Zhao J, et al. Reproducible copy number variation patterns among single circulating tumor cells of lung cancer patients. Proc Natl Acad Sci U S A. 2013;110:21083–8. [PubMed: 24324171]
59. Wang W, Zhu B, Wang X. Dynamic phenotypes: illustrating a single-cell odyssey. Cell Biol Toxicol. 2017. https://doi.org/10.1007/s10565-017-9400-2. [PubMed: 28638956]
60. Wang W, Wang X. Single-cell CRISPR screening in drug resistance. Cell Biol Toxicol. 2017;33:207–10. [PubMed: 28474250]
61. Wang Y, Navin NE. Advances and applications of single cell sequencing technologies. Mol Cell. 2015;58:598–609. [PubMed: 26000845]
62. Matzov D, Bashan A, Yonath A. A bright future for antibiotics? Annu Rev Biochem. 2017;86:567–83. [PubMed: 28654325]
63. Yan G, Yan X. Ribosomal proteomics: strategies, approaches, and perspectives. Biochimie. 2015;113:69–77. [PubMed: 25869001]

Linlin Zhang, MD, is a graduate student co-educated by Wenzhou Medical University and Fudan University. Her main research is focused on lung cancer, mitochondrial diseases, and metabolism.

Bijun Zhu, BS, is a project leader of electron microscope technique, Clinical Science Institute of Fudan University Zhongshan Hospital, Shanghai, China. She has published ten articles. She is interested in the field of oncology and iconography.

Xiangdong Wang, MD, PhD, is a distinguished professor of Medicine, director of Shanghai Institute of Clinical Bioinformatics, executive director of Clinical Science Institute of Fudan University Zhongshan Hospital, director of Fudan University Center of Clinical Bioinformatics, deputy director of Shanghai Respiratory Research Institute, and visiting professor of King's College of London. His main research is focused on clinical bioinformatics, disease-specific biomarkers, lung chronic diseases, cancer immunology, and molecular and cellular therapies. He is the author of more than 200 scientific publications with the impact factor about 900, citation number about 6918, h-index 46, i10-index 181, and cited journal impact factor about 7000.

Chapter 14
Epithelial Mitochondrial Dysfunction in Lung Disease

Linlin Zhang, William Wang, Bijun Zhu, and Xiangdong Wang

Abstract Since the twentieth century, scientists have studied the functions and mechanisms of the mitochondria. The mitochondrion plays many important roles in cell functioning and contributes to apoptosis, embryonic and tissue development, aging, etc. Consequently, mitochondrial dysfunction often has a direct impact on health such as aging, tumorigenesis, lung injury and COPD, etc. Recent evidence indicates that the mitochondria could also be a crucial contributor to immunity with functions such as biogenesis, fusion, and fission impacting various areas in initializing immunity. In this review, we will describe both the structure and various functions of the mitochondria with an emphasis on functions such ATP production which is crucial for a multitude of processes such as apoptosis, biosynthesis of Fe/S clusters, steroid synthesis, and, more fundamentally, cell survival. In addition, this review aims to investigate the relationship of epithelial mitochondria and lung disease. Cigarette smoke is known to induce structural and functional mutations in airway epithelial mitochondria often acting as an indicator for diseases such as COPD. Further evidence to support this speculation is the presence of reactive oxygen species (ROS) within cigarette smoke which is a factor in the development of COPD.

Keywords Mitochondria • Structure • Apoptosis • Reactive oxygen species • COPD

Linlin Zhang, William Wang and Bijun Zhu contributed equally to this work.

L. Zhang • W. Wang • B. Zhu • X. Wang (✉)
Zhongshan Hospital Institute of Clinical Science, Fudan University,
Shanghai Medical College, Shanghai, China
e-mail: xiangdong.wang@clintransmed.org

H. Sun, X. Wang (eds.), *Mitochondrial DNA and Diseases*, Advances in Experimental Medicine and Biology 1038,
https://doi.org/10.1007/978-981-10-6674-0_14

1 Introduction

Mitochondria are crucial in all cell types as they are the primary components of ATP production in cells via a complex set of chemical reactions. Additionally, they are also involved in functions such as programmed cell death, lipid synthesis, calcium signaling, and metabolism. The genes responsible for ribosome production are present within the mitochondria, which produce proteins as well as transfer RNA (tRNA) genes. It also provides a lock-and-key system that aids in deciphering the genetic code into an amino acid protein code. Mitochondria are crucial for continuous cell function and protein production (with a steady supply of tRNA keys) [1]. These main functions are represented in the Fig. 14.1. The mitochondrion is an organelle which is bound by two membranes, an inner and an outer. The dense matrix is enwrapped by the inner membrane, which consists of intermediary metabolism enzymes and different copies for genome enzymes. The genome within this matrix code does not only code for RNAs responsible for translating proteins, but it also codes for inner membrane proteins.

Recently tremendous progress has been made in understanding mitochondrial functions and mechanisms due to technological advancement in structural biology that span the resolution spectrum [2]. Recent discoveries in this field suggest an additional possible explanation for endosymbiosis: the mitochondria might represent a source of molecules that are responsible for host defense in the early eukaryotic

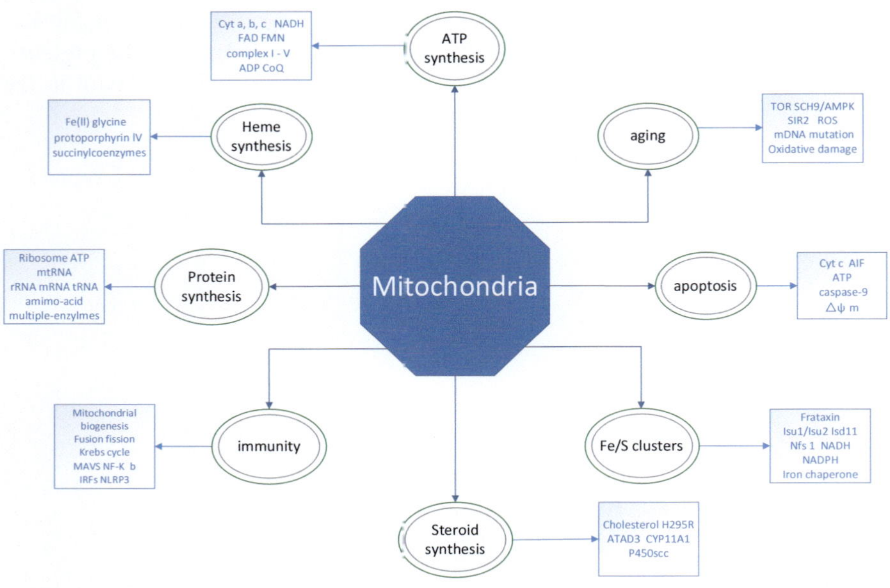

Fig. 14.1 The most important mitochondrial functions. In the oval shape, macro-effects of mitochondria on the cellular activity. And in the rectangle, there are some of the significant factors involved in the corresponding function [19, 24, 31, 62, 63]

cell. An example of this is that reactive oxygen species (ROS) could be involved in the digestion of phagocytosed pathogens; it could also send signals responsible for gene expression. These discoveries have revealed potential mitochondrial targets that could be exploited for therapeutic gain in inflammation and cancer [3].

2 Biology of Mitochondria

2.1 Structure

The development of biological sample techniques for electron microscopy has been pivotal in aiding our understanding of mitochondrial structures; likewise, advancement within the field of biochemical studies has greatly contributed to our understanding of mitochondrial function. Sjöstrand and Palades' [4] work has been important in discovering the uses of electron microscopy for mitochondria. Their model was used as a basis for the current models in textbooks with the inner mitochondrion membrane having a complex morphology and a continuous closed surface. The cristae that are formed from folds are like the bellows of an accordion. The model, known as the baffle model, depicts large openings from the cristae into the intermembrane space and protrudes to the opposite side. However, 3D images have illustrated that instead of the cristae having baffles and wide openings into the intermembrane space, it has more of a pleomorphic and tubular nature. The cristae are made of swollen cisterns or sacs that have narrow and frequently multiple tubular connections that are connected to the peripheral surface of the inner membrane (known as the inner boundary membrane). Orthodox rat-liver mitochondria which were reconstructed, both isolated [5] and in situ [6], demonstrated that the cisternal cristae are flattened and had less interconnections between each other. Furthermore, some orthodox mitochondria had cristae which were almost totally tubular shaped; this discovery coincided with what the microscope observations of liver mitochondria in situ were [7].

A common design theme was noted in the mitochondria in situ in multiple types of tissue when Perkins, Frey, and their coworkers utilized EM tomography to observe the mitochondria. The cristae that were observed were connected to the inner boundary membrane through a tubular structure with a diameter of approximately 28 nm, known as the crista junction. This discovery was consistent with what another group of scientists, Daems and Wisse, for *pediculi crista* recorded [8]. Although there are differences in the morphology and arrangement in cristae, when measured, it was discovered that the mitochondria that were reconstructed in situ in neural, brown adipose tissue (BAT) and *Neurospora* cell types all had uniform structural features. They all possessed crista with a junction diameter of 28 nm. The average distance across the outer membrane and inner boundary membrane is 20 nm. However, at contact sites, it is only 14 nm and has no space between as the thickness is double compared to single membranes. It has been suggested and supported that the inner membrane of the mitochondria is a dynamic structure; this means that it is able to adapt and quickly respond to changes within metabolic and osmotic environment [9]. Although unclear whether the cristae can regulate chemiosmosis, it is debatable whether they could potentially be part of the feedback mechanisms that the mitochondria utilize to respond to environmental alterations instead of just passively controlling volume adjustments. The baffle-like cristae structure originally suggested was discovered to not be an accurate representation of the organelle through the use of 3D tomographic images on a large variety of organisms. A narrower tubular connection to the inner boundary membrane was noted in situ and isolated in cristae that were slightly condensed or orthodox mitochondria. This design suggested that ATP production rates could potentially be regulated by internal metabolite gradients which could have functional implications [2].

2.2 Genome

Mitochondria regulation is mainly directed via a dual genome system, endogenous mitochondrial genes and mitochondrial genes, which transfer to the nucleus during evolution [10]. The mitochondrial "life cycle" begins with the coordinated synthesis of nuclear DNA (nDNA)- and mitochondrial DNA (mtDNA)-encoded proteins, membrane biosynthesis, and the targeting and folding of respiratory chain subunits. There are fundamental differences between the nuclear and mitochondrial genomes. An example of this is size; the mitochondrial genome is significantly smaller when compared to the nuclear genome. In addition, mtDNA molecules with high mitochondrial densities are small in contrast to the size of the nuclear genome. Furthermore, most mitochondrial proteins are encoded by nuclear genes and synthesized on cytosolic ribosomes, whereas mtDNA encodes 13 essential respiratory chain complex subunits [11]. Structurally, the mitochondrial genome is less complex and more easily regulated. The state of compaction for the nuclear DNA is when there are different transcription factors and coactivators that are interacting with each other and when the gene expression is accurately controlled via epigenetic determinant regulation [12]. In contrast, only a few mtDNA-binding proteins, particularly the mitochondrial transcription factor A (TFAM), regulate the transcription for the mitochondrial genome which expresses itself as one single unit [13]. Additionally, mtDNA is susceptible to damage as it is linked to the electron transport system. These components can rapidly mutate and cause harm to the mtDNA in comparison to the nuclear DNA which is less susceptible to damage [14]. Although the cause is not completely understood, it has been suggested to be related to how mtDNA is repaired. The mtDNA has few repair pathways in contrast to nuclear repair which requires a variety of pathways [15]. Additionally, it could be linked to the open structure of the mtDNA being more easily exposed to DNA-damaging ROS [13].

3 Function

3.1 Generate ATP

Mitochondria are integrated components of the cell, and their actions are undoubtedly linked to cellular activities. Energy must be supplied for cellular processes to function, and thus the mitochondria must be functional for cell survival [16].

The major function of mitochondria is to generate ATP through the process of oxidative phosphorylation (OXPHOS). Mitochondria provide 90% of the cell energy via OXPHOS, which is catalyzed by five membrane-bound protein complexes (complex I to V); [17] this multistep pathway results in the generation of 32 ATP molecules per glucose molecule and is used by quiescent cells in normoxia. The process where glucose is converted to pyruvate with lactate as a by-product is known as glycolysis. The production of lactate was originally thought to be produced during hypoxia. ATP production via glycolysis is faster than that by OXPHOS; however, this procedure only produces two molecules of ATP per glucose molecule [3]. Therefore, OXPHOS is much more efficient at producing energy than, for example, glycolysis or the Krebs cycle. Therefore, organisms capable of using this pathway must have received an enormous selective advantage [10]. Both glycolysis and mitochondrial metabolism influence the immune response [3].

3.2 Cell Death

There are currently three death models when classifying according to morphological criteria: autophagy, apoptosis, and necrosis. Autophagy is when degradation of proteins and organelles occur in the lysosomal pathway and the residues are able to be utilized by the cell as energy. There has been discussion about whether autophagy should be considered a death model as some argue that it would be more accurate to classify it as a degradation model. On the other hand, necrosis is the premature death of cells caused by autolysis. The exact mechanism on how necrosis is carried out is still not fully understood; however, there have been studies carried out recently [18] that illustrated that there are specific cellular pathways that carry out necrosis. This could suggest that necrosis does not only occur in relation to cell injury but could be programmed.

Programmed cell death is crucial for an organism and acts as one of the major mechanisms in regulating and maintaining cell numbers. It is also important in aiding immunity and destroying hostile cells. Apoptosis can be defined as a programmed cell death that occurs due to changes, both morphologically or biochemically [19]. Throughout evolution, this process has been refined and passed on as it is crucial in a variety of biological functions such as embryonic development, homeostasis, surveillance, remodeling, and immunity. Additionally, the pathways that regulate and enable apoptosis are crucial in carrying out basic biological processes such as differentiation, proliferation, inflammation, growth, and immunity resulting in apoptosis being crucial for an organism survival. Thus, any dysregulation related to apoptosis, an overexpression or under-expression, could be fatal to cells [20]. An example is carcinogenesis which is caused by insufficient apoptosis. On the other hand, excess apoptosis could potentially cause fatal conditions such as heart failure or stroke [21].

Mitochondria have a vital role in the regulation of apoptosis by the release of proapoptotic mediators (e.g., cytochrome *c*, apoptosis-inducing factor) in response to specific stimuli. However, the mitochondria are also important in producing ATP (linked to electron transfer chain) which is required for apoptosis to occur. Thus, changes in cellular energy (ATP) levels could potentially cause cells to die through necrosis or apoptosis [22]. There are two central pathways that mediate apoptosis: an extrinsic and intrinsic pathway. Figure 14.2 summarizes the intrinsic pathway

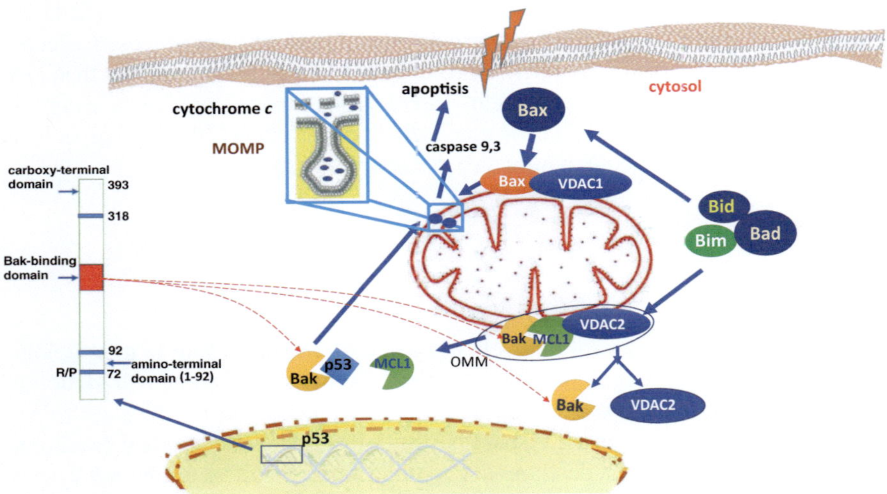

Fig. 14.2 The intrinsic pathway of apoptosis. The cells are stimulated by factors (e.g., stress signals); p53 is activated which contains the Bak-binding domain and a common coding region an arginine (R72) or proline (P72) at amino acid 72 in human p53 [64]. Compared to the P72 form of p53, the R72 which is located at amino acid 72 possesses a greater ability to localize to mitochondria. The mitochondrial pathway of apoptosis is relying on the BCL-2 family (proapoptotics (e.g., Bax and Bak)) for the release of proapoptotic factors (e.g., cytochrome c) from the mitochondrial intermembrane space (IMS), through the procedure of mitochondrial outer-membrane permeabilization (MOMP). On initiation of apoptosis, VDAC2 is displaced from Bak by activated (truncated) Bid, Bim, and Bad, whereas Mcl1 binding to Bak is disrupted by p53. The formation of a p53-Bak complex induces a conformational change in Bak which exists in the mitochondrial outer membrane initially, Bak oligomerization, and the release of cyto c from mitochondrial to cytosol. p53 interacts directly with Bak via a pocket structure formed by the BH1, BH2, and BH3 domains. Bax translocation to mitochondria, which initially located in cytosol, oligomerization, and insertion into the outer mitochondrial membrane combined with VDAC1 [65] via its C-terminal tail, which includes α9. Bax also stimulates the release of cyto c and other apoptogenic mitochondrial proteins and the activation of caspases

[23]. The exact mechanism in how these two pathways are linked to the central mechanism is not completely understood. However, it has been confirmed that the proapoptotic Bcl-2 protein is involved in conveying apoptotic stimuli.

3.3 *Biosynthesis of Fe/S Clusters*

In non-plant eukaryotes, mitochondria are the only compartment where iron/sulfur clusters are assembled. These clusters retain iron in its reduced form and function as cofactors in enzymatic redox reactions. Iron/sulfur proteins are of vital importance for many biological processes; an example of these is the mitochondrial electron transport chain which produces a redox reaction. Because of the toxicity of free ferrous ions as well as sulfide, the synthesis of Fe_2-S_2 as well as Fe_4-S_4 clusters is highly regulated. Iron/sulfur clusters prevent the presence of soluble, reduced iron within the cell, necessitating a complex transport pathway [24].

Besides their important biochemical role, Fe/S clusters are – in concert with the mitochondrial electron transport chain – a source of reactive oxygen species. Superoxide produced by the respiratory chain can, if accumulated above normal cellular levels, damages redox-sensitive proteins, such as the tricarboxylic acid cycle (TCA) enzyme aconitase, Aco1, by oxidizing its Fe_4-S_4 cluster, with the consequence that Fe^{2+} is released [25]. This release of Fe^{2+} is an additional source of ROS by producing hydroxyl radicals via the Fenton reaction [26].

3.4 *Steroid Synthesis*

Steroids aid in regulating a multitude of processes such as development and reproduction which means that steroid synthesis is a pivotal process for living organisms. Steroid synthesis is a multistep mechanism, and due to the wide variety of steroids, an effective and regulated mechanism is crucial in responding to physiological changes [27]. The mitochondria are the main site involved in regulating steroids synthesis. Cholesterol transfer from intracellular stores to the outer mitochondrial membrane (OMM) initializes protein biosynthesis. This transfer occurs via a complex, known as transduceosome, which is the site of interaction between OMM proteins and cytosolic proteins [28]. The cholesterol is then transported to the cytochrome P450 CYP11A1 enzyme located within the inner mitochondrial membrane (IMM). This cytochrome causes the cholesterol to be converted into pregnenolone, a known precursor for steroids. An important complex for this mechanism is the 800 kDa protein complex, known as the steroidogenic metabolon which can be identified at OMM-IMM contact sites.

Organelle plasticity and organelle-organelle interactions are key factors in regulating the formation of steroids, of importance are the endoplasmic reticulum (ER) and mitochondria [29]. Regions that are near appositions situated at the borders between ER membranes and OMM are known as mitochondria-associated membranes (MAMs). These were detected using biochemical and electron microscopy (EM) and are one of the main sites for the transport of calcium from ER to mitochondria. Their function also includes the regulation of mitochondrial homeostasis and cell apoptosis [30]. It has been suggested that an elevated concentration of cholesterol situated in the lipid raft microdomains could be an important source of free cholesterol for mitochondria to use during steroidogenesis [31].

4 Disease

4.1 Chronic Obstructive Pulmonary Disease

Chronic obstructive pulmonary disease (COPD) is a well-known chronic respiratory disease. It is characterized by an increased rate in decline of lung function resulting in chronic inflammation within the lungs. Another characteristic of COPD is limitation within irresistible airflow [32, 33] which is commonly related to conditions such as pulmonary arterial hypertension and hypoxemia which increases the patient susceptibility of having cardiopulmonary complications. A survey showed that about 65 million people have COPD globally, and almost 3 million people, majorly from low- and middle-income countries, died of complications associated with COPD in 2005 [34]. Cigarette smoking is the major etiological factor in this condition. Although COPD could be developed in nonsmokers, 90% of patients with COPD are smokers [35].

Due to the process of oxidative phosphorylation, there will be certain by-products produced that could potentially be harmful to the host, such as oxygen-based radicals. In response to this, the mitochondria have developed different systems, an enzymatic and a nonenzymatic system. In the presence of an elevated level of reactive oxygen species (ROS), the oxidative stress system will also be increased. If the mitochondria sustain damage, there is a likelihood that it could cause a feed-forward process; thus, damage could be further propagated and cause further harm [36].

Cigarette smoke is known to have harmful effects to the body; with over 4000 chemicals, there are a wide variety of potential effects. An example of this is the ROS that is present at a higher concentration in a gaseous state and the lipophilic components. These components are able to pass the lipid bilayer, and reports have shown that the cigarette smoke extract (where no ROS are present) is able to diffuse through the cell membranes and then ultimately affect mitochondrial function. Consequences of this could be an increased generation of ROS intracellularly [37]. Additionally, the function of mitochondria is impacted in the primary bronchial epithelial cells in the presence of cigarette smoke extract [22]. There are multiple factors and functions that are involved in the pathogenesis of COPD; one of these mechanisms is the oxidant-antioxidant imbalance mechanism [38]. Due to the elevated levels of ROS secreted from leukocytes and macrophages in response to the

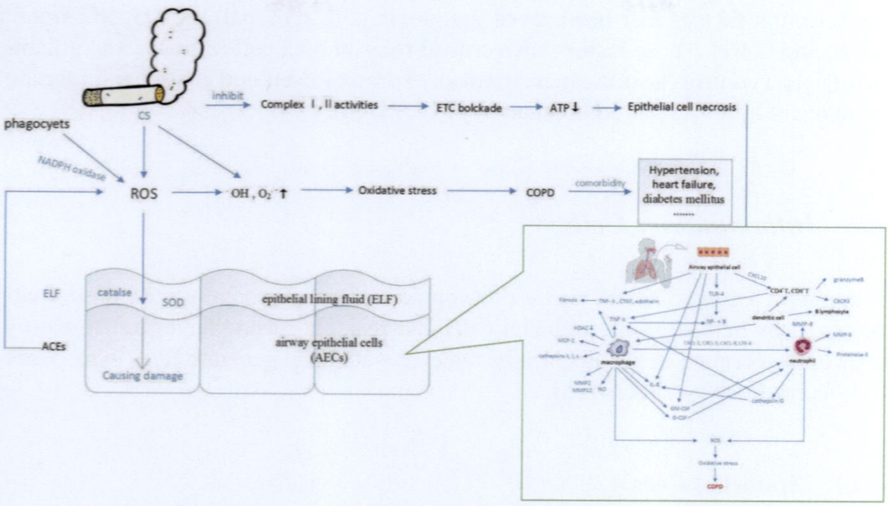

Fig. 14.3 The Role of Cigarette Smoke. The first barrier in lungs, which is achieved by cigarette smoke, is ELF, which covers AECs [66]. Cigarette smoke interacts with antioxidants in ELF. The most significant antioxidant of the normal human lower respiratory tract is catalase in ELF. Others include superoxide dismutase (SOD), glutathione reductase, peroxidase, and ceruloplasmin. After overcoming the ELF barrier, ROS in gaseous phase of cigarette smoke reach AECs plasma cell membranes, causing further pathological effects. However, these ROS are not capable of diffusing through the plasma membranes of these cells and therefore are not capable of entering the circulation. The ROS in cigarette smoke react with AECs plasma cell membranes, resulting in direct injury. Increased level or extended exposure to ROS also develops an inflammatory reaction which may activate epithelial cells and resident macrophages and the recruitment and activation of neutrophils, eosinophils, monocytes, and lymphocytes. This disturbs mitochondrial function, thereby decreasing the ability of mitochondria for ATP synthesis, leading to cellular necrosis

oxidants, an increased burden will be placed on the oxidative system resulting in the inflammatory response observed in COPD (Fig. 14.3). Additionally, due to continuous production of ROS within the body, even post-cessation, the oxidative stress is not relieved [39]. This could have a variety of consequences such as abnormal or pathological alterations to proteins, carbohydrates, lipids, etc. and result in alterations in cellular metabolism [40]. The initial barrier protecting the lungs and airway from harmful pathogens is the epithelial lining fluid (ELF) coating the airway epithelial cells (AECs). Catalase, an antioxidant present in the lower respiratory tract in the ELF, reacts to cigarette smoke [40]. ROS within the cigarette smoke are also able to react to the plasma cell membranes of the AECs and cause direct damage. A lipid class that is vital in maintaining the cellular membranes is the glycerophospholipids. The biological activities of the glycerophospholipids are impacted by the oxidation of esterified unsaturated fatty acids. An increased level of oxidized phospholipids can both directly or indirectly interact with the Toll-like receptor (TLR) signaling, resulting in lung injury. The trigger to activating this pathway is TLR4 stimulation, which subsequently activates NF-κB [41]. NF-κB is a family of transcription factors that have seven members: p50, p52, p100, p105, RelA/p65, RelB, and C-Rel. These factors have crucial roles in both cell response and inflammation as it controls how the gene network expresses itself and is also an important component in the airway inflammation of COPD [42].

4.2 Inflammatory Cells

A common feature of COPD is the development of an inflammatory response, characterized by activation of epithelial cells and resident macrophage. Inflammatory cells once recruited in the airspace become activated and generate ROS in response to inflammatory mediators [33].

4.2.1 Epithelial Cells

Cigarette smoke and other irritants inhaled (biomass fuel smoke) activate the epithelial cells causing a variety of inflammatory mediators such as TNF-α, IL-1β, GM-CSF, and IL-8 and IL-6 to be secreted. Additionally, within the smaller airways, epithelial cells could potentially also secrete TGF-β, known to be a contributing factor for local fibrosis [43]. A key factor required to maintain and regulate the integrity of the alveolar cell is the vascular endothelial growth factor (VEGF). Studies have shown that a blockage of these receptors, VEGF receptors (VEGFR2) in rats, could result in the apoptosis of alveolar cells and emphysema-like disorders [44]. Although characteristics of growth factors that participate in vital mechanisms such as epithelial cell proliferation, cell cycle, and differentiation within COPD are not fully understood, a study showed that squamous metaplasia within the airway epithelium in chronic bronchitis and COPD is frequent [45].

4.2.2 Macrophages

Macrophage is an important player in the pathophysiology of COPD and is also involved in the inflammatory response [46]. An elevated level (fivefold to tenfold) of macrophages can be observed in the airways, lung parenchyma, and bronchoalveolar lavage (BAL) fluid. Macrophages are concentrated where there is destruction in the alveolar walls in emphysema patients, and a correlation was discovered between the parenchyma macrophage numbers and the severity of the emphysema [47]. In addition, cigarette smoke could potentially cause macrophages to be activated causing inflammatory mediators secretion, such as TNF-α, IL-8, chemokine (C-C motif) ligand 2 (monocyte chemoattractant protein 1), leukotriene B4, and ROS [48].

4.2.3 Neutrophil

The stimulatory effect of cigarette smoke impacts the production of granulocytes and its survival within the respiratory tract. This could be due to interactions between GM-CSF and G-CSF secreted from lung macrophages. Additionally, neutrophils secrete factors such as NE, cathepsin G, and proteinase-3, which could potentially cause alveolar destruction. The factors mentioned are part of serine proteases which are also an important stimulator for mucus secretion [48]. A key factor for the recruitment of neutrophils in both the parenchyma and the airway is the adhesion to endothelial cells through E-selectin. This interaction is commonly elevated in COPD patients. The migration throughout the respiratory tract is directed by neutrophil chemotactic factors, such as leukotriene B4, CXCL1, CXCL5 (ENA-78), and CXCL8, which are elevated in COPD patients [49]. The activation of neutrophils within the airways in COPD patients is due to the increased levels of granule proteins, such as myeloperoxidase (MPO) and human neutrophil lipocalin.

4.2.4 Eosinophils

Although eosinophils are frequently discussed in the context of asthma, their exact role in COPD is less well known. There have been studies that suggest an elevated level of eosinophils can be observed in patients with COPD and bronchoalveolar lavage [50]; however, there are also studies which contradict this information [51]. It is speculated that COPD patients could potentially benefit from a corticosteroid therapy due to the presence of eosinophils; however, this requires further investigation [52, 53].

4.2.5 Dendritic Cells

For the innate and the adaptive system to function and relay information to each other, there needs to be a mediator; this mediator is known as the dendritic cell. The dendritic cell is an important cell that relays information and can activate a variety of the cells involved in immunity such as macrophages, neutrophils, and lymphocytes. Thus, these cells could also be important in how the body responses to cigarette smoke. An increased level of dendritic cell activation seems to be observed in COPD patients [54] and could also be related to how severe the disease is [55].

5 Single-Cell Analyses

Single-cell analysis is a rapidly emerging research field that has fascinated researchers from various disciplines such as biology, pharmacy, biotechnology, etc. [56]. Within recent years, it was discovered that each cell could vary greatly from one another causing a multitude of consequences to health [57]. Single-cell analysis is the most reliable approach for studying tumor heterogeneity. Advances in whole-genome analysis (WGA), next-generation sequence (NGS), fluorescence-activated cell sorting (FACS), and other techniques have made it possible to analyze multiple markers in single tumor cells isolated from fresh or fixed primary tumors and metastases [58]. This technique enables us to study each cell and how they vary within a specified population. This population is not strictly defined and could be organs, tissues, or a cell culture. Single-cell analysis is currently the most suitable approach to study biological processes such as cancer, physiological functions, and stem cell differentiation at a deeper level. Thus, it could also be said that single-cell analysis is crucial for drug development and disease studies [59]. Single-cell analysis has made tremendous progress over the recent years as it could potentially affect a wide variety of fields. The lack of cross-disciplinary collaborations in studying single cells coupled with difficulties in producing algorithms suitable for larger databases has resulted in an ineffective approach. Due to it being a new technique, there is not an established standard technique, and in comparison to other experimental platforms, single-cell analysis still requires time to mature [57].

6 Conclusion

Many of the ROS, both exogenous and endogenous involved in the pathophysiology of COPD, are associated with cigarette smoke. Numerous things have an impact on this; an example is the lipophilic components or how the inflammatory processes dynamically interact with each other. Even post-smoking cessation the inflammation could persist indicating the presence of an autoregulated mechanism; however, this only remains as a speculation as the mechanism is not yet understood [60]. One source that endogenous ROS could stem from is the mitochondria. Due to its natural mechanism, it is common that electrons are leaked from the electron transfer chain resulting in ROS. There are however intra- or extracellular factors that could result in excess amount of ROS causing an oxidative-antioxidative imbalance. One of endogenous sources of ROS is mitochondria. In addition, with impaired antioxidant defense, there is a large possibility of it contributing to the development of oxidative stress [40]. On this occasion, multifactors may lead to the disruption of the function and structure of airway epithelial mitochondria and eventually develop into COPD.

7 Future Outlooks

The early pioneers of mitochondria research would be fascinated by the discoveries that have been made concerning mitochondria and many of its functions. For the study of scholars in different fields, mitochondria have a diverse variety of functions. Mitochondria can be viewed as the powerhouse of immunity, in addition to their role as the powerhouse of the cell [61]. With the current technology and understanding, it is possible to investigate a variety of pathways responsible for COPD pathophysiology. However, these are mostly superficial, and more complex interactions such as how the mitochondria are linked to COPD remain unknown [40]. Thus, understanding the underlying mechanisms of the signaling pathway may provide valuable information regarding the therapeutic treatment of COPD. With the continuous innovation of technology and meticulous research in future, a large array of therapeutic possibilities for inflammatory diseases and cancer exist. For instance, single-cell analysis is very significant for the early diagnosis and prevention of disease and has great potential for future clinical application.

Acknowledgments The work was supported by Zhongshan Distinguished Professor Grant (XDW), the National Nature Science Foundation of China (91230204, 81270099, 81320108001, 81270131, 81300010), the Shanghai Committee of Science and Technology (12JC1402200, 12431900207, 11410708600, 14431905100), the operation funding of Shanghai Institute of Clinical Bioinformatics, the Ministry of Education for Academic Special Science and Research Foundation for PhD Education (20130071110043), and the National Key Research and Development Program (2016YFC0902400, 2017YFSF090207).

References

1. Bao L, Diao H, Dong N, Su X, Wang B, Mo Q, Yu H, Wang XD, Chen C. Histone deacetylase inhibitor induces cell apoptosis and cycle arrest in lung cancer cells via mitochondrial injury and p53 up-acetylation. Cell Biol Toxicol. 2016;32:469–82. [PubMed: 27423454]
2. Torrealba N, Aranguiz P, Alonso C, Rothermel BA, Lavandero S. Mitochondria in structural and functional cardiac remodeling. Adv Exp Med Biol. 2017;982:277–306. [PubMed: 28551793]
3. Mills EF, Kelly B, O'Neill LAJ. Mitochondria are the powerhouses of immunity. Nat Immunol. 2017;18:488–98. [PubMed: 28418387]
4. Palade GE. The fine structure of mitochondria. Anat Rec. 1952;114:427–51. [PubMed: 12996882]
5. Kharbangar A, Khynriam D, Prasad SB. Effect of cisplatin on mitochondrial protein, glutathione, and succinate dehydrogenase in Dalton lymphoma-bearing mice. Cell Biol Toxicol. 2000;16:363–73. [PubMed: 11254162]
6. Kakkar P, Mehrotra S, Viswanathan PN. Influence of antioxidants on the peroxidative swelling of mitochondria in vitro. Cell Biol Toxicol. 1998;14:313–21. [PubMed: 9808359]
7. Lea PJ, Hollenberg MJ. Mitochondrial structure revealed by high resolution scanning electron microscopy. Am J Anat. 1989;184:245–57. [PubMed: 2750680]
8. Sardão VA, Oliveira PJ, Holy J, Oliveira CR, Wallace KB. Morphological alterations induced by doxorubicin on H9c2 myoblasts: nuclear, mitochondrial, and cytoskeletal targets. Cell Biol Toxicol. 2009;25:227–43. [PubMed: 18386138]
9. Arismendi-Morillo G, Castellano-Ramírez A, Seyfried TN. Ultrastructural characterization of the mitochondria-associated membranes abnormalities in human astrocytomas: functional and therapeutics implications. Ultrastruct Patho. 2017;41:234–44. [PubMed: 28375672]
10. Wang M, Huang YP, Wu H, Song K, Wan C, Chi AN, Xiao YM, Zhao XY. Mitochondrial complex I deficiency leads to the retardation of early embryonic development in Ndufs4 knockout mice. PeerJ. 2017;5:e3339. [PubMed: 28533980]

11. Zhu J, Wang KZ, Chu CT. After the banquet mitochondrial biogenesis, mitophagy, and cell survival. Autophagy. 2013;9:1663–76. [PubMed: 23787782]
12. Bogenhagen DF. Mitochondrial DNA nucleoid structure. Biochim Biophys Acta. 1819;2012:914–20. [PubMed: 22142616]
13. Hu HL, Zhang ZX, Chen CS, Cai C, Zhao JP, Wang XD. Effects of mitochondrial K+ channel and membrane potential on hypoxic human pulmonary artery smooth muscle cells. Am J Respir Cell Mol Biol. 2010;42:661–6. [PubMed: 19617400]
14. Pesole G, Gissi C, De Chirico A, Saccone C. Nucleotide substitution rate of mammalian mitochondrial genome. J Mol Evol. 1999;48:427–34. [PubMed: 10079281]
15. Patenge N. Quantification of DNA damage and repair in mitochondrial, nuclear, and bacterial genomes by real-time PCR. Methods Mol Biol. 1644;2017:159–66. [PubMed: 28710762]
16. McBride HM, Neuspiel M, Wasiak S. Mitochondria: more than just a powerhouse. Curr Biol. 2006;16:R551–60. [PubMed: 16860735]
17. Arumugam P, Samson A, Ki J, Song JM. Knockdown of clusterin alters mitochondrial dynamics, facilitates necrosis in camptothecin-induced cancer stem cells. Cell Biol Toxicol. 2017;33:307–21. [PubMed: 28064403]
18. Kaarniranta K, Tokarz P, Koskela A, Paterno J, Blasiak J. Autophagy regulates death of retinal pigment epithelium cells in age-related macular degeneration. Cell Biol Toxicol. 2017;33:113–28. [PubMed: 27900566]
19. Shukla GS, Shukla A, Potts RJ, Osier M, Hart BA, Chiu JF. Cadmium-mediated oxidative stress in alveolar epithelial cells induces the expression of gamma-glutamylcysteine synthetase catalytic subunit and glutathione S-transferase alpha and pi isoforms: potential role of activator protein-1. Cell Biol Toxicol. 2000;16:347–62. [PubMed: 11254161]
20. Velentzas PD, Velentzas AD, Mpakou VE, Antonelou MH, Margaritis LH, Papassideri IS, Stravopodis DJ. Detrimental effects of proteasome inhibition activity in Drosophila melanogaster: implication of ER stress, autophagy, and apoptosis. Cell Biol Toxicol. 2013;29:13–37. [PubMed: 23161111]
21. Rincheval V, Bergeaud M, Mathieu L, Leroy J, Guillaume A, Mignotte B, Le Floch N, Vayssière JL. Differential effects of Bcl-2 and caspases on mitochondrial permeabilization during endogenous or exogenous reactive oxygen species-induced cell death: a comparative study of H_2O_2, paraquat, t-BHP, etoposide and TNF-α-induced cell death. Cell Biol Toxicol. 2012;28:239–53. [PubMed: 22491967]
22. Van der Toorn M, Slebos DJ, de Bruin HG, Leuvenink HG, Bakker SJ, Gans RO, Koëter GH, van Oosterhout AJ, Kauffman HF. Cigarette smoke-induced blockade of the mitochondrial respiratory chain switches lung epithelial cell apoptosis into necrosis. Am J Physiol Lung Cell Mol Physiol. 2007;292:L1211–8. [PubMed: 17209140]
23. Renvoizé C, Biola A, Pallardy M, Bréard J. Apoptosis: identification of dying cells. Cell Biol Toxicol. 1998;14:111–20. [PubMed: 9553722]
24. Vukotic M, Nolte H, König T, Saita S, Ananjew M, Krüger M, Tatsuta T, Langer T. Acylglycerol kinase mutated in Sengers syndrome is a subunit of the TIM22 protein translocase in mitochondria. Mol Cell. 2017;67:471–83. [PubMed: 28712724]
25. Zerin T, Kim JS, Gil HW, Song HY, Hong SY. Effects of formaldehyde on mitochondrial dysfunction and apoptosis in SK-N-SH neuroblastoma cells. Cell Biol Toxicol. 2015;31:261–72. [PubMed: 26728267]
26. Larsen FJ, Schiffer TA, Ørtenblad N, Zinner C, Morales-Alamo D, Willis SJ, Calbet JA, Holmberg HC, Boushel R. High-intensity sprint training inhibits mitochondrial respiration through aconitase inactivation. FASEB J. 2016;30:417–27. [PubMed: 26452378]
27. Kalayou S, Hamre AG, Ndossi D, Connolly L, Sørlie M, Ropstad E, Verhaegen S. Using SILAC proteomics to investigate the effect of the mycotoxin, alternariol, in the human H295R steroidogenesis model. Cell Biol Toxicol. 2014;30:361–76. [PubMed: 25416481]
28. Papadopoulos V, Miller WL. Role of mitochondria in steroidogenesis. Best Pract Res Clin Endocrinol Metab. 2012;26:771–90. [PubMed: 23168279]

29. Issop L, Rone MB, Papadopoulos V. Organelle plasticity and interactions in cholesterol transport and steroid biosynthesis. Mol Cell Endocrinol. 2013;371:34–46. [PubMed: 23246788]
30. Medvedev R, Hildt E, Paloen D. Look who's talking-the crosstalk between oxidative stress and autophagy supports exosomal-dependent release of HCV particles. Cell Biol Toxicol. 2017;33:211–31. [PubMed: 27987184]
31. Issop L, Fan J, Lee S, Rone MB, Basu K, Mui J, Papadopoulos V. Mitochondria-associated membrane formation in hormone-stimulated Leydig cell steroidogenesis: role of ATAD3. Endocrinology. 2015;156:334–45. [PubMed: 25375035]
32. Fang XC, Wang XD, Danderson M, Bai CX. Chronic obstructive pulmonary disease in China: the burden and the importance of proper management. Chest. 2011;139:920–9. [PubMed: 21467059]
33. Wang DC, Shi L, Zhu Z, Gao D, Zhang Y. Genomic mechanisms of transformation from chronic obstructive pulmonary disease to lung cancer. Semin Cancer Biol. 2017;42:52–9. [PubMed: 27840275]
34. Wu X, Yuan B, López E, Bai C, Wang XD. Gene polymorphisms and chronic obstructive pulmonary disease. J Cell Mol Med. 2014;18:15–26. [PubMed: 24256364]
35. Wang X. New biomarkers and therapeutics can be discovered during COPD-lung cancer transition. Cell Biol Toxicol. 2016;32:359–61. [PubMed: 27405768]
36. Farfán Labonne BE, Gutiérrez M, Gómez-Quiroz LE, Konigsberg Fainstein M, Bucio L, Souza V, Flores O, Ortíz V, Hernández E, Kershenobich D. Acetaldehyde-induced mitochondrial dysfunction sensitizes hepatocytes to oxidative damage. Cell Biol Toxicol. 2009;25:599–609. [PubMed: 19137438]
37. Kennedy CH, Catallo WJ, Wilson VL, Mitchell JB. Combustion products of 1,3-butadiene inhibit catalase activity and induce expression of oxidative DNA damage repair enzymes in human bronchial epithelial cells. Cell Biol Toxicol. 2009;25:457–70. [PubMed: 18685817]
38. Fischer BM, Voynow JA, Ghio AJ. COPD: balancing oxidants and antioxidants. Int J Chron Obstruct Pulmon Dis. 2015;10:261–76. [PubMed: 25673984]
39. Wu X, Sun X, Chen C, Bai C, Wang XD. Dynamic gene expressions of peripheral blood mononuclear cells in patients with acute exacerbation of chronic obstructive pulmonary disease: a preliminary study. Crit Care. 2014;18:508. [PubMed: 25407108]
40. Białas AJ, Sitarek P, Miłkowska-Dymanowska J, Piotrowski WJ, Górski P. The role of mitochondria and oxidative/antioxidative imbalance in pathobiology of chronic obstructive pulmonary disease. Oxid Med Cell Longev. 2016;2016:7808576. [PubMed: 28105251]
41. Chen C, Shi L, Li Y, Wang X, Yang S. Disease-specific dynamic biomarkers selected by integrating inflammatory mediators with clinical informatics in ARDS patients with severe pneumonia. Cell Biol Toxicol. 2016;32:169–84. [PubMed: 27095254]
42. Schuliga M. NF-κB signaling in chronic inflammatory airway disease. Biomolecules. 2015;5:1266–83. [PubMed: 26131974]
43. Chen H, Song ZJ, Qian MJ, Bai CX, Wang XD. Selection of disease-specific biomarkers by integrating inflammatory mediators with clinical informatics in AECOPD patients: a preliminary study. J Cell Mol Med. 2012;16:1286–97. [PubMed: 21883889]
44. Chen H, Wang YL, Bai CX, Wang XD. Alterations of serum inflammatory biomarkers in the healthy and chronic obstructive pulmonary disease patients with or without acute exacerbation. J Cell Mol Med. 2012;16:1286–97. [PubMed: 22343073]
45. Xu M, Wang X. Critical roles of mucin-1 in sensitivity of lung cancer cells to tumor necrosis factor-alpha and dexamethasone. Cell Biol Toxicol. 2017;33:361–71. [PubMed: 28470556]
46. Fang XC, Li SQ, Gao L, Zhao NQ, Wang XD, Bai CX. A short-term educational program improved physicians' adherence to guidelines for COPD and asthma in Shanghai. Clin Transl Med. 2012;1:13. [PubMed: 23369324]
47. Fang X, Netzer M, Baumgartner C, Bai C, Wang XD. Genetic network and gene set enrichment analysis to identify biomarkers related to cigarette smoking and lung cancer. Cancer Treat Rev. 2013;39:77–88. [PubMed: 22789435]

48. Fang X, Li K, Tao X, Chen C, Wang X, Wang L, Wang DC, Zhang Y, Bai C, Wang XD. Effects of phosphoinositide 3-kinase on protease-induced acute and chronic lung inflammation, remodeling, and emphysema in rats. Chest. 2013;143:1025–35. [PubMed: 23188423]
49. Truedsson M, Malm J, Barbara Sahlin K, Bugge M, Wieslander E, Dahlbäck M, Appelqvist R, Fehniger TE, Marko-Varga G. Biomarkers of early chronic obstructive pulmonary disease (COPD) in smokers and former smokers. Protocol of a longitudinal study. Clin Transl Med. 2016;5:9. [PubMed: 26951192]
50. Wang YL, Bai CX, Li K, Adler KB, Wang XD. Role of airway epithelial cells in development of asthma and allergic rhinitis. Respir Med. 2008;102:949–55. [PubMed: 18339528]
51. Sohal SS, Mahmood MQ, Walters EH. Clinical significance of epithelial mesenchymal transition (EMT) in chronic obstructive pulmonary disease (COPD): potential target for prevention of airway fibrosis and lung cancer. Clin Transl Med. 2014;3:33. [PubMed: 26932377]
52. King PT. Inflammation in chronic obstructive pulmonary disease and its role in cardiovascular disease and lung cancer. Clin Transl Med. 2015;4:68. [PubMed: 26220864]
53. Yan FG, Chen CS, Jing JY, Li W, Shen HB, Wang XD. Association between polymorphism of glutathione S-transferase P1 and chronic obstructive pulmonary disease: a meta-analysis. Respir Med. 2010;104:473–80. [PubMed: 20117922]
54. Freeman CM, Curtis JL. Lung dendritic cells: shaping immune responses throughout chronic obstructive pulmonary disease progression. Am J Respir Cell Mol Biol. 2017;56:152–9. [PubMed: 27767327]
55. Upham JW, Xi Y. Dendritic cells in human lung disease: recent advances. Chest. 2017;151:668–73. [PubMed: 27729261]
56. Wang W, Zhu B, Wang X. Dynamic phenotypes: illustrating a single-cell odyssey. Cell Biol Toxicol. 2017. https://doi.org/10.1007/s10565-017-9400-2. [PubMed: 28638956
57. Fessenden M. Metabolomics: small molecules, single cells. Nature. 2016;540:153–5. [PubMed: 27905420]
58. Wang W, Wang X. Single-cell CRISPR screening in drug resistance. Cell Biol Toxicol. 2017;33:207–10. [PubMed: 28474250]
59. Niu F, Wang DC, Lu JP, Wu W, Wang XD. Potentials of single-cell biology in identification and validation of disease biomarkers. J Cell Mol Med. 2016;20:1789–95. [PubMed: 27113384]
60. Barnes PJ. Cellular and molecular mechanisms of chronic obstructive pulmonary disease. Clin Chest Med. 2014;35:71–86. [PubMed: 24507838]
61. Jiang Y, Wang X, Hu D. Mitochondrial alterations during oxidative stress in chronic obstructive pulmonary disease. Int J Chron Obstruct Pulmon Dis. 2017;12:1153–62. [PubMed: 28458526]
62. Papa S, Martino PL, Capitanio G, Gaballo A, De Rasmo D, Signorile A, Petruzzella V. The oxidative phosphorylation system in mammalian mitochondria. Adv Exp Med Biol. 2012;942:3–37. [PubMed: 22399416]
63. Reddi AR, Hamza I. Heme mobilization in animals: a Metallolipid's journey. Acc Chem Res. 2016;49:1104–10. [PubMed: 27254265]
64. Leu JI, Dumont P, Hafey M, Murphy ME, George DL. Mitochondrial p53 activates Bak and causes disruption of a Bak-Mcl1 complex. Nat Cell Biol. 2004;6:443–50. [PubMed: 15077116]
65. Huang L, Han J, Ben-Hail D, He L, Li B, Chen Z, Wang Y, Yang Y, Liu L, Zhu Y, et al. A new fungal Diterpene induces VDAC1-dependent apoptosis in Bax/Bak-deficient cells. J Biol Chem. 2015;290:23563–78. [PubMed: 26253170]
66. Roland FH, Sina Z, Simone MB, Arjan K, Harold GB. Prolonged cigarette smoke exposure alters mitochondrial structure and function in airway epithelial cells. Respir Res. 2013;14:97. [PubMed: 24088173]

Linlin Zhang, MD, is a graduate student coeducated by Wenzhou Medical University and Fudan University. Her main research is focused on lung cancer, mitochondrial diseases, and metabolism.

Bijun Zhu, BS, is a project leader of electron microscope technique, Clinical Science Institute of Fudan University Zhongshan Hospital, Shanghai, China. She has published ten articles. She is interested in the field of oncology and iconography.

Xiangdong Wang, MD, PhD, is a distinguished professor of Medicine, director of Shanghai Institute of Clinical Bioinformatics, executive director of Clinical Science Institute of Fudan University Zhongshan Hospital, director of Fudan University Center of Clinical Bioinformatics, deputy director of Shanghai Respiratory Research Institute, and visiting professor of King's College of London. His main research is focused on clinical bioinformatics, disease-specific biomarkers, lung chronic diseases, cancer immunology, and molecular and cellular therapies. He is the author of more than 200 scientific publications with the impact factor about 900, citation number about 6918, h-index 46, i10-index 181, and cited journal impact factor about 7000.

Chapter 15
Significance of Mitochondria DNA Mutations in Diseases

Zhenhua Zhu and Xiangdong Wang

Abstract Mitochondria are essential double-membraned cytoplasmic organelles to support aerobic respiration and produce cellular energy by oxidative phosphorylation (OXPHOS). Mitochondrial functions are controlled by mitochondrial (mtDNA) and nuclear genomes (nDNA). Mutations of mtDNA result in mitochondrial dysfunction and multisystem diseases through compromising OXPHOS function directly by a point mutation or a large-scale mtDNA rearrangement. One or more of OXPHOS complexes are impaired and dysfunctional to affect tissues with high energy demands. mtDNA is more susceptible to oxidative damage and has more mutations than nDNA. Unlike diploid nDNA, mtDNA is a multi-copy genome transmitted and maternally inherited through oocyte. The multi-copy nature of mtDNA easily causes the heteroplasmy as a unique aspect of mtDNA, making mitochondrial diseases more complex and heterogeneous. mtDNA-associated mitochondrial dysfunction plays the important role in the development of multisystemic primary mitochondrial disease, neurodegeneration, and cancer. The present article overviews the occurrence of mtDNA mutation, interactions with other factors, and molecular mechanisms of mtDNA-associated diseases.

Keywords Mitochondria • Mutation • Oxidative phosphorylation • Heteroplasmy

1 Introduction

Mitochondria are one of the crucial cytoplasmic organelles with double membrane in cells, to sustain aerobic respiration and construct cellular energy in the form of adenosine triphosphate (ATP) through oxidative phosphorylation (OXPHOS), producing endogenous reactive oxygen species (ROS) as an accessory substance. The

Z. Zhu • X. Wang (✉)
Zhongshan Hospital Institute of Clinical Science, Fudan University,
Shanghai Medical College, Shanghai, China
e-mail: xiangdong.wang@clintransmed.org

H. Sun, X. Wang (eds.), *Mitochondrial DNA and Diseases*, Advances in Experimental Medicine and Biology 1038,
https://doi.org/10.1007/978-981-10-6674-0_15

mitochondria also play critical roles in the maintenance of cytosolic calcium homeostasis, regulation of initiating and executing cell apoptosis, biosynthesis of heme and steroid hormone, as well as control of multiple biochemical pathways, such as urea cycle and tricarboxylic acid (TCA) cycle [1]. The mitochondria have dual genetic impact to mitochondrial DNA (mtDNA) and nuclear DNA (nDNA), of which the mutation may cause mitochondrial dysfunction and multisystem syndrome. The present article aims to overview the important molecular mechanisms by which mtDNA mutations occur and change cell biological functions, leading to the development of diseases.

2 Comparison of Biological Characteristics Between mtDNA and nDNA

mtDNA, different from nDNA, is a multi-copy genome with a large number of copies per cell [2–5]. The number varies obviously among persons and tissues, on basis of the bioenergetic needs and cell types. nDNA is diploid and adheres to Mendelian laws genetically, while mtDNAs are transmitted by oocyte [6] and therefore maternally inherited [7]. Mitochondrial diseases are caused by mtDNA mutations and multi-copy nature changes also called heteroplasmy as a unique feature of mtDNA mutation [8]. The homoplasmy exists when all of the mtDNA molecules possess the equivalent genotype. Two kinds of mtDNAs are dispensed into the daughter cells at random in the separation of a heteroplasmic cell. The occurrence and repair of mtDNA mutations change with the process of repeated cell divisions [9], to form the homoplasmy. The replication of mtDNA is hardly linked to the cell cycle, even though a particular mtDNA may be duplicated repeatedly or not even once as a cell splits up. In such replication, a single mutation may prosper clonally or be thoroughly lost during the cell divisions. mtDNA is more vulnerable to oxidative impairment and exhibits a much higher mutation rate than nDNA [10]. mtDNA mending systems are insufficient to counteract the oxidative impairment, owning to a shortage of protective histones, limited DNA mending actions, proximity to respiratory chain complexes in the inner mitochondrial membrane, and the high rate of mitochondria ROS production.

3 mtDNA and OXPHOS Complexes

The human mitochondrial genome comprises about 1500 genes, among which 37 are encoded by mtDNA and the rest by nDNA [11]. mtDNA is a multi-copy, double-stranded, closed round molecule, with 16,569 nucleotides. It encodes 13 subunits of the crucial OXPHOS components, 22 transfer RNAs (tRNAs), and 2 ribosomal RNAs (rRNAs) (12S and 16S), for the protein synthesis of mitochondria [8]. In

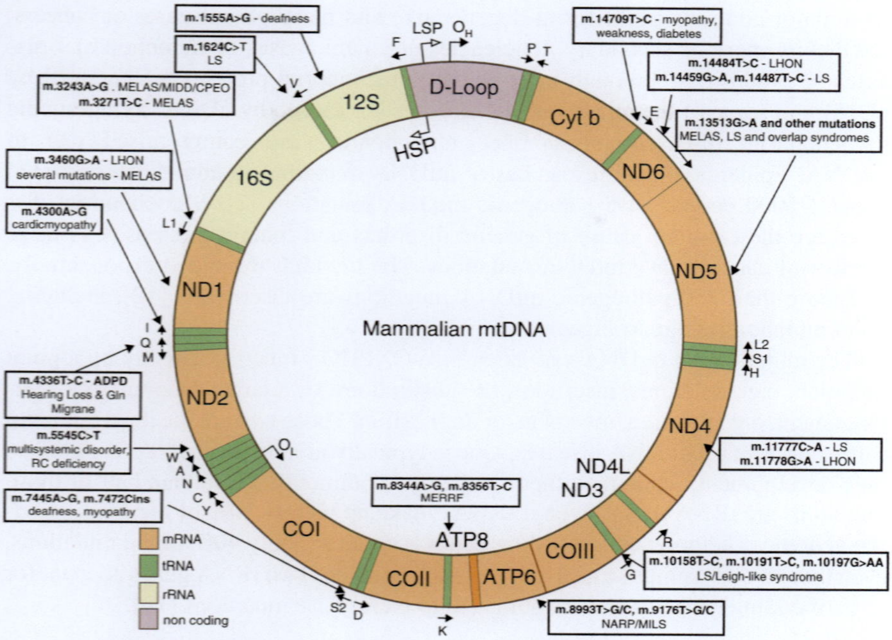

Fig. 15.1 The human mitochondrial genome and correlated mitochondrial diseases
The human mitochondrial genome is a double-stranded, circular molecule, on which the common mtDNA mutations, as well as their associated clinical presentations, are labeled at corresponding sites. LHON, Leber hereditary optic neuropathy; MELAS, mitochondrial encephalopathy, lactic acidosis, and stroke-like episodes; MERRF, myoclonic epilepsy and ragged red fibers; MIDD, maternally inherited diabetes and deafness; NARP, neurogenic weakness, ataxia, and retinitis pigmentosa; LS, Leigh syndrome; MILS, maternally inherited Leigh syndrome; CPEO, progressive external ophthalmoplegia; RC, respiratory chain

addition to the coding region for structural genes, mtDNA comprises about 1000 base pairs of control region, i.e., the D-loop region, consisted of the replication origins and transcription promoters. However, 300 nuclear-encoded proteins are required by the mitoproteome, for the production, assemblement, or sustainment of the five multimeric OXPHOS complexes (I–V) and other accessory mitochondrial processes (Fig. 15.1).

4 mtDNA-Dominant Diseases

mtDNA plays an important role in maintenance of functionally competent organelles. mtDNA mutations can influence the production of energy, oxidative stress, and cell living, resulting in development of diseases, by damage of mitochondrial function. Mitochondrial dysfunction resulted from mtDNA or nDNA changes can be a major contributing factor to degenerative and metabolic diseases or cancers. mtDNA primary or secondary deficiencies are main causes of mitochondrial diseases or syndromes, although most of the mitochondrial proteins are encoded by nDNA and some of the mitochondrial diseases are caused by nDNA. Symptomatic mitochondrial diseases appear once mitochondria are compromised due to mtDNA. Epidemiological researches of mtDNA mutations in adults demonstrated about 1/5000 people have pathogenic mtDNA mutations [2]. Mitochondrial diseases are the common cause of genetic disorders and inherited diseases. A large number of cancers have mtDNA mutations. The field has expanded exponentially, and more than 250 pathogenic mtDNA mutations are discovered [12], including point mutations and rearrangements.

The mutations in mtDNA can jeopardize OXPHOS function directly by a point mutation, e.g., deletions, insertions, or substitutions, or a large-scale mtDNA rearrangement, e.g., deletion, inversion, or duplication. Those two are the most common causes of primary mtDNA diseases. One is typically arising de novo during embryonic development, while the other is inherited commonly. More than half of those mutations are tRNA genes, even if tRNAs make up merely 10% of the overall coding genome. The polypeptide-coding genes account for only 40% of the mutations, even though they comprise nearly 70% of the genome. Two rRNA genes account for 15% of coding capacity and are obligated to merely 2% mutations [13].

5 Features of mtDNA Diseases

Threshold Effect The prevalence of mutational mtDNA is high, while it does not often cause obvious disease due to low heteroplasmy of mtDNA mutations. Heteroplasmic mutations frequently possess varying thresholds, a limit the cell can tolerate the faulty of mtDNA molecules. Increased amounts of mtDNA mutations can reduce the mitochondrial energetic function, resulting in the insufficiency of energy production in tissue. It was evidenced by the fact that the mutation burden is more than the limit, and metabolic disorder and corresponding clinical symptoms appear [5].

However, the disease is not developed only when the mutant load exceeds the definite threshold in a heteroplasmic person. The "threshold effect" varies among different tissues and mutation types. The ratio of mutated to inviolate mtDNA plays decisive roles in the formation of the disease severity. The high thresholds for pathogenicity make mtDNA mutations recessive and actual frequencies of mtDNA mutations underestimated in the population. Top ten of population frequency in pathogenic mtDNA mutations actually is much higher than we expected, i.e., approximately 1/200 of the population carry potentially harmful mtDNA mutations at subthreshold levels [14].

Heterogeneity The complexity and heterogeneity of mtDNA play important roles in the regulation of mitochondrial and cellular function. Mutations in different mitochondrial genes can show similar phenotypes, while the same gene can be distinct phenotypes. The heteroplasmy of the same mtDNA mutation is responsible for the formation of highly different phenotypes, e.g., variations in onset age, affected organs and tissues, or disease severity. It, however, stays unsettled why mitochondrial diseases reveal a broad variety of clinical phenotypes, even among patients who carry the equivalent mtDNA mutations. For instance, the mtDNA mutations present stereotypic phenotypes in Leber hereditary optic neuropathy [9], while other ones cause myoclonic epilepsy and ragged red fiber, mitochondrial encephalomyopathy, lactic acidosis, or stroke-like episode syndromes, leading to highly variable multisystem diseases [15]. Another example is that patients with over 90% of the m.8993 T $\rightarrow$ G mutation in the ATP synthase 6 (ATP6) gene present with maternally derived Leigh syndrome, as a fatal encephalopathy [16]. When mutations are between 70 and 90%, patients suffered from neuropathy, ataxia, and retinitis pigmentosa as a late-onset, recessionary but nonfatal illness [12].

6 Clinical Phenotypes of mtDNA Diseases

Major categories of diseases caused by mitochondrial disorder include primary mitochondrial disease, neurodegeneration, and cancer, of which mitochondrial diseases are heterogeneous and involved in multiple systems. The mitochondrial diseases present at any age and affect any organ as lower cellular respiration, compromised respiratory chain complex activity, or reduced ATP synthesis due to hypo-production of ATP through the OXPHOS pathway. Mitochondrial diseases can occur in any of impaired and dysfunctional OXPHOS complexes. Though any organ can be affected, tissues with high energy demands are more vulnerable, such as the heart, brain, nerves, muscles, and endocrine systems. Energetic deficiencies lead to the development of blindness, deafness, dementias, motion dysfunctions, myopathy, cardiomyopathy, or renal dysfunction. Respiratory chain malfunctions and OXPHOS defects caused by mtDNA mutations are present in neurodegenerative statuses, including Parkinson's disease [17], Alzheimer's disease, or multiple sclerosis (Fig. 15.2).

7 Mutations in mtDNA Diseases

Point Mutations mtDNA point mutations are an important cause of mitochondria disease with the prevalence about 1/200 population [18]. The highly recessive and heteroplasmic point mutations display considerable clinical heterogeneity of symptoms from plain sensorineural deafness to overwhelming syndromic neurological disease. Symptoms may appear in maturity or childhood, although mutations are maternally inherited (~75%) or can occur de novo (~25%) [3].

mtDNA point mutations may occur within every mtDNA gene, including protein genes (polypeptide mutations), rRNA, or tRNA genes (protein synthesis mutations). mt-tRNA genes represent about 10% of the mitochondrial genome and contribute to

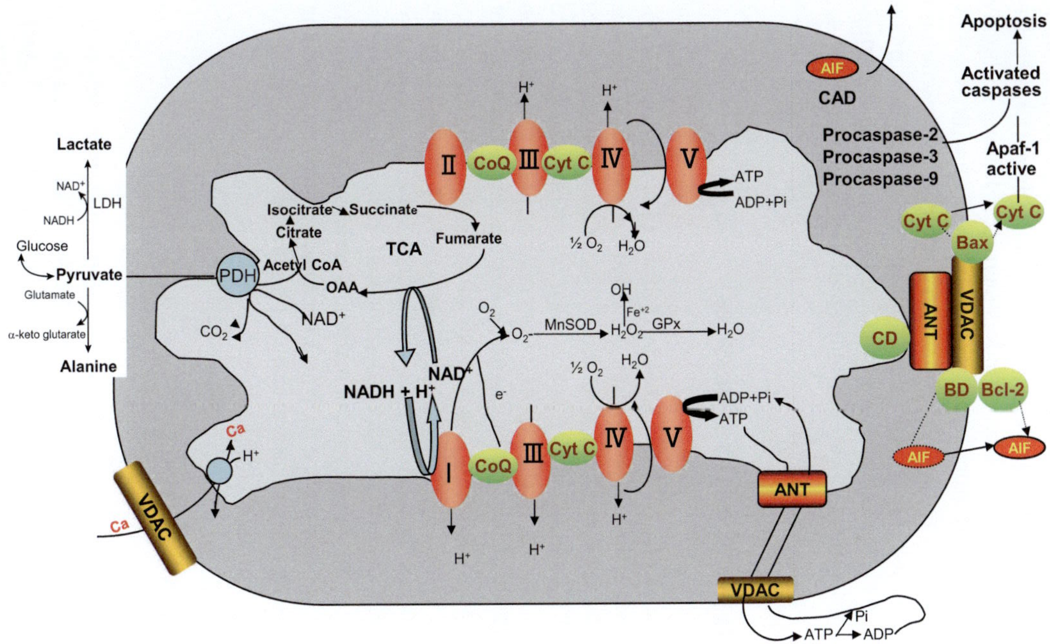

Fig. 15.2 The structure of OXPHOS complexes, partly encoded by mtDNA, and related diseases as a result of their dysfunction
Among the 37 mtDNA-encoded genes, 13 are subunits of the crucial OXPHOS components (red parts), including seven subunits of complex I (ND1, 2, 3, 4, 4 L, 5 and 6), one subunit of complex III (cytochrome b, Cyt b), three subunits of complex IV (Cyt c oxidase, COX I, II, and III), and two subunits of complex V (ATP8 and ATP6). These multimeric protein complexes shuttle electrons along the respiratory chain, which is facilitated by the decrease of the cofactors coenzyme Q10 (Q) and cytochrome c (cyt c). Electron transfer is coupled to the transfer of protons (H+) across the inner mitochondrial membrane from inter-membrane space (IMS) to mitochondrial matrix to produce a proton motive force, which is utilized by complex V (ATP synthase) to synthesize ATP. The major clinical diseases or manifestations caused by those mutations are also shown. ADP, adenosine diphosphate; ATP, adenosine triphosphate; TCA, tricarboxylic acid cycle

more than half of the mutations. Pathogenic point mutations in mitochondrial protein-coding genes impact specific respiratory chain complexes by reducing the availability of functional mt-tRNAs; mt-tRNA mutations may impair overall mitochondrial translation and protein synthesis. The clinical features of tRNA mutations are multiplex. Among the point mutations, the A → G transition at location 3243 in the tRNALeu (UUR) gene is the most prevalent one with high specificity and high prevalence about 5.7 per 100,000 [19]. The A → G transition at location 8344 in tRNALys leads to myoclonus epilepsy and ragged red fiber syndrome [13]. Different phenotypes can be led to by mutations in other tRNA genes, for instance, deafness due to tRNA-Ser (UCN) mutations or cardiomyopathy due to tRNA-Ile mutations [20].

Mutations in the mitochondrial rRNA genes are manifested as either aminoglycoside-induced deafness or non-syndromic sensorineural deafness. The most frequent mutation is A1555G in the 12S rRNA gene. The mutation changes the constitution of the rRNA, similar to the bacterial 16S rRNA. This interacts with aminoglycosides, resulting in translation fault. Deafness is not developed by all people carrying this mutation. As for polypeptide mutations, the well-known mutation is T → C and T → G transversion at location 8993 in ATP6 [21]. Different diseases may load with various mutants. A small load makes asymptomatic mutation carrier; an intermediate load leads to neurogenic weakness, ataxia, or retinitis pigmentosa; and a big load gives rise to maternally inherited Leigh syndrome. Multisystem diseases are caused by pathogenic polypeptide mutations, such as Leber hereditary optic neuropathy [9]. Other protein synthesis mutations are involved in encephalomyopathy, cardiomyopathy, mitochondrial myopathy, dystonia, gastrointestinal diseases, diabetes, renal dysfunction, Alzheimer's disease, or Parkinson's disease (Table 15.1).

Table 15.1 The common mitochondrial syndromes and their main affected organs

Acronym	Full name	Affected organs
Point mutation		
MELAS	Mitochondrial encephalopathy with lactic acidosis and stroke-like episodes	Skeletal muscle, brain, heart, endocrine pancreas, ear, gut
MERRF	Myoclonic epilepsy with ragged red fibers	Skeletal muscle, brain, heart
NARP	Neurogenic weakness, ataxia, and retinitis pigmentosa	Brain, eye
LHON	Leber hereditary optic neuropathy	Eye, heart
Single deletion		
KSS	Kearns–Sayre syndrome	Skeletal muscle, brain, heart, eye
PS	Pearson syndrome	Bone marrow, exocrine pancreas, skeletal muscle, brain
PEO	Progressive external ophthalmoplegia	Skeletal muscle (especially extraocular), heart
Other mutation		
LS	Leigh syndrome	Brain, eye, skeletal muscle, peripheral nerve

Deletion Mutations The large-scale deletions are the most common type in rearrangement mutations, of which over 120 different mtDNA deletions are associated with diseases. Among the mtDNA deletions detected, the most evident one is 4977 bp deletion with a 13-bp direct repetition flanking the 5′- and 3′-end breakpoints at nucleotide positions 8470/8482 and 13,447/13459, respectively. This mtDNA deletion was initially discovered in the muscle of patients with mitochondrial diseases in 1988 [22]. Point mutations are in tRNA and protein-encoding rRNA genes, while the deletion mutations occur at random and can be situated nearly anywhere in the mitochondrial genome. Deletion mutations of mtDNA play a decisive role in the severity of Kearns–Sayre syndrome associated with ophthalmoplegia, chronic progressive external ophthalmoplegia, or mitochondrial myopathy with ragged red fibers. The patients with Pearson syndrome, which is the most severe one, suffered acquired pan-hematopenia early in life and had transfusion-dependent living. The syndrome can be developed after the patients survive from the pancytopenia [23]. The pathology occurs in the substantia nigra of patients with Parkinson's disease and has the close association with deletion mutations of mtDNA [24].

Other Rearrangement Mutations Other than de novo deletion mutations, rearrangement mutations can also be insertion mutations, maternally transmitted, and instable. mtDNA rearrangement syndromes are inherited of mtDNA duplication mutation from mother and have the invariable heteroplasmy of clinical manifestations and seriousness. The mildest phenotypes of mtDNA rearrangement are deafness or type II diabetes [25]. Different phenotypes of mtDNA rearrangement are dependent upon distinctions between mtDNA mutation types (e.g., deletions or insertions), diversity of tissues where the rearrangement occurs, and severity of the mutation load in the tissue.

8 Cancers

The point mutations of mtDNA were initially noticed in human colorectal cancer [26]. After then, a large number of mtDNA change types were discovered in numerous human cancers where mtDNA modifications range from mild point and missense mutations to severe chain termination and insertion-deletion mutations. mtDNA mutations in cancer include oncogenic germ line mutations in the female germ line to predispose to cancer or tumor-specific somatic mutations in the tissues responsible for the tumor progression procedure. Somatic tumor mutations are strongly correlated with the polymorphisms. About 72% point mutations of mtDNA in human tumors were previously reported as polymorphic variants in the general population, while about 28% have tumorigenic mutations as the severe form to inhibit OXPHOS, produce ROS, and accelerate tumor cell proliferation and adaptive mutations as the milder form to assist tumors in the survival within new environments [7] (Fig. 15.3).

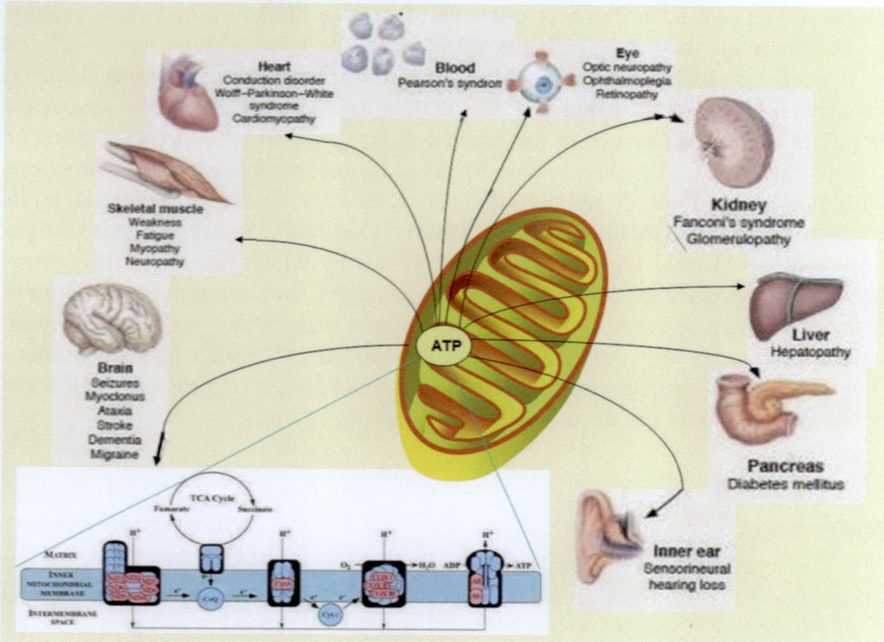

Fig. 15.3 The major functions of mitochondria
Mitochondria support aerobic respiration and produce adenosine triphosphate (ATP) by oxidative phosphorylation (OXPHOS), generating endogenous reactive oxygen species (ROS) as a by-product. Besides, mitochondria also control cytosolic calcium homeostasis, regulate apoptotic cell death, and host other important biochemical pathways, such as tricarboxylic acid (TCA) cycle and alanine metabolism
NAD, nicotinamide adenine dinucleotide; NADH, reduced form of nicotinamide adenine dinucleotide; FAD, flavin adenine dinucleotide; FADH2, reduced form of flavin adenine dinucleotide; Cyt C, cytochrome c; CoQ, coenzyme Q10; SOD, superoxide dismutase; LDH, lactate dehydrogenase; PDH, pyruvate dehydrogenase; VDAC, voltage-dependent anion channel; ANT, adenosine nucleotide translocator; AIF, apoptotic induction factor

9 Challenges and Outlooks

With increased knowledge of mtDNA mutations in mitochondrial disease, more alternatives of new therapies will be discovered and developed, although there are still a number of challenges to be faced and solved. There is an urgent need of more accurate and repeatable methodologies to ensure the accurate diagnosis of mtDNA mutation. It is critical to identify mutation-specific biomarkers to monitor the occurrence of mtDNA mutations and have disease-specific biomarkers [27–32]. There are still lacks of well-established in vitro and in vivo models of mtDNA-mutation diseases by means of introducing aimed mutations into mitochondria. Such models with targeted mtDNA mutations can be used to define elementary biochemical events in mitochondrial biogenesis and the exact mechanisms of diseases. The significance of heterogeneity of mtDNA mutations in clinical diagnosis and therapies should be furthermore clarified.

It is also a challenge to define the association of mtDNA mutations with the severity and form of mitochondrial dysfunction, since a huge number of mutations in mtDNA can be generated in the grade mix, mix, and link tie in combination of various mutations. Mitochondria are involved in uncountable biological functions within the cell, influenced by multiple factors, associated with various signal pathways, and stimulated by intra- and extracellular elements [33–40]. Future investigations will be focused on molecular mechanisms by which phenotype diversities occur, heterogeneity of mtDNA mutations develops, and mutations alter protein expression and activation. Single cell biology and gene editing will provide more hopes to overcome those challenges [41–45]. We need to further investigate functions of mtDNA mutations in diseases and regulations of crosstalk pathways between the nucleus and the mitochondria genomes as well.

References

1. Stehling O, Lill R. The role of mitochondria in cellular iron-sulfur protein biogenesis: mechanisms, connected processes, and diseases. Cold Spring Harb Perspect Biol. 2013;5(8):a011312. PMID:23906713
2. Wallace DC, et al. Mitochondrial DNA mutation associated with Leber's hereditary optic neuropathy. Science. 1988;242(4884):1427–30. PMID:3201231
3. Sallevelt SC, de Die-Smulders CE, Hendrickx AT, et al. De novo mtDNA point mutations are common and have a low recurrence risk. J Med Genet. 2017;54(2):73–83. PMID:27450679
4. Ylikallio E, Suomalainen A. Mechanisms of mitochondrial diseases. Ann Med. 2012;44(1):41–59. PMID:21806499
5. Alston CL, et al. The genetics and pathology of mitochondrial disease. J Pathol. 2017;241(2):236–50. PMID:27659608
6. Calvo SE, Clauser KR, Mootha VK. MitoCarta2.0: an updated inventory of mammalian mitochondrial proteins. Nucleic Acids Res. 2016;44(D1):D1251–7. PMID:26450961
7. Giles RE, et al. Maternal inheritance of human mitochondrial DNA. Proc Natl Acad Sci U S A. 1980;77(11):6715–9. PMID:6256757
8. Stewart JB, Chinnery PF. The dynamics of mitochondrial DNA heteroplasmy: implications for human health and disease. Nat Rev Genet. 2015;16(9):530–42. PMID:26281784
9. Brandon M, Baldi P, Wallace DC. Mitochondrial mutations in cancer. Oncogene. 2006;25(34):4647–62. PMID:16892079
10. Lee HC, Chang CM, Chi CW. Somatic mutations of mitochondrial DNA in aging and cancer progression. Ageing Res Rev. 2010;9(Suppl 1):S47–58. PMID:20816876
11. Anderson S, et al. Sequence and organization of the human mitochondrial genome. Nature. 1981;290(5806):457–65. PMID:7219534
12. Tuppen HA, et al. Mitochondrial DNA mutations and human disease. Biochim Biophys Acta. 2010;1797(2):113–28. PMID:19761752
13. Schon EA, DiMauro S, Hirano M. Human mitochondrial DNA: roles of inherited and somatic mutations. Nat Rev Genet. 2012;13(12):878–90. PMID:23154810
14. Elliott HR, et al. Pathogenic mitochondrial DNA mutations are common in the general population. Am J Hum Genet. 2008;83(2):254–60. PMID:18674747
15. Shoffner JM, et al. Myoclonic epilepsy and ragged-red fiber disease (MERRF) is associated with a mitochondrial DNA tRNA(Lys) mutation. Cell. 1990;61(6):931–7. PMID:2112427

16. Santorelli FM, et al. The mutation at nt 8993 of mitochondrial DNA is a common cause of Leigh's syndrome. Ann Neurol. 1993;34(6):827–34. PMID:8250532
17. Schapira AH, et al. Mitochondrial complex I deficiency in Parkinson's disease. J Neurochem. 1990;54(3):823–7. PMID:2154550
18. Chinnery PF, et al. Epigenetics, epidemiology and mitochondrial DNA diseases. Int J Epidemiol. 2012;41(1):177–87. PMID:22287136
19. Goto Y, Nonaka I, Horai S. A mutation in the tRNA(Leu)(UUR) gene associated with the MELAS subgroup of mitochondrial encephalomyopathies. Nature. 1990;348(6302):651–3. PMID:2102678
20. Scaglia F, Wong LJ. Human mitochondrial transfer RNAs: role of pathogenic mutation in disease. Muscle Nerve. 2008;37(2):150–71. PMID:17999409
21. Holt IJ, et al. A new mitochondrial disease associated with mitochondrial DNA heteroplasmy. Am J Hum Genet. 1990;46(3):428–33. PMID:2137962
22. Holt IJ, Harding AE, Morgan-Hughes JA. Deletions of muscle mitochondrial DNA in patients with mitochondrial myopathies. Nature. 1988;331(6158):717–9. PMID:2830540
23. Rotig A, et al. Spectrum of mitochondrial DNA rearrangements in the Pearson marrow-pancreas syndrome. Hum Mol Genet. 1995;4(8):1327–30. PMID:7581370
24. Kraytsberg Y, et al. Mitochondrial DNA deletions are abundant and cause functional impairment in aged human substantia nigra neurons. Nat Genet. 2006;38(5):518–20. PMID:16604072
25. Wallace DC. Mitochondrial DNA mutations in disease and aging. Environ Mol Mutagen. 2010;51(5):440–50. PMID:20544884
26. Polyak K, et al. Somatic mutations of the mitochondrial genome in human colorectal tumours. Nat Genet. 1998;20(3):291–3. PMID:9806551
27. Wang X. New biomarkers and therapeutics can be discovered during COPD-lung cancer transition. Cell Biol Toxicol. 2016;32(5):359–61. PMID:27405768
28. Chen C, Shi L, Li Y, Wang X, Yang S. Disease-specific dynamic biomarkers selected by integrating inflammatory mediators with clinical informatics in ARDS patients with severe pneumonia. Cell Biol Toxicol. 2016;32(3):169–84. PMID:27095254
29. Shi L, Zhu B, Xu M, Wang X. Selection of AECOPD-specific immunomodulatory biomarkers by integrating genomics and proteomics with clinical informatics. Cell Biol Toxicol. 2017. https://doi.org/10.1007/s10565-017-9405-x. (Epub ahead of print). PMID: 28779230
30. Gu J, Wang X. New future of cell biology and toxicology: thinking deeper. Cell Biol Toxicol. 2016;32(1):1–3. PMID:26874518
31. Wu Y, Geng XC, Wang JF, Miao YF, YL L, Li B. The HepaRG cell line, a superior in vitro model to L-02, HepG2 and hiHeps cell lines for assessing drug-induced liver injury. Cell Biol Toxicol. 2016;32(1):37–59. PMID:27027780
32. Zhu D, Liu Z, Pan Z, Qian M, Wang L, Zhu T, Xue Y, Wu DA. New method for classifying different phenotypes of kidney transplantation. Cell Biol Toxicol. 2016;32(4):323–32. PMID:27278387
33. Kikuchi S, Ninomiya T, Kohno T, Kojima T, Tatsumi H. Cobalt inhibits motility of axonal mitochondria and induces axonal degeneration in cultured dorsal root ganglion cells of rat. Cell Biol Toxicol. 2017. https://doi.org/10.1007/s10565-017-9402-0. (Epub ahead of print). PMID:28656345
34. Hsu HC, Li SJ, Chen CY, Chen MF. Eicosapentaenoic acid protects cardiomyoblasts from lipotoxicity in an autophagy-dependent manner. Cell Biol Toxicol. 2017. https://doi.org/10.1007/s10565-017-9406-9. (Epub ahead of print). PMID:28741157
35. Zhu LZ, Hou YJ, Zhao M, Yang MF, XT F, Sun JY, XY F, Shao LR, Zhang HF, Fan CD, Gao HL, Sun BL. Caudatin induces caspase-dependent apoptosis in human glioma cells with involvement of mitochondrial dysfunction and reactive oxygen species generation. Cell Biol Toxicol. 2016;32(4):333–45. PMID:27184666
36. Bao L, Diao H, Dong N, Su X, Wang B, Mo Q, Yu H, Wang X, Chen C. Histone deacetylase inhibitor induces cell apoptosis and cycle arrest in lung cancer cells via mitochondrial injury and p53 up-acetylation. Cell Biol Toxicol. 2016;32(6):469–82. PMID: 27423454

37. Vedi M, Sabina EP. Assessment of hepatoprotective and nephroprotective potential of withaferin A on bromobenzene-induced injury in Swiss albino mice: possible involvement of mitochondrial dysfunction and inflammation. Cell Biol Toxicol. 2016;32(5):373–90. PMID: 27250656
38. Sanyal S, Das P, Law S. Effect of chronic pesticide exposure on murine cornea: a histopathological, cytological and flow cytometric approach to study ocular damage by xenobiotics. Cell Biol Toxicol. 2016;32(1):7–22. PMID: 26897134
39. Soltani B, Ghaemi N, Sadeghizadeh M, Najafi F. Curcumin confers protection to irradiated THP-1 cells while its nanoformulation sensitizes these cells via apoptosis induction. Cell Biol Toxicol. 2016;32(6):543–61. PMID: 27473378
40. Keta O, Bulat T, Golić I, Incerti S, Korać A, Petrović I, Ristić-Fira A. The impact of autophagy on cell death modalities in CRL-5876 lung adenocarcinoma cells after their exposure to γ-rays and/or erlotinib. Cell Biol Toxicol. 2016;32(2):83–101. PMID: 27026538
41. Fang H, Wang W. Could CRISPR be the solution for gene editing's Gordian knot? Cell Biol Toxicol. 2016;32(6):465–7. PMID: 27614448
42. Wang W, Wang X, Single-cell CRISPR. Screening in drug resistance. Cell Biol Toxicol. 2017;33(3):207–10. PMID: 28474250
43. Sakuma T, Yamamoto T. Magic wands of CRISPR-lots of choices for gene knock-in. Cell Biol Toxicol. 2017. https://doi.org/10.1007/s10565-017-9409-6. PMID: 28828704
44. Wang W, Gao D, Wang X. Can single-cell RNA sequencing crack the mystery of cells? Cell Biol Toxicol. 2017. https://doi.org/10.1007/s10565-017-9404-y. (Epub ahead of print). PMID: 28733864
45. Wang W, Zhu B, Wang X. Dynamic phenotypes: illustrating a single-cell odyssey. Cell Biol Toxicol. 2017. https://doi.org/10.1007/s10565-017-9400-2. (Epub ahead of print). PMID: 28638956

Zhenhua Zhu, MD, an attending doctor of Department of Respiratory Medicine, Zhongshan Hospital, Fudan University. She was graduated from Shanghai Medical School, Fudan University. Her main research is focused on chronic lung diseases. She owns the honor of "rich experience in clinical trials." She has accomplished one "Youth Fund" of Zhongshan Hospital related to COPD.

Xiangdong Wang, MD, PhD, is a Distinguished Professor of Medicine, Director of Shanghai Institute of Clinical Bioinformatics, Executive Director of Clinical Science Institute of Fudan University Zhongshan Hospital, Director of Fudan University Center of Clinical Bioinformatics, Deputy Director of Shanghai Respiratory Research Institute, and visiting Professor of King's College of London. His main research is focused on clinical bioinformatics, disease-specific biomarkers, lung chronic diseases, cancer immunology, and molecular and cellular therapies. He is the author of more than 200 scientific publications with the impact factor about 900, citation number about 6918, h-index 46, i10-index 181, and cited journal impact factor about 7000.